FUNDAMENTALS OF MOLECULAR SPECTROSCOPY

ectroscopy

C. N. BANWELL

Lecturer in Chemistry
University of Sussex

McGRAW-HILL Book Company (UK) Limited

London · New York · St Louis · San Francisco · Auckland · Bogotá
Guatemala · Hamburg · Johannesburg · Lisbon · Madrid · Mexico
Montreal · New Delhi · Panama · Paris · San Juan · São Paulo
Singapore · Sydney · Tokyo · Toronto

Published by
McGRAW-HILL Book Company (UK) Limited
MAIDENHEAD · BERKSHIRE · ENGLAND

British Library Cataloguing in Publication Data

Banwell, C. N.
  Fundamentals of molecular spectroscopy.—3rd ed.
  1. Molecular Spectra
  I. Title
  535.8'4   QC454.M6

  ISBN 0–07–084139–X

Library of Congress Cataloging in Publication Data

Banwell, C. N.
  Fundamentals of molecular spectroscopy.
  Bibliography: p.
  Includes index.
  1. Molecular spectra. I. Title.
QD96.M65B36 1983   535.8'4   82–23949
ISBN 0–07–084139–X

Typeset by Santype International Ltd., Salisbury, Wilts.
Printed and bound in Great Britain by
William Clowes Limited, Beccles and London

CONTENTS

LIST OF TABLES

PREFACE

During the writing of this, the third edition of *Fundamentals of Molecular Spectroscopy*, I have borne very firmly in mind the aims of the original book—to emphasize the overall unity of the subject, and to offer a pictorial rather than a mathematical description of the principles of spectroscopy. The latter aim has received some criticism (although, it must be said, from teachers of the subject rather than from those trying to learn it); but, while it is true that universities and polytechnics now offer excellent courses in subjects such as quantum mechanics, a grasp of which can illuminate many areas of spectroscopy, experience shows that most students approaching spectroscopy for the first time prefer the simpler treatment offered here.

The general organization of the book remains unchanged, each chapter after the first presenting the fundamentals of a particular spectroscopic technique. Although each chapter is thus essentially self-contained, the intention is that the book shall be read as a whole, since concepts introduced in early chapters are often used later without further discussion.

In fact, although it is an almost unbelievable ten years since the second edition was published, there has been relatively little change in the basic ideas of spectroscopy, and much of the book is essentially unaltered. What changes have occurred have been mainly in the extension of existing techniques, and their increasing availability due to improvements in instrumentation. In this respect the two main areas of interest are the use of lasers as radiation sources, and the very rapid expansion in the use of Fourier transform techniques. Accordingly I have added a section on lasers to Chapter 1 and expanded the references to them in later chapters, particularly in Chapter 3. Equally, a general description of the Fourier transform method has been included in Chapter 1, and Chapter 7 has been extensively rewritten to take account of the wealth of additional information accessible from the application of this technique to nuclear magnetic resonance spectroscopy.

When a book has been in use for 15 years it becomes impossible to acknowledge the help received from the very many people, students and teachers, who have commented on it. My greatest debt is, of course, to Professor Sheppard of the University of East Anglia, who first introduced me to the beauty and fascination of spectroscopy, but I hope that all those who see their comments and suggestions incorporated into this edition will accept this general acknowledgement of my indebtedness, and will not think me churlish for leaving them unnamed.

C. N. BANWELL
University of Sussex

ONE

INTRODUCTION

1.1 CHARACTERIZATION OF ELECTROMAGNETIC RADIATION

Molecular spectroscopy may be defined as the study of the interaction of electromagnetic waves and matter. Throughout this book we shall be concerned with what spectroscopy can tell us of the structure of matter, so it is essential in this first chapter to discuss briefly the nature of electromagnetic radiation and the sort of interactions which may occur; we shall also consider, in outline, the experimental methods of spectroscopy.

Electromagnetic radiation, of which visible light forms an obvious but very small part, may be considered as a simple harmonic wave propagated from a source and travelling in straight lines except when refracted or reflected. The properties which undulate—corresponding to the physical displacement of a stretched string vibrating, or the alternate compressions and rarefactions of the atmosphere during the passage of a sound wave—are interconnected electric and magnetic fields. We shall see later that it is these undulatory fields which interact with matter giving rise to a spectrum.

It is trivial to show that any simple harmonic wave has properties of the sine wave, defined by $y = A \sin \theta$, which is plotted in Fig. 1.1. Here y is the displacement with a maximum value A, and θ is an angle varying between 0 and 360° (or 0 and 2π radians). The relevance of this representation to a travelling wave is best seen by considering the left-hand side of

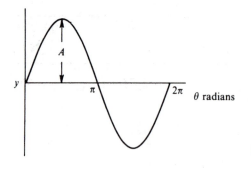

Figure 1.1 The curve of $y = A \sin \theta$.

Fig. 1.2. A point P travels with uniform angular velocity ω rad s^{-1} in a circular path of radius A; we measure the time from the instant when P passes O' and then, after a time t seconds, we imagine P to have described an angle $\theta = \omega t$ radians. Its vertical displacement is then $y = A \sin \theta = A \sin \omega t$, and we can plot this displacement against time as on the right-hand side of Fig. 1.2. After a time of $2\pi/\omega$ seconds, P will return to O', completing a 'cycle'. Further cycles of P will repeat the pattern and we can describe the displacement as a continuous function of time by the graph of Fig. 1.2.

In one second the pattern will repeat itself $\omega/2\pi$ times, and this is referred to as the *frequency* (v) of the wave. The SI unit of frequency is called the hertz (abbreviated to Hz) and has the dimensions of reciprocal seconds (abbreviated s^{-1}). We may then write:

$$y = A \sin \omega t = A \sin 2\pi v t \qquad (1.1)$$

as a basic equation of wave motion.

So far we have discussed the variation of displacement with time, but in order to consider the nature of a *travelling* wave, we are more interested in

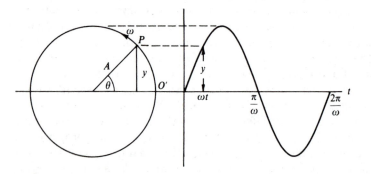

Figure 1.2 The description of a sine curve in terms of the circular motion of a point P at a uniform angular velocity of ω rad s^{-1}.

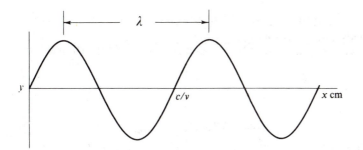

Figure 1.3 The concept of a travelling wave with a wavelength λ.

the distance variation of the displacement. For this we need the fundamental distance–time relationship:

$$x = ct \tag{1.2}$$

where x is the distance covered in time t at a speed c. Combining (1.1) and (1.2) we have:

$$y = A \sin 2\pi vt = A \sin \frac{2\pi vx}{c}$$

and the wave is shown in Fig. 1.3. Besides the frequency v, we now have another property by which we can characterize the wave—its *wavelength* λ, which is the distance travelled during a complete cycle. When the velocity is c metres per second and there are v cycles per second, there are evidently v waves in c metres, or

$$v\lambda = c \qquad \lambda = c/v \quad \text{metres} \tag{1.3}$$

so we have:

$$y = A \sin \frac{2\pi x}{\lambda} \tag{1.4}$$

In spectroscopy wavelengths are expressed in a variety of units, chosen so that in any particular range (see Fig. 1.4) the wavelength does not involve large powers of ten. Thus, in the microwave region, λ is measured in centimetres or millimetres, while in the infra-red it is usually given in micrometres (μm)—formerly called the *micron*—where

$$1 \ \mu\text{m} = 10^{-6} \text{ m} \tag{1.5}$$

In the visible and ultra-violet region, λ is still often expressed in Ångstrom units (Å) where $1 \text{ Å} = 10^{-10}$ m, although the proper SI unit for this region is the nanometre:

$$1 \text{ nm} = 10^{-9} \text{ m} = 10 \text{ Å} \tag{1.6}$$

There is yet a third way in which electromagnetic radiation can be usefully characterized, and this is in terms of the *wavenumber* $\bar{v}$. Formally this is defined as the reciprocal of the wavelength expressed in *centimetres*:

$$\bar{v} = 1/\lambda \quad \text{cm}^{-1} \tag{1.7}$$

and hence

$$y = A \sin 2\pi\bar{v}x \tag{1.8}$$

It is more useful to think of the wavenumber, however, as the number of compete waves or cycles contained in each centimetre length of radiation. Since the formal definition is based on the centimetre rather than the metre, the wavenumber is, of course, a non-SI unit; it is, however, so convenient a unit for the discussion of infra-red spectra that—like the Ångstrom—it will be many years before it falls into disuse.

It is unfortunate that the conventional symbols of wavenumber ($\bar{v}$) and frequency (v) are similar; confusion should not arise, however, if the units of any expression are kept in mind, since wavenumber is invariably expressed in reciprocal centimetres (cm^{-1}) and frequency in cycles per second (s^{-1} or Hz). The two are, in fact, proportional: $v = c\bar{v}$, where the proportionality constant is the velocity of radiation expressed in *centimetres* per second (that is, 3×10^{10} cm s^{-1}); the velocity in SI units is, of course, 3×10^{8} m s^{-1}.

1.2 THE QUANTIZATION OF ENERGY

Towards the end of the last century experimental data were observed which were quite incompatible with the previously accepted view that matter could take up energy continuously. In 1900 Max Planck published the revolutionary idea that the energy of an oscillator is discontinuous and that any change in its energy content can occur only by means of a jump between two distinct energy states. The idea was later extended to cover many other forms of the energy of matter.

A molecule in space can have many sorts of energy; e.g., it may possess rotational energy by virtue of bodily rotation about its centre of gravity; it will have vibrational energy due to the periodic displacement of its atoms from their equilibrium positions; it will have electronic energy since the electrons associated with each atom or bond are in unceasing motion, etc. The chemist or physicist is early familiar with the electronic energy states of an atom or molecule and accepts the idea that an electron can exist in one of several discrete energy levels: he learns to speak of the energy as being *quantized*. In much the same way the rotational, vibrational, and other energies of a molecule are also quantized—a particular molecule can exist in a variety of rotational, vibrational, etc., energy levels and can move from one level to another only by a sudden jump involving a finite amount of energy.

Consider two possible energy states of a system—two rotational energy levels of a molecule, for example—labelled E_1 and E_2 in the following diagram. The suffixes 1 and 2 used to distinguish these levels are, in fact,

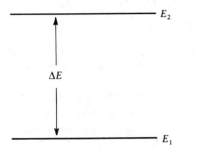

quantum numbers. The actual significance of quantum numbers goes far deeper than their use as a convenient label—in particular, we shall later see that analytical expressions for energy levels usually involve an algebraic function of one or more quantum numbers. Transitions can take place between the levels E_1 and E_2 provided the appropriate amount of energy, $\Delta E = E_2 - E_1$, can be either absorbed or emitted by the system. Planck suggested that such absorbed or emitted energy can take the form of electromagnetic radiation and that the frequency of the radiation has the simple form:

$$v = \Delta E/h \quad \text{Hz}$$

i.e.,

$$\Delta E = hv \quad \text{joules} \tag{1.9}$$

where we express our energies E in terms of the joule, and h is a universal constant—Planck's constant. This suggestion has been more than amply confirmed by experiment.

The significance of this is that if we take a molecule in state 1 and direct on to it a beam of radiation of a single frequency v (monochromatic radiation), where $v = \Delta E/h$, energy will be absorbed from the beam and the molecule will jump to state 2. A detector placed to collect the radiation after its interaction with the molecule will show that its intensity has decreased. Also if we use a beam containing a wide range of frequencies ('white' radiation), the detector will show that energy has been absorbed *only* from that frequency $v = \Delta E/h$, all other frequencies being undiminished in intensity. In this way we have produced a spectrum—an *absorption* spectrum.

Alternatively the molecule may already be in state 2 and may revert to state 1 with the consequent emission of radiation. A detector would show this radiation to have frequency $v = \Delta E/h$ only, and the *emission* spectrum so found is plainly complementary to the absorption spectrum of the previous paragraph.

The actual energy differences between the rotational, vibrational, and electronic energy levels are very small and may be measured in joules per molecule (or atom). In these units Planck's constant has the value:

$$h = 6{\cdot}63 \times 10^{-34} \text{ joules s molecule}^{-1}$$

Often we are interested in the total energy involved when a gram-molecule of a substance changes its energy state: for this we multiply by the Avogadro number $N = 6{\cdot}02 \times 10^{23}$.

However, the spectroscopist measures the various characteristics of the absorbed or emitted radiation during transitions between energy states and he often, rather loosely, uses frequency, wavelength, and wavenumber as if they were energy units. Thus in referring to 'an energy of 10 cm^{-1}' he means 'a separation between two energy states such that the associated radiation has a wavenumber value of 10 cm^{-1}'. The first expression is so simple and convenient that it is essential to become familiar with wavenumber and frequency energy units if one is to understand the spectroscopist's language. Throughout this book we shall use the symbol ε to represent energy in cm^{-1}.

It cannot be too firmly stressed at this point that the frequency of radiation associated with an energy change does *not* imply that the transition between energy levels occurs a certain number of times each second. Thus an electronic transition in an atom or molecule may absorb or emit radiation of frequency some 10^{15} Hz, but the electronic transition does not itself *occur* 10^{15} times per second. It may occur once or many times and on each occurrence it will absorb or emit an energy quantum of the appropriate frequency.

1.3 REGIONS OF THE SPECTRUM

Figure 1.4 illustrates in pictorial fashion the various, rather arbitrary, regions into which electromagnetic radiation has been divided. The boundaries between the regions are by no means precise, although the molecular processes associated with each region are quite different. Each succeeding chapter in this book deals essentially with one of these processes.

In increasing frequency the regions are:

1. Radiofrequency region: 3×10^6–3×10^{10} Hz; 10 m–1 cm wavelength. Nuclear magnetic resonance (n.m.r.) and electron spin resonance (e.s.r.) spectroscopy. The energy change involved is that arising from the reversal of spin of a nucleus or electron, and is of the order 0·001–10 joules/mole (Chapter 7).

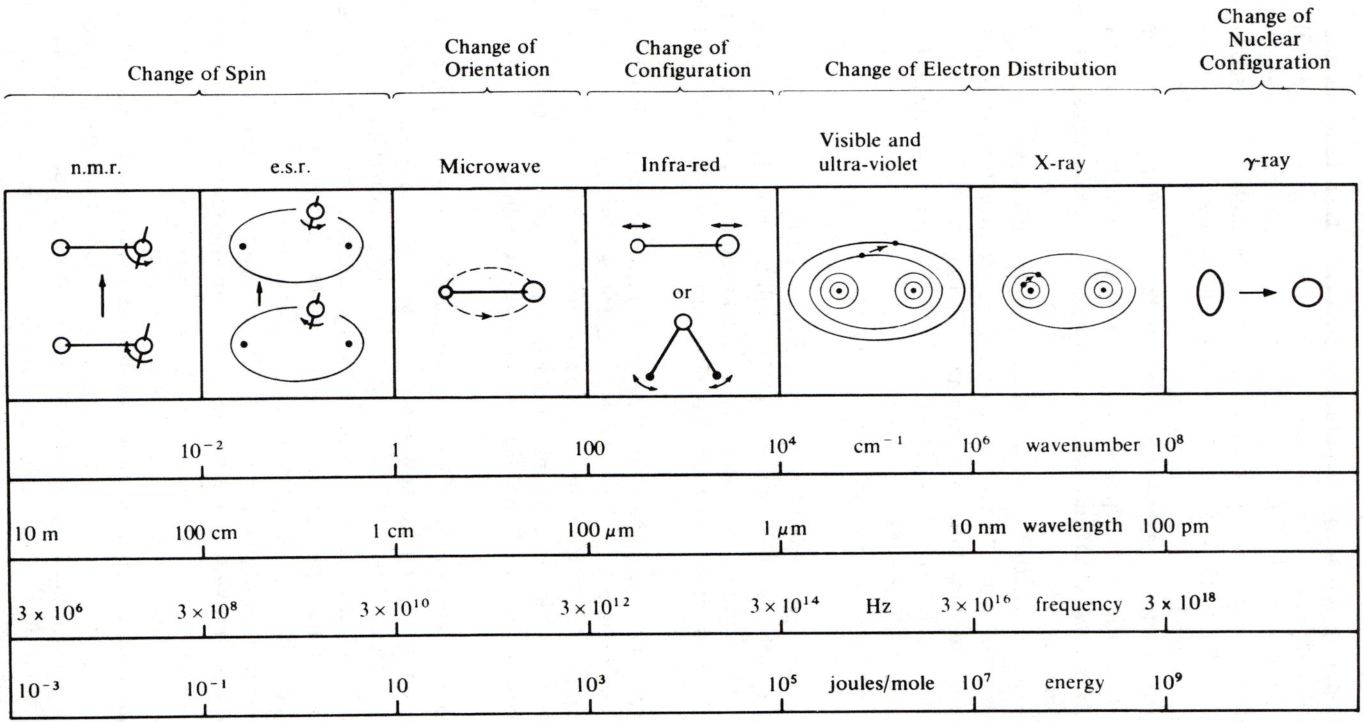

		Change of Spin		Change of Orientation	Change of Configuration	Change of Electron Distribution		Change of Nuclear Configuration
		n.m.r.	e.s.r.	Microwave	Infra-red	Visible and ultra-violet	X-ray	γ-ray

wavenumber cm^{-1}: 10^{-2} 1   100 10^4 10^6 10^8

wavelength: 10 m   100 cm   1 cm   100 μm   1 μm   10 nm   100 pm

frequency Hz: 3×10^6 3×10^8 3×10^{10} 3×10^{12} 3×10^{14} 3×10^{16} 3×10^{18}

energy joules/mole: 10^{-3} 10^{-1} 10 10^3 10^5 10^7 10^9

Figure 1.4 The regions of the electromagnetic spectrum.

2. Microwave region: 3×10^{10}–3×10^{12} Hz; 1 cm–100 μm wavelength. Rotational spectroscopy. Separations between the rotational levels of molecules are of the order of hundreds of joules per mole (Chapter 2).

3. Infra-red region: 3×10^{12}–3×10^{14} Hz; 100 μm–1 μm wavelength. Vibrational spectroscopy. One of the most valuable spectroscopic regions for the chemist. Separations between levels are some 10^4 joules/ mole (Chapter 3).

4. Visible and ultra-violet regions: 3×10^{14}–3×10^{16} Hz; 1 μm–10 nm wavelength. Electronic spectroscopy. The separations between the energies of valence electrons are some hundreds of kilojoules per mole (Chapters 5 and 6).

5. X-ray region: 3×10^{16}–3×10^{18} Hz; 10 nm–100 pm wavelength. Energy changes involving the inner electrons of an atom or a molecule, which may be of order ten thousand kilojoules (Chapter 5).

6. γ-ray region: 3×10^{18}–3×10^{20} Hz; 100 pm–1 pm wavelength. Energy changes involve the rearrangement of nuclear particles, having energies of 10^9–10^{11} joules per gram atom (Chapter 8).

One other type of spectroscopy, that discovered by Raman and bearing his name, is discussed in Chapter 4. This, it will be seen, yields information similar to that obtained in the microwave and infra-red regions, although the experimental method is such that observations are made in the visible region.

In order that there shall be some mechanism for interaction between the incident radiation and the nuclear, molecular, or electronic changes depicted in Fig. 1.4, there must be some electric or magnetic effect produced by the change which can be influenced by the electric or magnetic fields associated with the radiation. There are several possibilities:

1. The radiofrequency region. We may consider the nucleus and electron to be tiny charged particles, and it follows that their spin is associated with a tiny magnetic dipole. The reversal of this dipole consequent upon the spin reversal can interact with the magnetic field of electromagnetic radiation at the appropriate frequency. Consequently all such spin reversals produce an absorption or emission spectrum.

2. The visible and ultra-violet region. The excitation of a valence electron involves the moving of electronic charges in the molecule. The consequent change in the electric dipole gives rise to a spectrum by its interaction with the undulatory electric field of radiation.

3. The microwave region. A molecule such as hydrogen chloride, HCl, in which one atom (the hydrogen) carries a permanent net positive charge and the other a net negative charge, is said to have a permanent electric dipole moment. H_2 or Cl_2, on the other hand, in which there is no such charge separation, have a zero dipole. If we consider the rotation of HCl

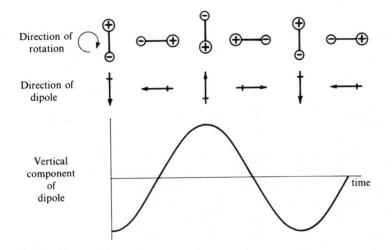

Figure 1.5 The rotation of a diatomic molecule, HCl, showing the fluctuation in the dipole moment measured in a particular direction.

(Fig. 1.5, where we notice that if only a pure rotation takes place, the centre of gravity of the molecule must not move), we see that the plus and minus charges change places periodically, and the component dipole moment in a given direction (say upwards in the plane of the paper) fluctuates regularly. This fluctuation is plotted in the lower half of Fig. 1.5, and it is seen to be exactly similar in form to the fluctuating electric field of radiation (cf. Fig. 1.2). Thus interaction can occur, energy can be absorbed or emitted, and the rotation gives rise to a spectrum. All molecules having a permanent moment are said to be 'microwave active'. If there is no dipole, as in H_2 or Cl_2, no interaction can take place and the molecule is 'microwave inactive'. This imposes a limitation on the applicability of microwave spectroscopy.

4. The infra-red region. Here it is a vibration, rather than a rotation, which must give rise to a dipole change. Consider the carbon dioxide molecule as an example, in which the three atoms are arranged linearly with a small net positive charge on the carbon and small negative charges on the oxygens:

$$\overset{\delta -}{O} \rule{1cm}{0.4pt} \overset{2\delta +}{C} \rule{1cm}{0.4pt} \overset{\delta -}{O}$$

During the mode of vibration known as the 'symmetric stretch', the molecule is alternately stretched and compressed, both C—O bonds changing simultaneously, as in Fig. 1.6. Plainly the dipole moment remains zero throughout the whole of this motion, and this particular vibration is thus 'infra-red inactive'.

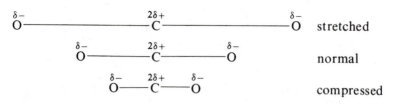

Figure 1.6 The symmetric stretching vibration of the carbon dioxide molecule.

However, there is another stretching vibration called the anti-symmetrical stretch, depicted in Fig. 1.7. Here one bond stretches while the other is compressed, and vice versa. As the figure shows, there is a periodic alteration in the dipole moment, and the vibration is thus 'infra-red active'. One further vibration is allowed to this molecule (see Chapter 3 for a more detailed discussion), known as the bending mode. This, as shown in Fig. 1.8, is also infra-red active. In both these motions the centre of gravity does not move.

Although dipole change requirements do impose some limitation on the application of infra-red spectroscopy, the appearance or non-appearance of certain vibration frequencies can give valuable information about the structure of a particular molecule (see Chapter 3).

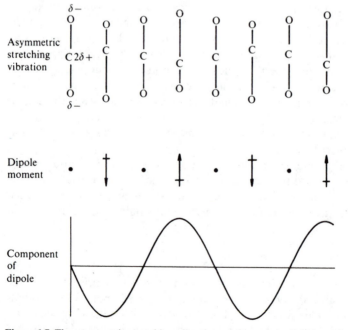

Figure 1.7 The asymmetric stretching vibration of the carbon dioxide molecule showing the fluctuation in the dipole moment.

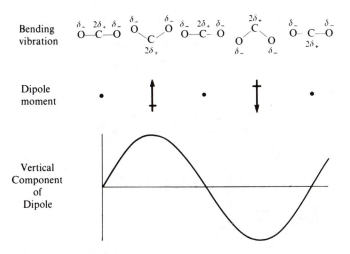

Figure 1.8 The bending motion of the carbon dioxide molecule and its associated dipole fluctuation.

5. There is a rather special requirement for a molecular motion to be 'Raman active'; this is that the electrical *polarizability* of the molecule must change during the motion. This will be discussed fully in Chapter 4.

1.4 REPRESENTATION OF SPECTRA

We show in Fig. 1.9 a highly schematic diagram of a spectrometer suitable for use in the visible and ultra-violet regions of the spectrum. A 'white' source is focused by lens 1 on to a narrow slit (arranged perpendicularly to the plane of the paper) and is then made into a parallel beam by lens 2. After passing through the sample it is separated into its constituent frequencies by a prism and is then focused on to a photographic plate by lens 3; the vertical image of the slit will thus appear on the plate. Rays have been drawn to show the points at which two frequencies, v_1 and v_2, are focused.

If the sample container is empty, the photographic plate, after development, should ideally show an even blackening over the whole range of frequencies covered (i.e., from A to B). The ideal situation is seldom realized, if only because the source does not usually radiate all frequencies with the same intensity, but in any case the blackening of the plate serves to indicate the relative intensities of the frequencies emitted by the source.

If we now imagine the sample space to be filled with a substance having only two possible energy levels, E_1 and E_2, the photographic plate, after development, will show a blackening at all points except at the frequency $v = (E_2 - E_1)/h$, since energy at this frequency will have been absorbed by

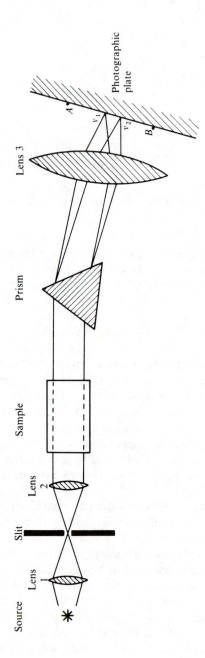

Figure 1.9 Schematic diagram of a spectrometer suitable for operation in the visible region.

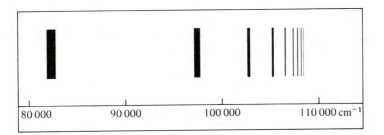

Figure 1.10 Schematic diagram of the absorption spectrum of atomic hydrogen recorded on a photographic plate.

the sample in raising each molecule from state 1 to state 2. Further if, as is almost always the case, there are many possible energy levels, E_1, E_2, ..., E_j, E_k, ... available to the sample, a series of absorption lines will appear on the photographic plate at frequencies given by $v = (E_j - E_k)/h$. A typical spectrum may then appear as in Fig. 1.10.

At this point it may be helpful to consider what happens to the energy absorbed in the sort of process described above. In the ultra-violet, visible, and infra-red regions it is an experimental fact that a given sample continues to show an absorption spectrum for as long as we care to irradiate it—in other words, a finite number of sample molecules appear to be capable of absorbing an infinite amount of energy. Plainly the molecules must be able to rid themselves of the absorbed energy.

A possible mechanism for this is by thermal collisions. An energized molecule collides with its neighbours and gradually loses its excess energy to them as kinetic energy—the sample as a whole becomes warm.

Another mechanism is that energy gained from radiation is lost as radiation once more. A molecule in the ground state absorbs energy at frequency v and its energy is raised an amount $\Delta E = hv$ above the ground state. It is thus in an excited, unstable condition, but by emitting radiation of frequency v again, it can revert to the ground state and is able to reabsorb from the radiation beam once more. In this case, it is often asked how an absorption spectrum can arise at all, since the absorbed energy is re-emitted by the sample. The answer is simply that the radiation is re-emitted in a random direction and the proportion of such radiation reaching the detector is minute—in fact re-emitted radiation has as much chance of reaching the source as the detector. The net effect, then, is an absorption from the directed beam and, when re-emission occurs, a scattering into the surroundings. The scattered radiation can, of course, be collected and observed as an emission spectrum which will be—with important reservations to be discussed in Chapter 4—the complement of the absorption spectrum. Under the right conditions much of the radiation emitted from a sample can be in a very coherent beam—so-called laser radiation. We discuss this in Sec. 1.10.

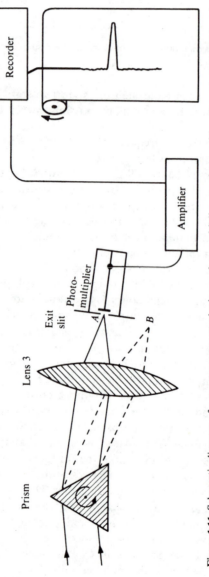

Figure 1.11 Schematic diagram of a spectrometer employing a photomultiplier or other sensitive element as detector and recording the spectrum graphically on chart paper.

14

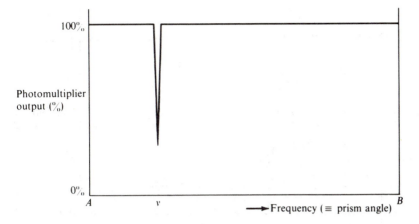

Figure 1.12 The idealized spectrum of a molecule undergoing a single transition.

In modern spectrometers the detector is rarely the simple photographic plate of Fig. 1.9. One of the most sensitive and useful devices in the visible and ultra-violet region is the photomultiplier tube, consisting of a light-sensitive surface which emits electrons when light falls upon it. The tiny electron current may be amplified and applied to an ammeter or pen recorder. The spectrometer would then appear somewhat as in Fig. 1.11, where the sensitive element of the photomultiplier is situated at the point A of Fig. 1.9. The physical width of the beam falling on the detector can be limited by the provision of an 'exit slit' just in front of the detector entrance.

The frequency of the light falling on the photomultiplier may be altered either by physically moving the latter from A to B or, more usually, by steady rotation of the prism. If, as before, we imagine the sample to contain a substance having just two energy levels, the photomultiplier output will, ideally, vary with the prism orientation as in Fig. 1.12. We say that the spectrum has been *scanned* between the frequencies represented by A and B, and such a picture is referred to, rather grandly, as a spectrum in the 'frequency domain', to indicate that it records the detector output against frequency. In Sec. 1.8 we shall discuss 'time domain' spectroscopy, where the detector output is recorded as a function of time.

Again, the ideal situation of Fig. 1.12 is seldom attained. Not only does the source emissivity vary with frequency, but often the sensitivity of the photomultiplier is also frequency-dependent. Thus the baseline—the 'sample-empty' condition—is never horizontal, although matters can usually be arranged so that it is approximately linear. Further, since it is impossible to make either of the slits infinitely narrow, a *range* of frequencies, rather than just a single frequency, falls on the photomultiplier at any given setting of the prism. This results in a broadening of the absorbance peak, and the final spectrum may appear rather as in Fig. 1.13. In this

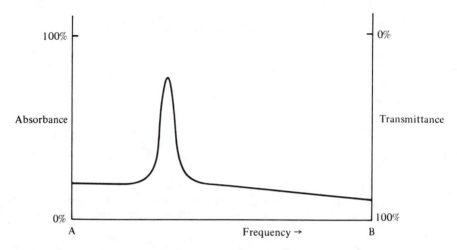

Figure 1.13 The usual appearance of the spectrum of a molecule undergoing a single transition (cf. Fig. 1.12); here the background is no longer constant and the absorption region is of finite width.

figure, too, we have plotted absorbance upwards from 0 to 100% and transmittance—its complement—downwards. This is the usual way in which such spectra are represented.

If there are again several energy levels available to the sample, it is very unlikely that there is the same probability of transition between the various levels. The question of transition probability will be discussed more fully in Sec. 1.7 but here we may note that differences in transition probability will mean that the absorbance (or transmittance) at each absorbing frequency will differ. This is shown by the varying intensities of the lines on the photographic plate of Fig. 1.10 and, more precisely, by the recorder trace of Fig. 1.14(*a*).

Figure 1.14(*a*) shows the sort of record which is produced by most modern spectrometers, whatever the region in which they operate. One other form of presentation is often adopted, however, particularly in the microwave and radiofrequency regions, and this is to record the *derivative* of the spectral trace instead of the trace itself. The derivative of a curve is simply its slope at a given point. In calculus notation, the derivative of the spectral trace is dA/dv, where A is the absorbance. The derivative record is thus a plot of the slope dA/dv against v; this is shown in Fig. 1.14(*b*) corresponding to the plot of Fig. 1.14(*a*).

Although at first sight more complex, the derivative trace has advantages over the direct record in some circumstances. Firstly, it indicates rather more precisely the centre of each absorbance peak: at the centre of a peak, the A curve is horizontal, hence dA/dv is zero, and the centres are marked by the intersection of the derivative curve with the axis. Further, for

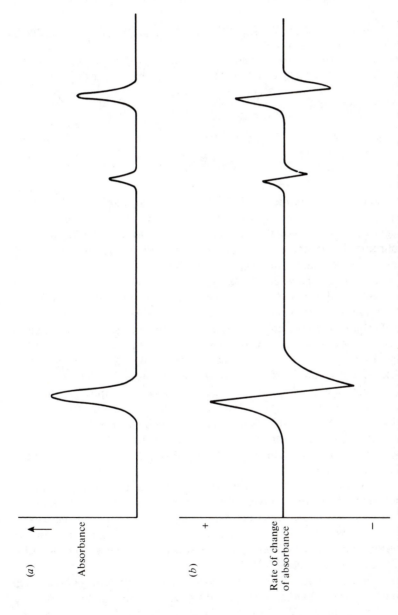

Figure 1.14 To illustrate the relation between absorption and derivative spectra: in (a) the absorption nuclear magnetic resonance spectrum of benzyl alcohol, $C_6H_5CH_2OH$, is shown, and in (b) the derivative (or dispersion) spectrum of the same molecule.

(a)

Absorbance

(b)

+

Rate of change of absorbance

−

17

instrumental reasons, it is often better to measure the relative intensities of absorbance peaks from the derivative curve than from the direct trace.

1.5 BASIC ELEMENTS OF PRACTICAL SPECTROSCOPY

Spectrometers used in various regions of the spectrum naturally differ widely from each other in construction. These differences will be discussed in more detail in the following chapters, but here it will probably be helpful to indicate the basic features which are common to all types of spectrometer. We may, for this purpose, consider absorption and emission spectrometers separately.

1. *Absorption instruments.* Figure 1.15(*a*) shows, in block diagram form, the components of an absorption spectrometer which might be used in the infra-red, visible, and ultra-violet regions. The radiation from a white source is directed by some guiding device (e.g., the lens of Fig. 1.9, or mirrors) on to the sample, from which it passes through an analyser (e.g., the prism of Fig. 1.9), which selects the frequency reaching the detector at any given time. The signal from the latter passes to a recorder which is synchronized with the analyser so as to produce a trace of the absorbance as the frequency varies.

   Placed, often, between the sample and the analyser is a *modulator*; this mechanical or electronic device interrupts the radiation beam a certain number of times per second, usually fixed somewhere between 10 and 1000 times, and its effect is to cause the detector to send an alternating current signal to the recorder, with a fixed frequency of 10–1000 Hz, rather than the direct current signal which would result from a steady, uninterrupted beam. This has two main advantages: (*a*) the amplifier in the recorder can be of a.c. type which is, in general, simpler to construct and more reliable in operation than a d.c. amplifier, and (*b*) the amplifier can be tuned to select only that frequency which the modulator imposes on the signal, thus ignoring all other signals. In this way stray radiation and other extraneous signals are removed from the spectral trace and a better, cleaner spectrum results.

   In the microwave and radiofrequency regions it is possible to construct monochromatic sources whose emission frequency can be varied over a range. In this case, as Fig. 1.15(*b*) shows, no analyser is necessary, the source being, in a sense, its own analyser. Now it is necessary for the recorder to be synchronized with the source-scanning device in order that a spectral trace be obtained.

2. *Emission instruments.* The layout now differs in that the sample, after excitation, is its own source, and it is necessary only to collect the emitted radiation, analyse, and record it in the usual way. Figure 1.16

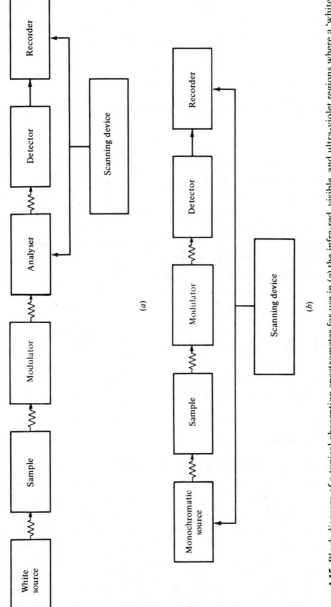

Figure 1.15 Block diagram of a typical absorption spectrometer for use in (*a*) the infra-red, visible, and ultra-violet regions where a 'white' source is available, and (*b*) the microwave and radiofrequency regions where the source can be tuned over a considerable range of frequencies.

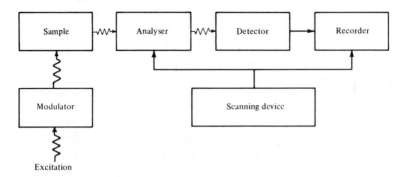

Figure 1.16 Block diagram of a typical emission spectrometer.

shows, schematically, a typical spectrometer. The excitation can be thermal or electrical, but often takes the form of electromagnetic radiation. In the latter case it is essential that the detector does not collect radiation directly from the exciting beam, and the two are placed at right angles as shown. A modulator placed between the source of excitation and the sample, together with a tuned detector-amplifier, ensures that the only emission recorded from the sample arises directly from excitation; any other spontaneous emission is ignored.

1.6 SIGNAL-TO-NOISE: RESOLVING POWER

Two other spectroscopic terms may be conveniently discussed at this point since they will recur in succeeding chapters.

Signal-to-Noise Ratio

Since almost all modern spectrometers use some form of electronic amplification to magnify the signal produced by the detector, every recorded spectrum has a background of random fluctuations caused by spurious electronic signals produced by the detector, or generated in the amplifying equipment. These fluctuations are usually referred to as 'noise'. In order that a real spectral peak should show itself as such and be sufficiently distinguished from the noise, it must have an intensity some three or four times that of the noise fluctuations (a signal-to-noise ratio of three or four). This requirement places a lower limit on the intensity of observable signals. In Sec. 1.9 we refer briefly to a computer-averaging technique by which it is possible to improve the effective signal-to-noise ratio.

Resolving Power

This is a somewhat imprecise concept which can, however, be defined rather arbitrarily and is often used as a measure of the performance of a spectrometer. We shall here consider it in general terms only.

No molecular absorption takes place at a single frequency only, but always over a spread of frequencies, usually very narrow but sometimes quite large (see Sec. 1.7); it is for this reason that we have up to now drawn spectra with broadened line shapes (cf. Fig. 1.14(a)).

Let us consider two such lines close together, as on the right of Fig. 1.17(a): the dotted curve represents the absorption due to each line separately, the full line their combined absorption. We shall first take the exit slit width to be larger than the separation between the lines. Scanning the

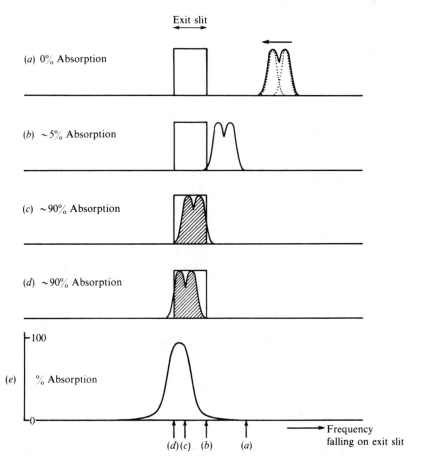

Figure 1.17 Illustrating the relation between slit width and resolving power: see text for discussion.

spectrum plainly involves moving the twin absorbance peaks steadily to the left so that they pass across the exit slit and into the detector; the situation at successive stages is shown in (b), (c), and (d) of Fig. 1.17, the shaded area showing the amount of absorbance which the detector would register. At (e) of this figure, the absorbance is plotted against frequency, together with the approximate positions of stages (a), (b), (c), and (d).

It is quite evident that the separation between the lines has disappeared under these conditions—the lines are not *resolved*. It is equally evident that the use of a much narrower slit would result in their resolution—the *resolving power* would be increased. In fact, provided the slit width is less than the separation between the lines, the detector output will show a minimum between them.

However, it must be remembered that a narrower slit allows less total energy from the beam to reach the detector and consequently the intrinsic signal strength will be less. There comes a point when decreasing the slit width results in such weak signals that they become indistinguishable from the background noise mentioned in the previous paragraph. Thus spectroscopy is a continual battle to find the minimum slit width consistent with acceptable signal-to-noise values. Improvements in resolving power may arise not only as a result of obtaining better dispersion of the radiation by the analyser (e.g., by the use of a diffraction grating rather than a prism for the ultra-violet and infra-red regions) but also by using a more sensitive detector.

1.7 THE WIDTH AND INTENSITY OF SPECTRAL TRANSITIONS

In the preceding sections we have seen that a spectral transition has the important property of *position*, measured in terms of its frequency, wavelength, or wavenumber; there are two other important properties, its *width* and its *intensity*, and we shall consider these briefly here.

1.7.1 The Width of Spectral Lines

Throughout this chapter we have drawn spectral absorptions and emissions not as infinitely sharp lines but as more or less broad peaks; we have seen that one reason for this is that the mechanical slits in spectrometers are not infinitely narrow and thus allow a *range* of frequencies, rather than a single frequency, to fall on the detector, hence blurring the pattern. While improvements in spectrometer design can improve the resolving power of an instrument, however, there is nonetheless a minimum width inherent in any atomic or molecular transition—the *natural line width*—beyond which no instrument, however superior, will show a sharpening. This width arises essentially because the energy levels of atomic and molecular systems are not precisely determined, but have a certain fuzziness or imprecision. Several factors contribute to this.

1. *Collision broadening.* Atoms or molecules in liquid and gaseous phases are in continual motion and collide frequently with each other. These collisions inevitably cause some deformation of the particles and hence perturb, to some extent, the energies of at least the outer electrons in each. This immediately gives a possible explanation for the width of visible and ultra-violet spectral lines, since these deal largely with transitions between outer electronic shells. Equally vibrational and rotational spectra are broadened since collisions interfere with these motions too. In general, molecular interactions are more severe in liquids than in gases, and gas-phase spectra usually exhibit sharper lines than those of the corresponding liquid.

    In the case of solids, the motions of the particles are more limited in extent and less random in direction, so that solid-phase spectra are often sharp but show evidence of interactions by the splitting of lines into two or more components.

2. *Doppler broadening.* Again in liquids and gases the motion of the particles causes their absorption and emission frequencies to show a Doppler shift; since the motion is random in a given sample, shifts to both high and low frequencies occur and hence the spectral line is broadened. In general, for liquids collision broadening is the most important factor, whereas for gases, where collision broadening is less pronounced, the Doppler effect often determines the natural line width.

3. *Heisenberg uncertainty principle.* Even in an isolated, stationary molecule or atom the energy levels are not infinitely sharp, due to the operation of a fundamental and very important principle, the Uncertainty Principle of Heisenberg. In effect this says that, if a system exists in an energy state for a limited time δt seconds, then the energy of that state will be uncertain (fuzzy) to an extent δE where

$$\delta E \times \delta t \approx h/2\pi \approx 10^{-34} \text{ J s} \qquad (1.10)$$

where h is again Planck's constant. Thus we see that the lowest energy state of a system *is* sharply defined since, left to itself, the system will remain in that state for an infinite time; thus $\delta t = \infty$, and $\delta E = 0$. But, for example, the lifetime of an excited electronic state is usually only about 10^{-8} s, which gives a value for δE of about $10^{-34}/10^{-8} = 10^{-26}$ J. A transition between this state and the ground state will thus have an energy uncertainty of δE, and a corresponding uncertainty in the associated radiation frequency of $\delta E/h$, which we can write as:

$$\delta v = \frac{\delta E}{h} \approx \frac{h}{2\pi h \delta t} \approx \frac{1}{2\pi \delta t} \qquad (1.11)$$

Thus for our example of an excited electronic state lifetime of 10^{-8} s, $\delta v \approx 10^8$ Hz. This apparently large uncertainty is, in fact, small compared with the usual radiation frequency of such transitions, 10^{14}–10^{16} Hz, and so the natural line width is said to be small; in fact, the apparent widths of electronic transitions are far more dependent on collision and Doppler broadening than on energy uncertainties.

On the other hand an excited electron *spin* state may exist for some 10^{-7} s which, from Eq. (1.11), leads to a frequency uncertainty of some 10^7 Hz for a transition. This, compared with the usual frequency of such transitions, 10^8–10^9 Hz, represents a very broad transition indeed, and here the Heisenberg uncertainty relation is by far the most important effect.

Further examples of the application of Heisenberg's principle will be given in later chapters.

1.7.2 The Intensity of Spectral Lines

When discussing spectral intensities there are three main factors to be considered: the likelihood of a system in one state changing to another state— the *transition probability*; the number of atoms or molecules initially in the state from which the transition occurs—the *population*; and the amount of material present giving rise to the spectrum—the *concentration* or *path length* of the sample.

1. *Transition probability.* The detailed calculation of absolute transition probabilities is basically a straightforward matter, but as it involves a knowledge of the precise quantum mechanical wave functions of the two states between which the transition occurs, it can seldom be done with accuracy and is, in any case, beyond the scope of this book. We shall generally content ourselves with qualitative statements about relative transition probabilities without attempting any detailed calculations.

    At a much lower level of sophistication, however, it is often possible to decide whether a particular transition is forbidden or allowed (i.e., whether the transition probability is zero or non-zero). This process is essentially the deduction of *selection rules*, which allow us to decide between which levels transitions will give rise to spectral lines, and it can often be carried out through pictorial arguments very like those we have already used in discussing the activity or otherwise of processes in Sec. 1.3.

2. *Population of states.* If we have two levels from which transitions to a third are equally probable, then obviously the most intense spectral line will arise from the level which initially has the greater population. There

is a simple statistical rule governing the population of a set of energy levels.

For example, if we have a total of N molecules distributed between two different energy states, a lower and an upper with energies E_{lower} and E_{upper}, respectively, we would intuitively expect most of the molecules to occupy the lower state. Proper statistical analysis bears this out and shows that, *at equilibrium*

$$\frac{N_{upper}}{N_{lower}} = \exp(-\Delta E/kT) \tag{1.12}$$

where $\Delta E = E_{upper} - E_{lower}$, T is the temperature in K, and k is a universal constant. The expression is known as the Boltzmann distribution, after its originator, and k, which has a value of $1\cdot38 \times 10^{-23}$ J K^{-1}, as Boltzmann's constant. Examples showing the use of this very important expression will recur throughout the remaining chapters.

3. *Path length of sample.* Clearly if a sample is absorbing energy from a beam of radiation, the more sample the beam traverses the more energy will be absorbed from it. We might expect that twice as much sample would give twice the absorption, but a very simple argument shows that this is not so. Consider two identical samples of the same material, S_1 and S_2, and assume that S_1 or S_2 alone absorb 50 per cent of the energy falling on them, allowing the remaining 50 per cent to pass through. If we pass a beam of initial intensity I_0 through S_1, 50 per cent of I_0 will be absorbed and the intensity of the beam leaving S_1 will be $\frac{1}{2}I_0$; if we then pass this beam through S_2 a further 50 per cent will be absorbed, and $\frac{1}{2} \times \frac{1}{2}I_0 = \frac{1}{4}I_0$ will leave S_2. Thus two 50 per cent absorptions in succession do not add up to 100 per cent but only to 75 per cent absorption. An exactly similar relationship exists between the *concentration* of a sample and the amount of energy absorption—a doubling of the concentration produces something less than a doubling of the absorption.

These relationships are best expressed in terms of the Beer–Lambert law, which is:

$$\frac{I}{I_0} = \exp(-\varepsilon cl) \tag{1.13}$$

where I_0 is the intensity of radiation falling on the sample, and I that part transmitted, c and l are the sample concentration and length, and ε is the extinction coefficient or absorption coefficient, which is a constant for a given type of transition (e.g., electronic, vibrational, etc.) occurring within a particular sample. Clearly ε is closely connected with the transition probability discussed above, a large probability being associated with a large ε, and vice versa.

1.8 FOURIER TRANSFORM SPECTROSCOPY

One of the major disadvantages of the conventional method of producing a spectrum, such as that of Fig. 1.13, is its inherent slowness. Essentially each point of the spectrum has to be recorded separately—the spectrometer is set to start reading at one end, the frequency is swept smoothly across the whole span of the spectrum, and the detector signal is monitored and recorded. The inefficiency of such a method is clear when one considers taking a spectrum with only one or two 'lines' in it—we have to sweep from one end to the other in order to find the lines, but most of the time is spent recording nothing but background noise. Until recently it was only in the visible and ultra-violet regions that the whole of a spectrum could be recorded simultaneously—on a photographic plate—but a new development, Fourier transform spectroscopy, is providing simultaneous and almost instantaneous recording of the whole spectrum in the magnetic resonance, microwave, and infra-red regions. In this section we shall briefly discuss the basic ideas of the technique, leaving to later chapters more detailed consideration of its methods and applications.

Although equally applicable to both emission and absorption spectroscopy, it is probably easier to visualize Fourier transform (FT) spectroscopy in terms of emission, so let us take the spectrum of Fig. 1.13 to represent the emission of radiation by a sample. Further we shall, for the moment, ignore the line-broadening discussed in the previous section, and think of this radiation as a pure sine wave at some quite precise frequency, v. If a detector capable of responding sufficiently rapidly receives this emitted radiation, its output will be an oscillating signal, again of frequency v. Note that we here consider the detector output as a function of *time* ('time domain' spectroscopy) rather than as the function of frequency ('frequency domain') plotted in Fig. 1.13.

Now consider a sample-emitting radiation at two different frequencies; a detector receiving the total radiation will 'see' the *sum* of the two sine waves. We illustrate, diagrammatically, the two separate but superimposed waves in Fig. 1.18(*a*), and their sum in (*b*), and we note that the detector output shows an oscillation due to the frequency of the two waves, but also a slow increase and decrease in overall amplitude. This latter is often called the 'beat' frequency, by analogy with a similar phenomenon for musical tones, and it arises because the two component waves are sometimes 'in step' (as at points *A* and *B* in Fig. 1.18 where their amplitudes reinforce each other), and sometimes out of step (as at *C* where they cancel). In fact the beat frequency is always equal to the *difference* in frequency of the component waves—if they differ in frequency by 10 cycles per second (Hz) then the beat oscillation is also at 10 Hz. This is illustrated by Fig. 1.18(*c*) and (*d*) where we show sine waves with half the frequency separation of those in Fig. 1.18(*a*) and (*b*), and we see that the beat frequency has also halved.

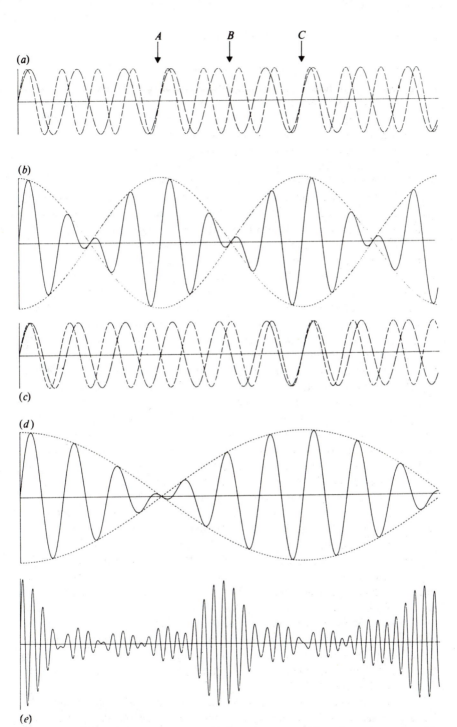

Figure 1.18 Adding of sine waves. (*a*) and (*c*) show the superposition of two sine waves with slightly different frequency, with (*b*) and (*d*) their sums; (*e*) shows the summation of five sine waves with different frequencies.

Mathematically it is simple, but tedious, to resolve a combined wave such as Fig. 1.18(b) into its components. Essentially each component wave has its own frequency and maximum amplitude, so two components require us to evaluate four unknowns from the composite curve. In principle, observation of the time domain signal at four points and solution of four simultaneous equations will yield the information we seek.

Adding more than two sine waves complicates the resultant combined wave and makes the resolution into components even more tedious, but does not change the principle. Figure 1.18(e) shows the result of superimposing five sine waves each of slightly different frequency. It would need 10 measured points and the solution of 10 simultaneous equations to determine the frequency and relative amplitude of each component. Fortunately there is a simple and quite general way to resolve a complex wave into its frequency components; this is the mathematical process known as the Fourier transform, named after the French mathematician Jean Baptiste Fourier who developed the method in the early 1800s. Even more fortunately we do not need to know how the process works: it suffices to say that it is essentially a matter of integration of the complex waveform, and that it may be carried out very conveniently nowadays by computer.

As an example of its operation let us consider the complex waveform of Fig. 1.18(b) and imagine that a suitable detector is responding to this waveform. A computer receiving the detector output might typically be set to sample it once every millisecond and to store, say, 2000 samplings in separate memory locations; it would thus need to collect the signal for just two seconds. The computer would then apply the Fourier transform process to the stored data, taking a further second or so, and the component sine wave frequencies and intensities would be displayed. Conventionally the display would not take the form of Fig. 1.18(a), where the actual periodic variation of the waves is shown, but would simply be the *spectrum* of the waves, i.e., two very sharp peaks of equal height plotted on a suitable frequency scale to show where the two frequencies occur. This is shown in Fig. 1.19, where the complex wave in Fig. 1.19(a) (taken from Fig. 1.18(b)) is seen to give rise to the spectrum of Fig. 1.19(b). Essentially the Fourier transform has converted the *time* domain plot of Fig. 1.19(a) into the *frequency* domain spectrum of Fig. 1.19(b). The process described above would have taken, perhaps, four or five seconds only. Essentially the detector collects *all* the spectral information simultaneously, and the computer 'decodes' that information into the conventional spectrum. It is in this way that the FT method speeds the collection of spectral data, typically, by factors of 10 to 1000.

There are one or two further points to consider before we leave this basic discussion of FT spectroscopy. Firstly we must recall that real samples do not emit radiation at precise frequencies; as we saw in the previous section, each emission is more or less broadened by various processes, and

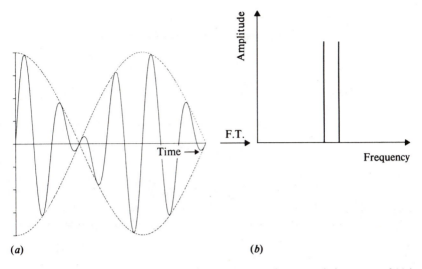

(a)   *(b)*

Figure 1.19 The use of the Fourier transform to convert the summed sine waves of (*a*) into their frequency spectrum, (*b*).

so each 'line' is really a small package of slightly different frequencies. We show a typical package in Fig. 1.20(*a*). In Fig. 1.20(*b*) we see that the package can be considered as arising from a large number of sample molecules radiating at exactly v_{max}, the frequency maximum of the package, with a smaller number of molecules radiating at frequencies away from that maximum, the number decreasing as the separation increases. Now, if we want to discover the total signal emitted by such a package we could, if we had the time, plot out a sine wave for each frequency using an intensity proportional to the number of molecules radiating at that frequency, and then add all the sine waves together. We are fortunately spared this task because it turns out that the Fourier transform is a reciprocal process; just as FT converts a time domain signal to a frequency domain spectrum, so it will carry out the reverse conversion. Thus if we supply the frequency curve of Fig. 1.20(*a*) to a computer and carry out the FT, the resultant display will be exactly the same as adding the component sine waves. The result is shown in Fig. 1.20(*c*).

We see that a detector receiving the total radiation from a single broad-line emission will show an oscillating signal whose overall amplitude decays smoothly to zero. The oscillation is the beat pattern set up by all the superimposed, but slightly different, sine waves emitted by the samples; and the signal decays because, if we imagine all the waves in the package to be 'in step' initially, after some time has elapsed the many different frequencies concerned will be very much out of step, and on average half will have their amplitudes in the positive sense and half in the negative, thus giving a

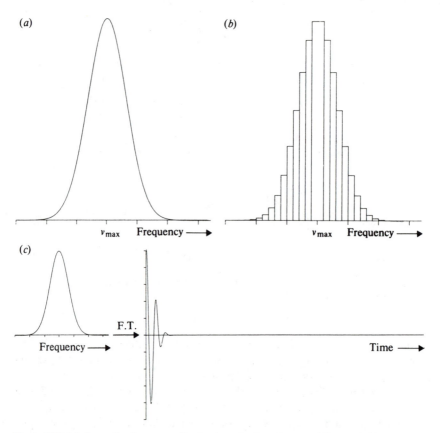

Figure 1.20 (*a*) shows the frequency distribution of a broad spectral line and (*b*) an approximate histogram of the frequencies; (*c*) is the Fourier transform of (*a*).

resultant sum of zero. Another way to think of this is to remember that two waves setting out in step with an infinitesimally small difference in frequency will take an infinite time to get back in step again, i.e., they will never do so. The distribution of Fig. 1.20(*a*) has many infinitely close frequencies within it and so, after a few cycles, none of the individual waves ever get back into step again. If the band had been *infinitely* broad, i.e., containing an infinite number of infinitesimally close neighbours, none would have ever been in step after the first instant, and the FT of such a 'white' source is a single decaying signal with no beats. We shall return to this in a moment.

The corollary of these arguments is that the *rate of decay* of the overall signal is dependent on the *width* of the original spectral peak. This is illustrated in Figs. 1.21(*a*) and (*b*) where (*a*) shows the FT of a relatively narrow signal and (*b*) that of a broader signal at the same central frequency.

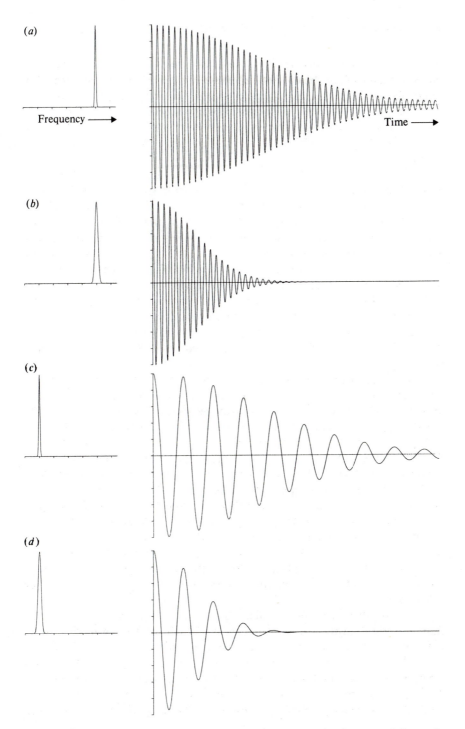

(a)

Frequency ⟶   Time ⟶

(b)

(c)

(d)

Figure 1.21 Showing the variation in the time domain signal from a spectral line as its position and breadth vary.

The beat frequency is identical, but the decay is more rapid in the latter. On the other hand, Figs. 1.21(c) and (d) show the FT of two more lines with peaks the same *width* as in Fig. 1.21(a) and (b) respectively, but at a different central *frequency*. We see that increasing the frequency of the line gives rise to an increased beat frequency. In general it is clear that the position and width of a frequency package can be recovered from the time domain signal by FT.

Next we must consider the situation when a sample emits radiation at more than one frequency, i.e., has more than one spectral peak. Not surprisingly the overall result is again the summation of the individual peaks, and beat patterns of varying complexity are built up. Thus Fig. 1.22(a) shows the time domain signal detected from two separate spectral peaks, choosing the frequencies used in Fig. 1.21(a) and (c). Figure 1.22(b) shows the effect of moving the two peaks close together. The beat pattern becomes more pronounced if three spectral lines are involved (Fig. 1.22(c)), and rather complex when several randomly spaced lines of different intensities are emitted, as in Fig. 1.22(d). As before, however, we should remember that brief observation of the signals whose patterns are shown on the right of Fig. 1.22 can be rapidly Fourier-transformed into their resultant spectra shown on the left.

Finally, although we stated initially that the FT process is most easily visualized in terms of emission of radiation, the technique is just as readily applied to absorption. We have already seen that a 'white' source would show a single decay signal with no beats; an approximation to this is given in Fig. 1.23(a), where a very broad emission line (which can be considered as a white source covering a limited region of the spectrum) and its Fourier transform are shown. We can now imagine an absorbing sample making a 'hole' in this radiation, as approximated by the left-hand side of Fig. 1.23(b), with its resulting FT shown on the right. Although we may find it difficult to imagine Fourier-transforming (or even just adding up) the *absence* of radiation at a particular frequency, in practice a detector will collect a perfectly sensible signal which can be stored by a computer, transformed, and displayed as the normal adsorption spectrum. The experimental technique for absorption FT spectroscopy in the infra-red region is described briefly in Sec. 3.8.

We see, then, that the FT method allows us to record spectra much more rapidly than the conventional sweep technique. This in itself is valuable; spectrometers are costly instruments, and the more work we can get from them in a given time, the more justified is the initial investment. But rapid data collection brings other benefits, for example, in being able to record the spectra of transient species such as unstable molecules or intermediates in a chemical reaction. Since the technique essentially reduces the

Figure 1.22 Time domain signals from several spectral lines.

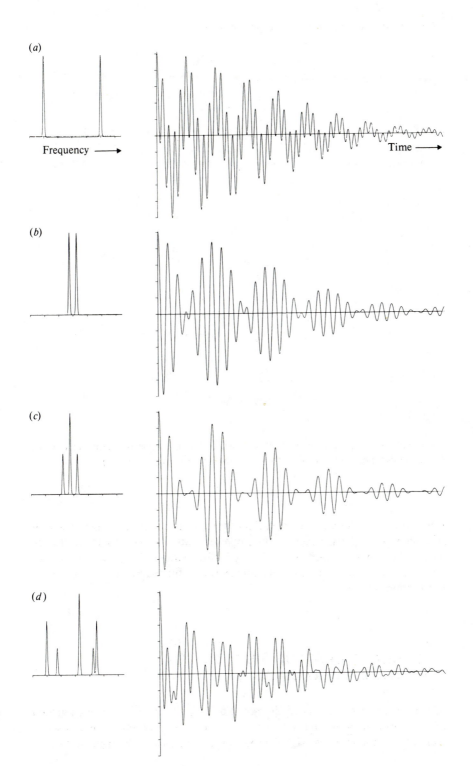

Frequency ⟶

Time ⟶

(b)

(c)

(d)

33

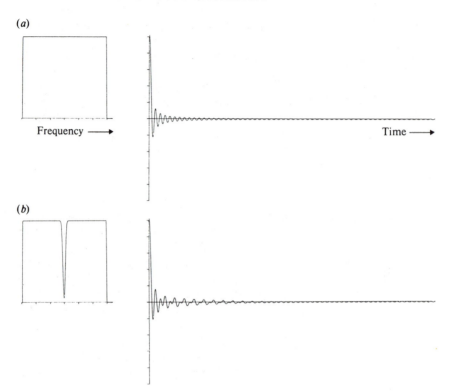

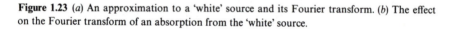

Figure 1.23 (*a*) An approximation to a 'white' source and its Fourier transform. (*b*) The effect on the Fourier transform of an absorption from the 'white' source.

time spent obtaining a spectrum from minutes to seconds or even fractions of a second, it vastly increases the range of materials which can be studied. There are other advantages in using FT instruments, but we shall leave discussion of them until the relevant chapters on magnetic resonance and infra-red spectroscopy.

1.9 ENHANCEMENT OF SPECTRA: COMPUTER AVERAGING

We have already mentioned, in Sec. 1.6, that the problem of background noise imposes a limitation on the sensitivity of any spectroscopic technique—unless a real signal peak stands out clearly from noise fluctuations it is impossible to be sure that it *is* a signal. A signal-to-noise (S/N) ratio of 3 or 4 is usually reckoned necessary for unambiguous recognition of a signal. There are several ways in which S/N can be improved for a given sample, but all require the expenditure of time. Thus it is possible electronically to damp out oscillations of the recorder pen so that it is less susceptible to high-frequency noise. The baseline of the spectrum will then be smoother, but, because the pen responds more slowly to any change (including changes in signal), one must sweep more slowly across the spectrum. Nor is FT spectroscopy immune from noise—detector and amplifier noises occur during the collection of data, and are transformed into spurious frequencies in the spectrum.

The advent of cheaper and more powerful computers offers another method of signal enhancement, known as 'computer averaging of transients' or the CAT technique, which involves recording the spectrum stepwise into a computer. Of course, this is already done if FT is intended, but it is just as easy to sample a frequency domain spectrum at, say, 2000 closely spaced points, and to store the intensity at each point in 2000 separate computer memory locations. This process may then be repeated as many times as we wish, but each time *adding* the new data into that already existing. Although in any one scan a weak signal may not be visible above the noise level, after n summed scans the signal will be n times larger in the store, whereas the noise, being random, will sometimes contribute to the store in a positive sense and sometimes negatively, so it will accumulate less rapidly. In fact it may be shown that n scans increase the noise level in the store by $n^{1/2}$, so the net gain in S/N is $n/n^{1/2} = n^{1/2}$.

If a single scan takes several minutes, as is usually the case in conventional frequency-sweep spectroscopy, the necessity to store 100 scans in order to give an improvement in S/N by a factor of 10 is rather costly in instrument time, so CAT techniques are not often applied to such measurements. However, the combination of CAT with FT is very powerful indeed. Here one time-domain scan can be completed in a second or two, and 100 scans will only occupy a couple of minutes; thus a tenfold gain in S/N can be achieved in a total time often less than that required for a non-enhanced spectrum by ordinary sweep methods. So useful is the combination of CAT with FT that virtually all FT spectrometers are routinely equipped with CAT facilities.

Other benefits follow from the addition of a computer to a spectrometer. The spectrum of a solvent or other background can be stored in the computer and subtracted from the observed spectrum in order to isolate the spectrum of the substance, or peak intensities can be automatically measured and converted to sample concentrations. Even the operation of the spectrometer itself can usefully be entrusted to the computer—samples can be changed automatically, and the optimum operating conditions can be determined and set for each new sample.

1.10 STIMULATED EMISSION: LASERS

We have already mentioned that, once radiation has been absorbed by a sample, the sample can lose its excess energy either by thermal collisions or by re-emission of radiation. In this section we shall consider the latter process in more detail, because it leads to the very important topic of laser radiation.

Radiation may be emitted by an excited molecule or atom either *spontaneously* or as the result of some stimulus acting on the molecule, called *stimulated emission*. Which of these two processes is most likely to occur in any given case depends on the energy jump involved, i.e., on the frequency of the radiation being emitted. For high-frequency transitions (infra-red, visible, and ultra-violet upwards) spontaneous emission is by far the most likely; conversely, for low-frequency changes (microwave and magnetic resonance) spontaneous emission is unlikely and, if the right conditions obtain, stimulated emission will occur.

Stimulated emission is a resonance phenomenon—an excited state drops to the ground state (emitting radiation of frequency $v = \Delta E/h$, where ΔE is the energy gap), only when a photon (i.e., radiation) of the same frequency v interacts with the system. We illustrate the situation in Fig. 1.24. On the left, in both (a) and (b), we show the excitation of a molecule by absorption of radiation of frequency $v_{ex.}$. At the right in (a) we show spon-

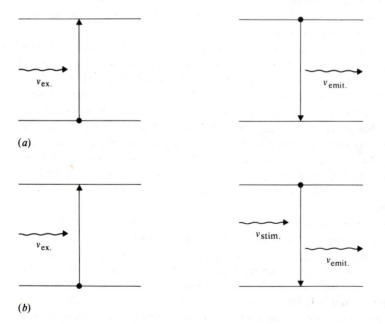

(a)

(b)

Figure 1.24 Showing (a) spontaneous and (b) stimulated emission from an excited energy state.

taneous emission, when radiation $v_{\text{emit.}}$ is spontaneously given out, and, in (b) stimulated emission where a photon of frequency $v_{\text{stim.}}$ interacts with the excited state and causes radiation of frequency $v_{\text{emit.}}$ to be released. Note particularly that, although we have given different subscripts to $v_{\text{ex.}}$, $v_{\text{emit.}}$, and $v_{\text{stim.}}$ in order to indicate their origins, they all represent *exactly the same frequency*, the frequency $\Delta E/h$.

Radiation emitted under stimulation of this sort has three very important qualities. Firstly it is of a very *precisely defined* frequency: the excited state does not spontaneously decay, so it is inherently long-lived, which implies (see the discussion of Heisenberg uncertainty in Sec. 1.7.1) a narrow energy level. Secondly the emitted radiation is *in phase* with the stimulating radiation: the excited state is stimulated to emit by interaction with the oscillating electromagnetic field of $v_{\text{stim.}}$, so it is not surprising that the maximum amplitude of the emitted wave coincides with that of $v_{\text{stim.}}$. And, since the waves are exactly the same frequency, they remain in phase as they leave the sample. Finally, the stimulating and emitted radiation are *coherent*, which means that they travel in precisely the same direction. In contrast *spontaneous* emission can occur at any time (so each emitted photon is not necessarily in phase with any other), in any direction, and within a more or less broad range of frequencies.

Of course the stimulating radiation of Fig. 1.24 is still present in the system after emission has occurred—it is in no way absorbed—so it can go on to interact with another excited molecule to induce more emission. Equally the emitted radiation has the right frequency to stimulate emission from yet another excited molecule. Clearly, all the time a supply of excited molecules exists, this process is likely to cascade and a great deal of radiation may be emitted coherently. This *amplification* of the original stimulating photon is reflected in the name of the process—*l*ight *a*mplification by *s*timulated *e*mission of *r*adiation, or *laser*.

In fact, as we have said, light (or, more properly, visible radiation) is far more likely to be emitted *spontaneously*, and so not to have the coherent properties of laser radiation. It was in the microwave region that the first successful amplification by stimulated emission was performed (and the process was therefore christened *maser*, standing for *m*icrowave *a*mplification by *s*timulated *e*mission of *r*adiation). For the process to be possible in higher-frequency regions it is necessary to find systems with long-lived excited states so that stimulated, rather than spontaneous, emission may predominate, and this may only be achieved if more than two energy levels are involved.

Consider the three energy levels of Fig. 1.25(a). Excitation from the ground state, level 1, to the normal excited state, level 2, can occur by absorption of radiation, as usual. Provided that, as well as emitting spontaneously, level 2 can transfer some molecules into a metastable state (level 3) which cannot easily revert spontaneously to the ground state, then the

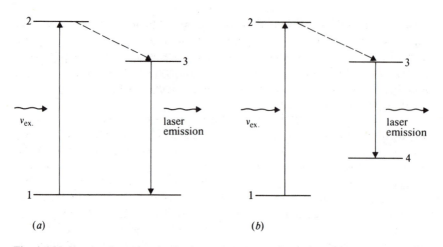

Figure 1.25 Showing the energy levels of (*a*) a three-level and (*b*) a four-level system operating as a laser.

population of level 3 builds up, and laser action becomes feasible. The ruby laser is an example of this type of three-level system. Ruby is basically aluminium oxide containing a trace (about 0·05 per cent) of chromium ions, which gives it its characteristic colour. A discharge tube wound round a rod of ruby is flashed very briefly to raise the chromium ions into an excited electronic state; they drop rapidly by thermal, non-radiative processes, into a metastable state some $14\,000$ cm^{-1} above the ground state, and they then revert to the ground state, by laser action, emitting radiation at about 690 nm wavelength. The decay from excited to metastable state releases quite large amounts of heat, so the ruby must be allowed to cool before another excitation cycle is commenced; it is thus operated as a pulsed laser.

In some cases the laser emission arises by reversion of level 3 to a lower state other than the original level 1, as shown in the four-level system of Fig. 1.25(*b*). This situation occurs particularly when, as is quite possible, levels 3 and 4 belong to an entirely *different* molecular species from levels 1 and 2. For example in the helium–neon laser it is the helium atoms which are initially excited (level 1 to 2), and which then transfer their excitation energy to neon atoms by collisions; this can happen only because neon has an excited state with almost exactly the same energy as the excited state of helium, so a resonance transfer of energy is possible. It also happens that the excited state of neon does not readily undergo an ordinary spectroscopic transition back to its ground state, so the conditions are ideal for laser action. Provided the exciting radiation for helium is maintained, so replenishing the population of excited helium atoms, this type of laser can operate continuously. We shall discuss the precise electronic energy levels involved in more detail in Chapter 5.

The extreme coherence of laser radiation makes it ideal in applications like communications, distance measurement, etc., but from the spectroscopic point of view its very narrow frequency spread makes it directly useful in only one area. This is Raman spectroscopy where, as we shall see in Chapter 4, the requirement is for an intense monochromatic source. For this the laser—almost *any* laser—is ideal.

For virtually all other spectroscopic measurements, however, either a wide-band or a tunable source is desirable. Lasers cannot be wide-band (although the CO_2 laser, to be described in Sec. 3.8.4, comes close to this), but recent developments have led to their becoming tunable. In order to change the emission frequency of a laser system it is necessary to be able to modify the energy levels between which transitions take place. Solids emitting laser radiation can be subjected to varying temperatures or pressures in order smoothly to change the relevant energies, but the extent of such changes is relatively small. More usefully, lasers made from coloured organic substances in solution—the so-called dye lasers—are widely tunable. In these the active material is usually a rare-earth ion held in the centre of an organic 'ligand' which complexes firmly to the ion. The tuning is brought about by changes in temperature, solvent, or concentration, and it is now possible to produce laser emission anywhere from the near infra-red to the ultra-violet. The possibilities for using such intense, sharply defined but variable-frequency sources routinely in spectroscopy are tremendous, and this technique will certainly become increasingly important in the near future.

PROBLEMS

(Useful constants: $N = 6 \cdot 023 \times 10^{23}$ mol^{-1}; $k = 1 \cdot 381 \times 10^{-23}$ J K^{-1}; $h = 6 \cdot 626 \times 10^{-34}$ J s; $c = 2 \cdot 998 \times 10^8$ m s^{-1}.)

1.1 The wavelength of the radiation absorbed during a particular spectroscopic transition is observed to be 10 μm. Express this in frequency (Hz) and in wavenumber (cm^{-1}), and calculate the energy change during the transition in both joules per molecule and joules per mole. If the energy change were twice as large, what would be the wavelength of the corresponding radiation?

1.2 Which of the following molecules would show (a) a microwave (rotational) spectrum, (b) an infra-red (vibrational) spectrum: Br_2, HBr, CS_2?

1.3 A particular molecule is known to undergo spectroscopic transitions between the ground state and two excited states, (a) and (b), its lifetime in (a) being about 10 s, and in (b) about 0·1 s. Calculate the approximate uncertainty in the excited state energy levels and the widths of the associated spectral 'lines' in hertz.

1.4 A certain transition involves an energy change of $4 \cdot 005 \times 10^{-22}$ J molecule^{-1}. If there are 1000 molecules in the ground state, what is the approximate equilibrium population of the excited state at temperatures of (a) 29 K, (b) 145 K, (c) 290 K and (d) 2900 K? What would your answer have been if the energy change were 10 times greater?

TWO

MICROWAVE SPECTROSCOPY

2.1 THE ROTATION OF MOLECULES

We saw in the previous chapter that spectroscopy in the microwave region is concerned with the study of rotating molecules. The rotation of a three-dimensional body may be quite complex and it is convenient to resolve it into rotational components about three mutually perpendicular directions through the centre of gravity—the principal axes of rotation. Thus a body has three principal *moments of inertia*, one about each axis, usually designated I_A, I_B, and I_C.

Molecules may be classified into groups according to the relative values of their three principal moments of inertia—which, it will be seen, is tantamount to classifying them according to their shapes. We shall describe this classification here before discussing the details of the rotational spectra arising from each group.

1. *Linear molecules.* These, as the name implies, are molecules in which all the atoms are arranged in a straight line, such as hydrogen chloride HCl, or carbon oxysulphide OCS, illustrated below. The three directions of rotation may be taken as (*a*) about the bond axis, (*b*) end-over-end

rotation in the plane of the paper, and (*c*) end-over-end rotation at right angles to the plane. It is self-evident that the moments of (*b*) and (*c*) are the same (i.e., $I_B = I_C$) while that of (*a*) is very small. As an approximation we may say that $I_A = 0$, although it should be noted that this *is* only an approximation (see p. 46).

Thus for linear molecules we have:

$$I_B = I_C \qquad I_A = 0 \tag{2.1}$$

2. *Symmetric tops.* Consider a molecule such as methyl fluoride, where the three hydrogen atoms are bonded tetrahedrally to the carbon, as shown below. As in the case of linear molecules, the end-over-end rotation in,

and out of, the plane of the paper are still identical and we have $I_B = I_C$. The moment of inertia about the C—F bond axis (chosen as the main rotational axis since the centre of gravity lies along it) is now not negligible, however, because it involves the rotation of three comparatively massive hydrogen atoms off this axis. Such a molecule spinning about this axis can be imagined as a top, and hence the name of the class. We have then:

$$\text{Symmetric tops:} \quad I_B = I_C \neq I_A \qquad I_A \neq 0 \tag{2.2}$$

There are two subdivisions of this class which we may mention: if, as in methyl fluoride above, $I_B = I_C > I_A$, then the molecule is called a *prolate* symmetric top; whereas if $I_B = I_C < I_A$, it is referred to as *oblate*. An example of the latter type is boron trichloride, which, as shown, is planar and symmetrical. In this case $I_A = 2I_B = 2I_C$

3. *Spherical tops.* When a molecule has all three moments of inertia identical, it is called a spherical top. A simple example is the tetrahedral molecule methane CH_4. We have then:

$$\text{Spherical tops:} \quad I_A = I_B = I_C \tag{2.3}$$

In fact these molecules are only of academic interest in this chapter. Since they can have no dipole moment owing to their symmetry, rotation alone can produce no dipole change and hence no rotational spectrum is observable.

4. *Asymmetric tops.* These molecules, to which the majority of substances belong, have all three moments of inertia different:

$$I_A \neq I_B \neq I_C \tag{2.4}$$

Simple examples are water H_2O, and vinyl chloride $CH_2{=}CHCl$.

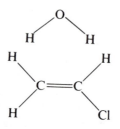

Perhaps it should be pointed out that one can (and often does) describe the classification of molecules into the four rotational classes in far more rigorous terms than have been used above (see, for example, Herzberg, *Molecular Spectra and Molecular Structure*, vol. II). However, for the purposes of this book the above description is adequate.

2.2 ROTATIONAL SPECTRA

We have seen that rotational energy, along with all other forms of molecular energy, is quantized: this means that a molecule cannot have any arbitrary amount of rotational energy (i.e., any arbitrary value of angular momentum) but its energy is limited to certain definite values depending on the shape and size of the molecule concerned. The permitted energy values—the so-called rotational energy *levels*—may in principle be calculated for any molecule by solving the Schrödinger equation for the system represented by that molecule. For simple molecules the mathematics involved is straightforward but tedious, while for complicated systems it is probably impossible without gross approximations. We shall not concern ourselves unduly with this, however, being content merely to accept the results of existing solutions and to point out where reasonable approximations may lead.

We shall consider each class of rotating molecule in turn, discussing the linear molecule in most detail, because much of its treatment can be directly extended to symmetrical and unsymmetrical molecules.

2.3 DIATOMIC MOLECULES

2.3.1 The Rigid Diatomic Molecule

We start with this, the simplest of all linear molecules, shown in Fig. 2.1. Masses m_1 and m_2 are joined by a rigid bar (the bond) whose length is

$$r_0 = r_1 + r_2 \tag{2.5}$$

The molecule rotates end-over-end about a point C, the centre of gravity: this is defined by the moment, or balancing, equation:

$$m_1 r_1 = m_2 r_2 \tag{2.6}$$

The moment of inertia about C is defined by:

$$\begin{aligned}
I &= m_1 r_1^2 + m_2 r_2^2 \\
&= m_2 r_2 r_1 + m_1 r_1 r_2 \quad \text{(from (2.6))} \\
&= r_1 r_2 (m_1 + m_2)
\end{aligned} \tag{2.7}$$

But, from (2.5) and (2.6):

$$m_1 r_1 = m_2 r_2 = m_2 (r_0 - r_1)$$

therefore,

$$r_1 = \frac{m_2 r_0}{m_1 + m_2} \quad \text{and} \quad r_2 = \frac{m_1 r_0}{m_1 + m_2} \tag{2.8}$$

Replacing (2.8) into (2.7):

$$I = \frac{m_1 m_2}{m_1 + m_2} r_0^2 = \mu r_0^2 \tag{2.9}$$

where we have written $\mu = m_1 m_2 / (m_1 + m_2)$, and μ is called the *reduced mass* of the system. Equation (2.9) defines the moment of inertia conveniently in terms of the atomic masses and the bond length.

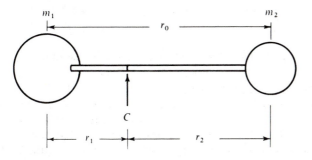

Figure 2.1 A rigid diatomic molecule treated as two masses, m_1 and m_2, joined by a rigid bar of length $r_0 = r_1 + r_2$.

By the use of the Schrödinger equation it may be shown that the rotational energy levels allowed to the rigid diatomic molecule are given by the expression:

$$E_J = \frac{h^2}{8\pi^2 I} J(J + 1) \text{ joules} \qquad \text{where } J = 0, 1, 2, \dots \qquad (2.10)$$

In this expression h is Planck's constant, and I is the moment of inertia, either I_B or I_C, since both are equal. The quantity J, which can take integral values from zero upwards, is called the *rotational quantum number*: its restriction to integral values arises directly out of the solution to the Schrödinger equation and is by no means arbitrary, and it is this restriction which effectively allows only certain discrete rotational energy levels to the molecule.

Equation (2.10) expresses the allowed energies in joules; we, however, are interested in differences between these energies, or, more particularly, in the corresponding frequency, $v = \Delta E/h$ Hz, or wavenumber, $\bar{v} = \Delta E/hc$ cm^{-1}, of the radiation emitted or absorbed as a consequence of changes between energy levels. In the rotational region spectra are usually discussed in terms of wavenumber, so it is useful to consider energies expressed in these units. We write:

$$\varepsilon_J = \frac{E_J}{hc} = \frac{h}{8\pi^2 Ic} J(J + 1) \text{ cm}^{-1} \qquad (J = 0, 1, 2, \dots) \qquad (2.11)$$

where c, the velocity of light, is here expressed in cm s^{-1}, since the unit of wavenumber is reciprocal *centimetres*.

Equation (2.11) is usually abbreviated to:

$$\varepsilon_J = BJ(J + 1) \text{ cm}^{-1} \qquad (J = 0, 1, 2, \dots) \qquad (2.12)$$

where B, the *rotational constant*, is given by

$$B = \frac{h}{8\pi^2 I_B c} \quad \text{cm}^{-1} \qquad (2.13)$$

in which we have used explicitly the moment of inertia I_B. We might equally well have used I_C and a rotational constant C, but the notation of (2.13) is conventional.

From Eq. (2.12) we can show the allowed energy levels diagrammatically as in Fig. 2.2. Plainly for $J = 0$ we have $\varepsilon_J = 0$ and we would say that the molecule is not rotating at all. For $J = 1$, the rotational energy is $\varepsilon_1 = 2B$ and a rotating molecule then has its lowest angular momentum. We may continue to calculate ε_J with increasing J values and, in principle, there is no limit to the rotational energy the molecule may have. In practice, of course, there comes a point at which the centrifugal force of a rapidly rotating diatomic molecule is greater than the strength of the bond, and the molecule is disrupted, but this point is not reached at normal temperatures.

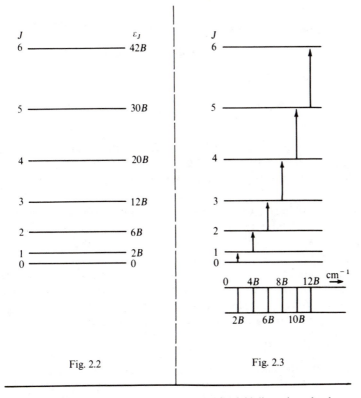

Fig. 2.2

Fig. 2.3

Figure 2.2 The allowed rotational energy levels of a rigid diatomic molecule.

Figure 2.3 Allowed transitions between the energy levels of a rigid diatomic molecule and the spectrum which arises from them.

We now need to consider *differences* between the levels in order to discuss the spectrum. If we imagine the molecule to be in the $J = 0$ state (the *ground rotational state*, in which no rotation occurs), we can let incident radiation be absorbed to raise it to the $J = 1$ state. Plainly the energy absorbed will be:

$$\varepsilon_{J=1} - \varepsilon_{J=0} = 2B - 0 = 2B \quad \text{cm}^{-1}$$

and, therefore,

$$\bar{v}_{J=0 \to J=1} = 2B \quad \text{cm}^{-1} \tag{2.14}$$

In other words, an absorption line will appear at $2B$ cm^{-1}. If now the molecule is raised from the $J = 1$ to the $J = 2$ level by the absorption of more energy, we see immediately:

$$\bar{v}_{J=1 \to J=2} = \varepsilon_{J=2} - \varepsilon_{J=1}$$
$$= 6B - 2B = 4B \quad \text{cm}^{-1} \tag{2.15}$$

In general, to raise the molecule from the state J to state $J + 1$, we would have:

$$\bar{v}_{J \to J+1} = B(J + 1)(J + 2) - BJ(J + 1)$$
$$= B[J^2 + 3J + 2 - (J^2 + J)]$$

or

$$\bar{v}_{J \to J+1} = 2B(J + 1) \text{ cm}^{-1} \tag{2.16}$$

Thus a stepwise raising of the rotational energy results in an absorption spectrum consisting of lines at $2B$, $4B$, $6B$, ..., cm^{-1}, while a similar lowering would result in an identical emission spectrum. This is shown at the foot of Fig. 2.3.

In deriving this pattern we have made the assumption that a transition can occur from a particular level only to its immediate neighbour, either above or below: we have not, for instance, considered the sequence of transitions $J = 0 \to J = 2 \to J = 4 \ldots$. In fact, a rather sophisticated application of the Schrödinger wave equation shows that, for this molecule, we need only consider transitions in which J changes by one unit—all other transitions being spectroscopically *forbidden*. Such a result is called a *selection rule*, and we may formulate it for the rigid diatomic rotator as:

$$\text{Selection rule:} \quad \Delta J = \pm 1 \tag{2.17}$$

Thus Eq. (2.16) gives the *whole* spectrum to be expected from such a molecule.

Of course, only if the molecule is asymmetric (heteronuclear) will this spectrum be observed, since if it is homonuclear there will be no dipole component change during the rotation, and hence no interaction with radiation. Thus molecules such as HCl and CO will show a rotational spectrum, while N_2 and O_2 will not. Remember also, that rotation about the bond axis was rejected in Sec. 2.1: we can now see that there are two reasons for this. Firstly, the moment of inertia is very small about the bond so, applying Eqs (2.10) or (2.11) we see that the energy levels would be extremely widely spaced: this means that a molecule requires a great deal of energy to be raised from the $J = 0$ to the $J = 1$ state, and such transitions do not occur under normal spectroscopic conditions. Thus diatomic (and all linear) molecules are in the $J = 0$ state for rotation about the bond axis, and they may be said to be not rotating. Secondly, even if such a transition should occur, there will be no dipole change and hence no spectrum.

To conclude this section we shall apply Eq. (2.16) to an observed spectrum in order to determine the moment of inertia and hence the bond length. Gilliam et al.† have measured the first line ($J = 0$) in the rotation

† Gilliam, Johnson, and Gordy, *Physical Review*, **78**, 140 (1950).

spectrum of carbon monoxide as 3.84235 cm^{-1}. Hence from Eq. (2.16):

$$\bar{v}_{0 \to 1} = 3.84235 = 2B \quad \text{cm}^{-1}$$

or,

$$B = 1.92118 \text{ cm}^{-1}$$

Rewriting Eq. (2.13) as: $I = h/8\pi^2 Bc$, we have

$$I_{\text{CO}} = \frac{6.626 \times 10^{-34}}{8\pi^2 \times 2.99793 \times 10^{10} \times B} = \frac{27.9907 \times 10^{-47}}{B} \text{ kg m}^2$$

$$= 14.5695_4 \times 10^{-47} \text{ kg m}^2$$

where we express the velocity of light in cm s^{-1}, since B is in cm^{-1}. But the moment of inertia is μr^2 (cf. Eq. (2.9)) and, knowing the relative atomic weights (H = 1.0080) to be C = 12.0000, O = 15.9994, and the absolute mass of the hydrogen atom to be 1.67343×10^{-27} kg, we can calculate the masses of carbon and oxygen, respectively, as 19.92168 and 26.56136×10^{-27} kg. The reduced mass is then:

$$\mu = \frac{19.92168 \times 26.56136 \times 10^{-54}}{46.48303 \times 10^{-27}} = 11.38365 \times 10^{-27} \text{ kg}$$

Hence:

$$r^2 = \frac{I}{\mu} = 1.2799 \times 10^{-20} \text{ m}^2$$

and

$$r_{\text{CO}} = 0.1131 \text{ nm (or } 1.131 \text{ Å)}$$

2.3.2 The Intensities of Spectral Lines

We want now to consider briefly the relative intensities of the spectral lines of Eq. (2.16); for this a prime requirement is plainly a knowledge of the relative probabilities of transition between the various energy levels. Does, for instance, a molecule have more or less chance of making the transition $J = 0 \to J = 1$ than the transition $J = 1 \to J = 2$? We mentioned above calculations which show that a change of $\Delta J = \pm 2$, ± 3, etc., was forbidden—in other words, the transition probability for all these changes is zero. Precisely similar calculations show that the probability of all changes with $\Delta J = \pm 1$ is almost the same—all, to a good approximation, are equally likely to occur.

This does not mean, however, that all spectral lines will be equally intense. Although the intrinsic probability that a single molecule in the $J = 0$ state, say, will move to $J = 1$ is the same as that of a single molecule moving from $J = 1$ to $J = 2$, in an assemblage of molecules, such as in a normal gas sample, there will be different numbers of molecules in each level to begin with, and therefore different total numbers of molecules will carry out transitions between the various levels. In fact, since the intrinsic probabilities are identical, the line intensities will be directly proportional to the initial numbers of molecules in each level.

The first factor governing the population of the levels is the Boltzmann distribution (cf. Sec. 1.7.2). Here we know that the rotational energy in the lowest level is zero, since $J = 0$, so, if we have N_0 molecules in this state, the number in any higher state is given by:

$$N_J/N_0 = \exp\left(-E_J/kT\right) = \exp\left\{-BhcJ(J + 1)/kT\right\} \qquad (2.18)$$

where, we must remember, c is the velocity of light in cm s^{-1} when B is in cm^{-1}. A very simple calculation shows how N_J varies with J; for example, taking a typical value of $B = 2$ cm^{-1}, and room temperature (say $T = 300$ K), the relative population in the $J = 1$ state is:

$$\frac{N_1}{N_0} = \exp\left\{-\frac{2 \times 6{\cdot}63 \times 10^{-34} \times 3 \times 10^{10} \times 1 \times 2}{1{\cdot}38 \times 10^{-23} \times 300}\right\}$$

$$= \exp\left(-0{\cdot}019\right) \approx 0{\cdot}98$$

and we see that there are almost as many molecules in the $J = 1$ state, at equilibrium, as in the $J = 0$. In a similar way the two graphs of Fig. 2.4 have been calculated, showing the more rapid decrease of N_J/N_0 with increasing J and with larger B.

A second factor is also required—the possibility of *degeneracy* in the energy states. Degeneracy is the existence of two or more energy states which have exactly the *same* energy. In the case of the diatomic rotator we may approach the problem in terms of its angular momentum.

The defining equations for the energy and angular momentum of a rotator are:

$$E = \tfrac{1}{2}I\omega^2 \qquad \mathbf{P} = I\omega$$

where I is the moment of inertia, ω the rotational frequency (in radians per second), and $\mathbf{P}$ the angular momentum. Rearrangement of these gives

$$\mathbf{P} = \sqrt{2EI}$$

The energy level expression of Eq. (2.10) can be rewritten:

$$2EI = J(J + 1)\frac{h^2}{4\pi^2}$$

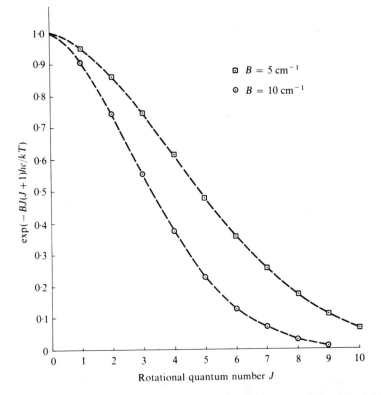

Figure 2.4 The Boltzmann populations of the rotational energy levels of Fig. 2.2. The diagram has been drawn taking values of $B = 5$ and 10 cm^{-1} and $T = 300$ K in Eq. (2.18).

and hence

$$\mathbf{P} = \sqrt{J(J + 1)}\,\frac{h}{2\pi} = \sqrt{J(J + 1)}\ \text{units} \qquad (2.19)$$

where, following convention, we take $h/2\pi$ as the fundamental unit of angular momentum. Thus we see that $\mathbf{P}$, like E, is quantized.

Throughout the above derivation $\mathbf{P}$ has been printed in bold face type to show that it is a *vector*—i.e., it has *direction* as well as *magnitude*. The direction of the angular momentum vector is conventionally taken to be along the axis about which rotation occurs and it is usually drawn as an arrow of length proportional to the magnitude of the momentum. The number of different directions which an angular momentum vector may take up is limited by a quantum mechanical law which may be stated:

'For integral values of the rotational quantum number (in this case J), the angular momentum vector may only take up directions such that its component along a given reference direction is zero or an integral multiple of angular momentum units.'

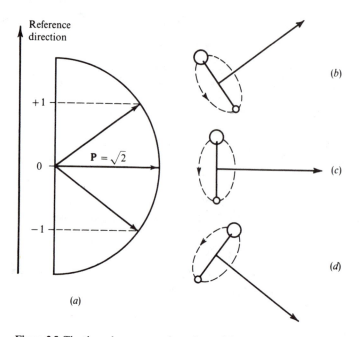

Figure 2.5 The three degenerate orientations of the rotational angular momentum vector for a molecule with $J = 1$.

We can see the implications of this most easily by means of a diagram. In Fig. 2.5 we show the case $J = 1$. Here $\mathbf{P} = \sqrt{1 \times 2}$ units $= \sqrt{2}$, and, as Fig. 2.5(a) shows, a vector of length $\sqrt{2}\,(= 1\cdot41)$ can have only *three* integral or zero components along a reference direction (here assumed to be from top to bottom in the plane of the paper): $+1$, 0, and -1. Thus the angular momentum vector in this instance can be oriented in only three different directions (Fig. 2.5(b)–(d)) with respect to the reference direction. All three rotational directions are, of course, associated with the same angular momentum and hence the same rotational energy: the $J = 1$ level is thus threefold degenerate.

Figure 2.6(a) and (b) shows the situation for $J = 2$ ($\mathbf{P} = \sqrt{6}$) and $J = 3$ ($\mathbf{P} = 2\sqrt{3}$) with fivefold and sevenfold degeneracy respectively. In general it may readily be seen that *each energy level is $2J + 1$-fold degenerate*.

Thus we see that, although the molecular population in each level decreases exponentially (Eq. (2.18)), the number of degenerate levels available increases rapidly with J. The total relative population at an energy E_J will plainly be:

$$\text{Population} \propto (2J + 1) \exp\left(-E_J/kT\right) \tag{2.20}$$

When this is plotted against J the points fall on a curve of the type shown in Fig. 2.7, indicating that the population rises to a maximum and then

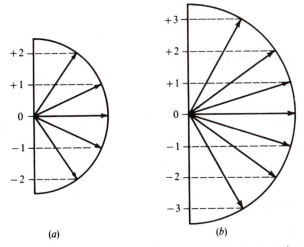

(a)   (b)

Figure 2.6 The five and seven degenerate rotational orientations for a molecule with $J = 2$ and $J = 3$ respectively.

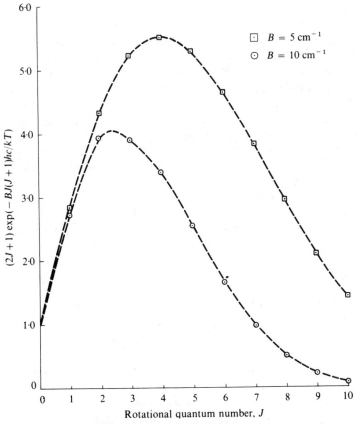

Figure 2.7 The total relative populations, including degeneracy, of the rotational energy levels of a diatomic molecule. The diagram has been drawn for the same conditions as Fig. 2.4.

diminishes. Differentiation of Eq. (2.20) shows that the population is a maximum at the nearest integral J value to:

$$\text{Maximum population:} \quad J = \sqrt{\frac{kT}{2hcB}} - \frac{1}{2} \qquad (2.21)$$

We have seen that line intensities are directly proportional to the populations of the rotational levels, hence it is plain that transitions between levels with very low or very high J values will have small intensities while the intensity will be a maximum at or near the J value given by Eq. (2.21).

2.3.3 The Effect of Isotopic Substitution

When a particular atom in a molecule is replaced by its isotope—an element identical in every way except for its atomic mass—the resulting substance is identical chemically with the original. In particular there is no appreciable change in internuclear distance on isotopic substitution. There is, however, a change in total mass and hence in the moment of inertia and B value for the molecule.

Considering carbon monoxide as an example, we see that on going from $^{12}C^{16}O$ to $^{13}C^{16}O$ there is a mass increase and hence a decrease in the B value. If we designate the ^{13}C molecule with a prime we have $B > B'$. This change will be reflected in the rotational energy levels of the molecule and Fig. 2.8 shows, much exaggerated, the relative lowering of the ^{13}C levels with respect to those of ^{12}C. Plainly, as shown by the diagram at the foot of Fig. 2.8, the spectrum of the heavier species will show a smaller separation between the lines ($2B'$) than that of the lighter one ($2B$). Again the effect has been much exaggerated for clarity, and the transitions due to the heavier molecule are shown dashed.

Observation of this decreased separation has led to the evaluation of precise atomic weights. Gilliam et al., as already stated, found the first rotational absorption of $^{12}C^{16}O$ to be at 3.84235 cm^{-1}, while that of $^{13}C^{16}O$ was at 3.67337 cm^{-1}. The values of B determined from these figures are:

$$B = 1.92118 \text{ cm}^{-1} \quad \text{and} \quad B' = 1.83669 \text{ cm}^{-1}$$

where the prime refers to the heavier molecule. We have immediately:

$$\frac{B}{B'} = \frac{h}{8\pi^2 I c} \cdot \frac{8\pi^2 I'c}{h} = \frac{I'}{I} = \frac{\mu'}{\mu} = 1.046$$

where μ is the reduced mass, and the internuclear distance is considered unchanged by isotopic substitution. Taking the mass of oxygen to be 15.9994 and that of carbon-12 to be 12.00, we have:

$$\frac{\mu'}{\mu} = 1.046 = \frac{15.9994 m'}{15.9994 + m'} \times \frac{12 + 15.9994}{12 \times 15.9994}$$

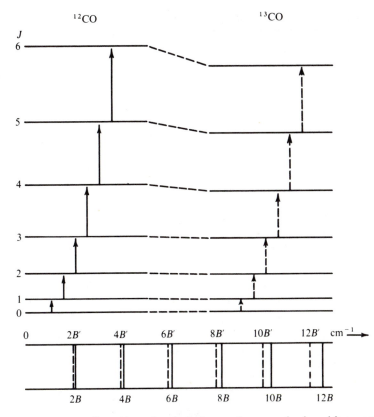

Figure 2.8 The effect of isotopic substitution on the energy levels and hence rotational spectrum of a diatomic molecule such as carbon monoxide.

from which m', the atomic weight of carbon-13, is found to be 13·0007. This is within 0·02 per cent of the best value obtained in other ways.

It is noteworthy that the data quoted above were obtained by Gilliam et al. from $^{13}C^{16}O$ molecules in natural abundance (i.e., about 1 per cent of ordinary carbon monoxide). Thus, besides allowing an extremely precise determination of atomic weights, microwave studies can give directly an estimate of the abundance of isotopes by comparison of absorption intensities.

2.3.4 The Non-Rigid Rotator

At the end of Sec. 2.3.1 we indicated how internuclear distances could be calculated from microwave spectra. It must be admitted that we selected our data carefully at this point—spectral lines for carbon monoxide, other than the first, would not have shown the constant $2B$ separation predicted

Table 2.1 Rotation spectrum of hydrogen fluoride

J	$\bar{\nu}_{obs.}$† (cm^{-1})	$\bar{\nu}_{calc.}$‡ (cm^{-1})	$\Delta\bar{\nu}_{obs.}$ (cm^{-1})	B $(=\frac{1}{2}\Delta\bar{\nu})$	r (nm)
0	41·08	41·11			
			41·11	20·56	0·0929
1	82·19	82·18			
			40·96	20·48	0·0931
2	123·15	123·14			
			40·85	20·43	0·0932
3	164·00	163·94			
			40·62	20·31	0·0935
4	204·62	204·55			
			40·31	20·16	0·0938
5	244·93	244·89			
			40·08	20·04	0·0941
6	285·01	284·93			
			39·64	19·82	0·0946
7	324·65	324·61			
			39·28	19·64	0·0951
8	363·93	363·89			
			38·89	19·45	0·0955
9	402·82	402·70			
			38·31	19·16	0·0963
10	441·13	441·00			
			37·81	18·91	0·0969
11	478·94	478·74			

† Lines numbered according to $\bar{\nu}_J = 2B(J + 1)$ cm^{-1}. Observed data from 'An Examination of the Far Infra-red Spectrum of Hydrogen Fluoride' by A. A. Mason and A. H. Nielsen, published as Scientific Report No. 5, August 1963, Contract No. AF 19(604)-7981, by kind permission of the authors.

‡ See Sec. 2.3.5 for details of the calculation.

by Eq. (2.16). This is shown by the spectrum of hydrogen fluoride given in Table 2.1; it is evident that the separation between successive lines (and hence the apparent B value) decreases steadily with increasing J.

The reason for this decrease may be seen if we calculate internuclear distances from the B values. The calculations are exactly similar to those of Sec. 2.3.1 and the results are shown in column 6 of Table 2.1. Plainly the bond length increases with J and we can see that our assumption of a *rigid* bond is only an approximation; in fact, of course, all bonds are elastic to some extent, and the increase in length with J merely reflects the fact that the more quickly a diatomic molecule rotates the greater is the centrifugal force tending to move the atoms apart.

Before showing how this elasticity may be quantitatively allowed for in rotational spectra, we shall consider briefly two of its consequences. First, when the bond is elastic, a molecule may have vibrational energy—i.e., the bond will stretch and compress periodically with a certain fundamental frequency dependent upon the masses of the atoms and the elasticity (or force constant) of the bond. If the motion is simple harmonic (which, we shall see in Chapter 3, is usually a very good approximation to the truth) the force constant is given by:

$$k = 4\pi^2 \bar{\omega}^2 c^2 \mu \qquad (2.22)$$

where $\bar{\omega}$ is the vibration frequency (expressed in cm^{-1}), and c and μ have their previous definitions. Plainly the variation of B with J is determined by the force constant—the weaker the bond, the more readily will it distort under centrifugal forces.

The second consequence of elasticity is that the quantities r and B vary during a vibration. When these quantities are measured by microwave techniques many hundreds of vibrations occur during a rotation, and hence the measured value is an average. However, from the defining equation of B we have:

$$B = \frac{h}{8\pi^2 Ic} = \frac{h}{8\pi^2 c \mu r^2}$$

or

$$B \propto 1/r^2 \qquad (2.23)$$

since all other quantities are independent of vibration. Now, although in simple harmonic motion a molecular bond is compressed and extended an equal amount on each side of the equilibrium distance and the average value of the distance is therefore unchanged, the average value of $1/r^2$ is *not* equal to $1/r_e^2$, where r_e is the equilibrium distance. We can see this most easily by an example. Consider a bond of equilibrium length 0·1 nm vibrating between the limits 0·09 and 0·11 nm. We have:

$$\langle r \rangle_{av.} = \frac{0·09 + 0·11}{2} = 0·1 = r_e$$

but

$$\left\langle \frac{1}{r^2} \right\rangle_{av.} = \frac{(1/0·09)^2 + (1/0·11)^2}{2} = 103·05 \text{ nm}^2$$

and therefore $\langle r \rangle_{av.} = \sqrt{1/103·5} = 0·0985$ nm. The difference, though small, is not negligible compared with the precision with which B can be measured spectroscopically. And in fact the real situation is rather worse. We shall see in Chapter 3 that real vibrations are not simple harmonic, since a real bond may be stretched more easily than it may be compressed, and this usually results in $r_{av.}$ being greater than $r_{eq.}$.

It is usual, then, to define three different sets of values for B and r. At the equilibrium separation, r_e, between the nuclei, the rotational constant is B_e; in the vibrational ground state the average internuclear separation is r_0 associated with a rotational constant B_0; while if the molecule has excess vibrational energy the quantities are r_v and B_v, where v is the vibrational quantum number.

During the remainder of this chapter we shall ignore the small differences between B_0, B_e, and B_v—the discrepancy is most important in the consideration of vibrational spectra in Chapter 3.

We should note, in passing that the rotational spectrum of hydrogen fluoride given in Table 2.1 extends from the microwave well into the infrared region (cf. Fig. 1.4). This underlines the comment made in Chapter 1 that there is no fundamental distinction between spectral regions, only differences in technique. Since hydrogen fluoride, together with other diatomic hydrides, has a small moment of inertia and hence a large B value, the spacings between rotational energy levels become large and fall into the infra-red region after only a few transitions. Historically, indeed, the moments of inertia and bond lengths of these molecules were first determined from spectral studies using infra-red techniques.

2.3.5 The Spectrum of a Non-Rigid Rotator

The Schrödinger wave equation may be set up for a non-rigid molecule, and the rotational energy levels are found to be:

$$E_J = \frac{h^2}{8\pi^2 I} J(J + 1) - \frac{h^4}{32\pi^4 I^2 r^2 k} J^2(J + 1)^2 \; J$$

or

$$\varepsilon_J = E_J/hc = BJ(J + 1) - DJ^2(J + 1)^2 \text{ cm}^{-1} \qquad (2.24)$$

where the rotational constant, B, is as defined previously, and the *centrifugal distortion constant* D, is given by:

$$D = \frac{h^3}{32\pi^4 I^2 r^2 kc} \quad \text{cm}^{-1} \qquad (2.25)$$

which is a positive quantity. Equation (2.24) applies for a simple harmonic force field only; if the force field is anharmonic, the expression becomes:

$$\varepsilon_J = BJ(J + 1) - DJ^2(J + 1)^2 + HJ^3(J + 1)^3 + KJ^4(J + 1)^4 \cdots \text{ cm}^{-1}$$

$$(2.26)$$

where H, K, etc., are small constants dependent upon the geometry of the molecule. They are, however, negligible compared with D and most modern spectroscopic data are adequately fitted by Eq. (2.24).

From the defining equations of B and D it may be shown directly that

$$D = \frac{16B^3\pi^2\mu c^2}{k} = \frac{4B^3}{\bar{\omega}^2} \qquad (2.27)$$

where $\bar{\omega}$ is the vibrational frequency of the bond, and k has been expressed according to Eq. (2.22). We shall see in Chapter 3 that vibrational frequencies are usually of the order of 10^3 cm^{-1}, while B we have found to be of the order of 10 cm^{-1}. Thus we see that D, being of the order 10^{-3} cm^{-1}, is very small compared with B. For small J, therefore, the correction term $DJ^2(J + 1)^2$ is almost negligible, while for J values of 10 or more it may become appreciable.

Figure 2.9 shows, much exaggerated, the lowering of rotational levels when passing from the rigid to the non-rigid diatomic molecule. The spectra are also compared, the dashed lines connecting corresponding energy levels and transitions of the rigid and the non-rigid molecules. It should be noted that the selection rule for the latter is still $\Delta J = \pm 1$.

We may easily write an analytical expression for the transitions:

$$\varepsilon_{J+1} - \varepsilon_J = \bar{\nu}_J = B[(J + 1)(J + 2) - J(J + 1)]$$
$$- D[(J + 1)^2(J + 2)^2 - J^2(J + 1)^2]$$
$$= 2B(J + 1) - 4D(J + 1)^3 \text{ cm}^{-1} \qquad (2.28)$$

where $\bar{\nu}_J$ represents equally the upward transition from J to $J + 1$, or the downward from $J + 1$ to J. Thus we see analytically, and from Fig. 2.9, that the spectrum of the elastic rotor is similar to that of the rigid molecule except that each line is displaced slightly to low frequency, the displacement increasing with $(J + 1)^3$.

A knowledge of D gives rise to two useful items of information. Firstly, it allows us to determine the J value of lines in an observed spectrum. If we have measured a few isolated transitions it is not always easy to determine from which J value they arise; however, fitting Eq. (2.28) to them— provided three consecutive lines have been measured—gives unique values for B, D, and J. The precision of such fitting is shown by Table 2.1 where the wavenumbers are calculated from the equation:

$$\bar{\nu}_J = 41\cdot122(J + 1) - 8\cdot52 \times 10^{-3}(J + 1)^3 \text{ cm}^{-1} \qquad (2.29)$$

Secondly, a knowledge of D enables us to determine—although rather inaccurately—the vibrational frequency of a diatomic molecule. From the above data for hydrogen fluoride and Eq. (2.27) we have:

$$\bar{\omega}^2 = \frac{4B^3}{D} = 16\cdot33 \times 10^6 \text{ (cm}^{-1})^2$$

i.e.,

$$\bar{\omega} \approx 4050 \text{ cm}^{-1}$$

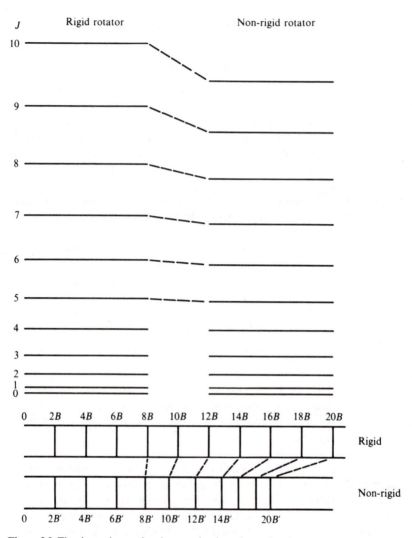

Figure 2.9 The change in rotational energy levels and rotational spectrum when passing from a rigid to a non-rigid diatomic molecule. Levels on the right calculated using $D = 10^{-3}B$.

In the next chapter we shall see that a more precise determination leads to the value $4138 \cdot 3$ cm^{-1}; the two per cent inaccuracy in the present calculation is due partly to the assumption of simple harmonic motion, and partly to the very small, and hence relatively inaccurate, value of D.

The force constant follows directly:

$$k = 4\pi^2 c^2 \bar{\omega}^2 \mu = 960 \text{ N m}^{-1}$$

which indicates, as expected, that H—F is a relatively strong bond.

2.4 POLYATOMIC MOLECULES

2.4.1 Linear Molecules

We consider first molecules such as carbon oxysulphide OCS, or chloro-acetylene HC≡CCl, where all the atoms lie on a straight line, since this type gives rise to a particularly simple spectra in the microwave region. Since $I_B = I_C$; $I_A = 0$, as for diatomic molecules, the energy levels are given by a formula identical with Eq. (2.26), i.e.,

$$\varepsilon_J = BJ(J + 1) - DJ^2(J + 1)^2 + \cdots \text{ cm}^{-1} \qquad (2.30)$$

and the spectrum will show the same $2B$ separation modified by the distortion constant. In fact, the whole of the discussion on diatomic molecules applies equally to all linear molecules; three points, however, should be underlined:

1. Since the moment of inertia for the end-over-end rotation of a polyatomic linear molecule is considerably greater than that of a diatomic molecule, the B value will be much smaller, and the spectral lines more closely spaced. Thus B values for diatomic molecules are about 10 cm^{-1}, while for triatomic molecules they can be 1 cm^{-1} or less, and for larger molecules smaller still.
2. The molecule must, as usual, possess a dipole moment if it is to exhibit a rotational spectrum. Thus OCS will be microwave active, while OCO (more usually written CO_2) will not. In particular, it should be noted that isotopic substitution does not lead to a dipole moment since the bond lengths and atomic charges are unaltered by the substitution. Thus $^{16}OC^{18}O$ is microwave inactive.
3. A non-cyclic polyatomic molecule containing N atoms has altogether $N - 1$ individual bond lengths to be determined. Thus in the triatomic molecule OCS there is the CO distance, r_{CO}, and the CS distance, r_{CS}. On the other hand, there is only *one* moment of inertia for the end-over-end rotation of OCS, and only this one value can be determined from the spectrum. Table 2.2 shows the data for this molecule. Over the four lines observed there is seen to be no appreciable centrifugal distortion, and, taking the value of B as 0.2027 cm^{-1}, we calculate:

$$I_B = \frac{h}{8\pi^2 Bc} = 137.95 \times 10^{-47} \text{ kg m}^2$$

From this one observation it is plainly impossible to deduce the two unknowns, r_{CO} and r_{CS}. The difficulty can be overcome, however, if we study a molecule with different atomic masses but the *same* bond lengths—i.e., an isotopically substituted molecule—since this will have a different moment of inertia.

Table 2.2 Microwave spectrum of carbon oxy-sulphide

$J \to J+1$	$\bar{v}_{obs.}$ (cm^{-1})	$\Delta\bar{v}$	B (cm^{-1})
$0 \to 1$	$\cdots$		
		2×0.4055	0.2027
$1 \to 2$	0.8109		
		0.4054	0.2027
$2 \to 3$	1.2163		
		0.4054	0.2027
$3 \to 4$	1.6217		
		0.4054	0.2027
$4 \to 5$	2.0271		

Let us consider the rotation of OCS in some detail. Figure 2.10 shows the molecule, where r_O, r_C, and r_S represent the distances of the atoms from the centre of gravity. Consideration of moments gives:

$$m_O r_O + m_C r_C = m_S r_S \tag{2.31}$$

where m_i is the mass of atom i. The moment of inertia is:

$$I = m_O r_O^2 + m_C r_C^2 + m_S r_S^2 \tag{2.32}$$

and we have the further equations:

$$r_O = r_{CO} + r_C \qquad r_S = r_{CS} - r_C \tag{2.33}$$

where r_{CO} and r_{CS} are the bond lengths of the molecule. It is these we wish to determine. Substituting (2.33) in (2.31) and collecting terms:

$$(m_C + m_O + m_S)r_C = m_S r_{CS} - m_O r_{CO}$$

or

$$Mr_C = m_S r_{CS} - m_O r_{CO} \tag{2.34}$$

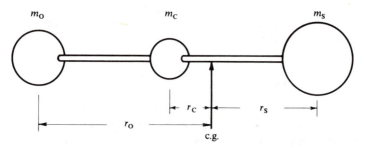

Figure 2.10 The molecule of carbon oxysulphide, OCS, showing the distances of each atom from the centre of gravity.

where we write M for the total mass of the molecule. Substituting (2.33) in (2.32):

$$I = m_O(r_{CO} + r_C)^2 + m_C r_C^2 + m_S(r_{CS} - r_C)^2$$
$$= M r_C^2 + 2r_C(m_O r_{CO} - m_S r_{CS}) + m_O r_{CO}^2 + m_S r_{CS}^2$$

and finally substituting for r_C from Eq. (2.34):

$$I = m_O r_{CO}^2 + m_S r_{CS}^2 - \frac{(m_O r_{CO} - m_S r_{CS})^2}{M} \tag{2.35}$$

Considering now the isotopic molecule, ^{18}OCS, we may write m_O' for m_O throughout Eq. (2.35):

$$I' = m_O' r_{CO}^2 + m_S r_{CS}^2 - \frac{(m_O' r_{CO} - m_S r_{CS})^2}{M'} \tag{2.36}$$

and we can now solve for r_{CO} and r_{CS}, provided we have extracted a value for I' from the microwave spectrum of the isotopic molecule. Note that we do *not* need to write r_{CO}', since we assume that the bond length is unaltered by isotopic substitution. This assumption may be checked by studying the molecules $^{16}OC^{34}S$ and $^{18}OC^{34}S$, since we would then have four moments of inertia. The bond distances found are quite consistent, and hence justify the assumption.

The extension of the above discussion to molecules with more than three atoms is straightforward; it suffices to say here that microwave studies have led to very precise determinations of many bond lengths in such molecules.

2.4.2 Symmetric Top Molecule

Although the rotational energy levels of this type of molecule are more complicated than those of linear molecules, we shall see that, because of their symmetry, their pure rotational spectra are still relatively simple. Choosing methyl fluoride again as our example we remember that

$$I_B = I_C \neq I_A \qquad I_A \neq 0$$

There are now two directions of rotation in which the molecule might absorb or emit energy—that about the main symmetry axis (the C—F bond in this case) and that perpendicular to this axis.

We thus need two quantum numbers to describe the degree of rotation, one for I_A and one for I_B or I_C. However, it turns out to be very convenient mathematically to have a quantum number to represent the *total* angular momentum of the molecule, which is the sum of the separate angular momenta about the two different axes. This is usually chosen to be the quantum number J. Reverting for a moment to *linear* molecules, remember that we there used J to represent the end-over-end rotation of a molecule:

however, this was the *only* sort of rotation allowed, so it is quite consistent to use J, in general, to represent the *total angular momentum*. It is then conventional to use K to represent the angular momentum about the top axis—i.e., about the C—F bond in this case.

Let us briefly consider what values are allowed to K and J. Both must, by the conditions of quantum mechanics, be integral or zero. The total angular momentum can be as large as we like, that is, J can be 0, 1, 2,..., ∞ (except, of course, for the theoretical possibility that a real molecule will be disrupted at very high rotational speeds). Once we have chosen J, however, K is rather more limited. Let us consider the case when $J = 3$. Plainly the rotational energy can be divided in several ways between motion about the main symmetry axis and motion perpendicular to this. If *all* the rotation is about the axis, $K = 3$; but note that K cannot be greater than J since J is the *total* angular momentum. Equally we could have $K = 2$, 1, or 0, in which case the motion perpendicular to the axis increases accordingly. Additionally, however, K can be negative—we can imagine positive and negative values of K to correspond with clockwise and anti-clockwise rotation about the symmetry axis—and so can have values -1, -2, or -3.

In general, then, for a total angular momentum J, we see that K can take values:

$$K = J, J - 1, J - 2, \ldots, 0, \ldots, -(J - 1), -J \qquad (2.37)$$

which is a total of $2J + 1$ values altogether. This figure of $2J + 1$ is important and will recur.

If we take first the case of a *rigid* symmetric top—i.e., one in which the bonds are supposed not to stretch under centrifugal forces—the Schrödinger equation may be solved to give the allowed energy levels for rotation as:

$$\varepsilon_{J, K} = E_{J, K}/hc = BJ(J + 1) + (A - B)K^2 \quad \text{cm}^{-1} \qquad (2.38)$$

where, as before,

$$B = \frac{h}{8\pi^2 I_B c} \quad \text{and} \quad A = \frac{h}{8\pi^2 I_A c}$$

Note that the energy depends on K^2, so that it is immaterial whether the top spins clockwise or anticlockwise: the energy is the same for a given angular momentum. For all $K > 0$, therefore, the rotational energy levels are *doubly degenerate*.

The selection rules for this molecule may be shown to be:

$$\Delta J = \pm 1 \text{ (as before)} \quad \text{and} \quad \Delta K = 0 \qquad (2.39)$$

and, when these are applied to Eq. (2.38), the spectrum is given by:

$$\varepsilon_{J+1, K} - \varepsilon_{J, K} = \bar{\nu}_{J, K} = B(J + 1)(J + 2) + (A - B)K^2$$
$$- [BJ(J + 1) + (A - B)K^2]$$
$$= 2B(J + 1) \text{ cm}^{-1} \qquad (2.40)$$

Thus the spectrum is independent of K, and hence rotational changes about the symmetry axis do not give rise to a rotational spectrum. The reason for this is quite evident—rotation about the symmetry axis does not change the dipole moment perpendicular to the axis (which always remains zero), and hence the rotation cannot interact with radiation. Equation (2.40) shows that the spectrum is just the same as for a linear molecule and that only one moment of inertia—that for end-over-end rotation—can be measured.

Both Eqs (2.38) and (2.40) are for a rigid molecule, however, and we have already seen that microwave spectroscopy is well able to detect the departure of real molecules from this idealized state. When centrifugal stretching is taken into account, the energy levels become:

$$\varepsilon_{J, K} = BJ(J + 1) + (A - B)K^2 - D_J J^2(J + 1)^2$$
$$- D_{JK} J(J + 1)K^2 - D_K K^4 \quad \text{cm}^{-1} \quad (2.41)$$

where, in an obvious notation, D_J, D_{JK}, and D_K are small correction terms for non-rigidity. The selection rules are unchanged (Eq. (2.39)), and so the spectrum is:

$$\bar{\nu}_{J, K} = \varepsilon_{J+1, K} - \varepsilon_{J, K}$$
$$= 2B(J + 1) - 4D_J(J + 1)^3 - 2D_{JK}(J + 1)K^2 \quad \text{cm}^{-1} \quad (2.42)$$

We see that the spectrum will be basically that of a linear molecule (including centrifugal stretching) with an additional term depending on K^2.

It is easy to see why this spectrum now depends on the axial rotation (i.e., depends on K), although such rotation produces no dipole change. Figure 2.11 illustrates methyl fluoride for (a) $K = 0$, no axial rotation, and (b) $K > 0$, the molecule rotating about the symmetry axis. We see, from the much exaggerated diagram, that axial rotation widens the HCH angles and stretches the C—H bonds. The distorted molecule (b) has a different moment of inertia for end-over-end rotation from (a). If we write Eq. (2.42) as:

$$\bar{\nu}_{JK} = 2(J + 1)[B - 2D_J(J + 1)^2 - D_{JK} K^2] \quad \text{cm}^{-1}$$

we can see more clearly that the centrifugal distortion constants D_J and D_{JK} can be considered as correction terms to the rotational constant B, and hence as perturbing the moment of inertia I_B.

Since each value of J is associated with $2J + 1$ values of K, we see that each line characterized by a certain J value must have $2J + 1$ components.

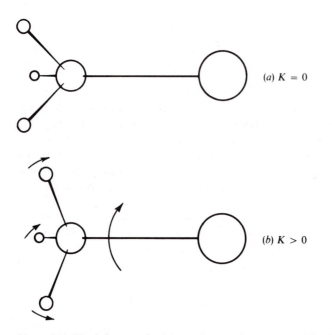

(a) K = 0

(b) K > 0

Figure 2.11 The influence of axial rotation on the moment of inertia of a symmetric top molecule, e.g., methyl fluoride, CH_3F. In (a) there is no axial rotation ($K = 0$), and in (b) $K > 0$.

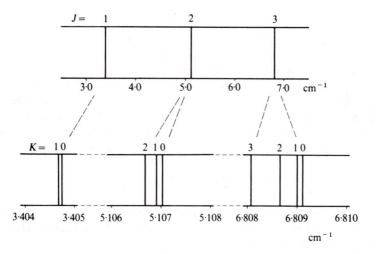

Figure 2.12 A diagrammatic representation of the rotational spectrum of the symmetric top molecule methyl fluoride, CH_3F.

However, since K only appears as K^2 in Eq. (2.42), there will be only $J + 1$ *different* frequencies, all those with $K > 0$ being doubly degenerate. We may tabulate a few lines as follows:

$$J = 0,\ K = 0 \qquad \bar{\nu}_{JK} = 2B - 4D_J \quad \text{cm}^{-1}$$

$$J = 1,\ K = 0 \qquad \bar{\nu}_{JK} = 4B - 32D_J$$

$$K = \pm 1 \qquad \bar{\nu}_{JK} = 4B - 32D_J - 4D_{JK} \qquad\qquad (2.43)$$

$$J = 2,\ K = 0 \qquad \bar{\nu}_{JK} = 6B - 108D_J$$

$$K = \pm 1 \qquad \bar{\nu}_{JK} = 6B - 108D_J - 6D_{JK}$$

$$K = \pm 2 \qquad \bar{\nu}_{JK} = 6B - 108D_J - 24D_{JK}, \quad \text{etc.}$$

Let us now compare these with the observed spectrum of methyl fluoride. This is shown as a line diagram in Fig. 2.12, and the frequencies are tabulated in Table 2.3. Fitting these data to equations such as (2.43) leads directly to:

$$B = 0.851\,204 \text{ cm}^{-1}$$

$$D_J = 2.00 \times 10^{-6} \text{ cm}^{-1}$$

$$D_{JK} = 1.47 \times 10^{-5} \text{ cm}^{-1}$$

The calculated frequencies of Table 2.3 show how precisely such measurements may now be made.

Once again each spectrum examined yields only one value of B, but the spectra of isotopic molecules can, in principle, give sufficient information for the calculation of all the bond lengths and angles of symmetric top molecules, together with estimates of the force constant of each bond.

Table 2.3 Microwave spectrum of methyl fluoride

J	K	$\bar{\nu}_{\text{obs.}}$ †(cm^{-1})	$\bar{\nu}_{\text{calc.}}$ (cm^{-1})
1	0	3·404 75	3·404 752
	1	3·404 70	3·404 693
2	0	5·107 01	5·107 008
	1	5·106 92	5·106 920
	2	5·106 65	5·106 655
3	0	6·809 12	6·809 120
	1	6·809 00	6·809 002
	2	6·808 65	6·808 649
	3	6·808 06	6·808 062

† Taken from W. Gordy, *Physical Review*, **93**, 406 (1954), by kind permission of the author.

Table 2.4 Some molecular data determined by microwave spectroscopy

Molecule	Type	Bond length (nm)	Bond angle (deg.)	Dipole moment† (debyes)
NaCl	Diatomic	0.23606 ± 0.00001	—	8.5 ± 0.2
COS	Linear	$\begin{cases} 0.1164 \pm 0.0001 \text{ (CO)} \\ 0.1559 \pm 0.0001 \text{ (CS)} \end{cases}$	—	0.712 ± 0.004
HCN	Linear	$\begin{cases} 0.106317 \pm 0.000005 \text{ (CH)} \\ 0.115535 \pm 0.000006 \text{ (CN)} \end{cases}$	—	2.986 ± 0.004
NH_3	Sym. Top	0.1008 ± 0.0004	107.3 ± 0.2	1.47 ± 0.01
CH_3Cl	Sym. Top	$\begin{cases} 0.10959 \pm 0.00005 \text{ (CH)} \\ 0.17812 \pm 0.00005 \text{ (CCl)} \end{cases}$	108.0 ± 0.2 (HCH)	1.871 ± 0.005
H_2O	Asym. Top	0.09584 ± 0.00005	104.5 ± 0.3	1.846 ± 0.005
O_3	Asym. Top	0.1278 ± 0.0002	116.8 ± 0.5	0.53 ± 0.02

† Measured from the Stark effect, cf. Sec. 2.5.2.

2.4.3 Asymmetric Top Molecules

Since spherical tops show no microwave spectrum (cf. Sec. 2.1(3)) the only other class of molecule of interest here is the asymmetric top which has (Sec. 2.1(4)) all three moments of inertia different. These molecules will not detain us long since their rotational energy levels and spectra are very complex—in fact, no analytical expressions can be written for them corresponding to Eqs (2.24) and (2.28) for linear or Eqs (2.41) and (2.42) for symmetric top molecules. Each molecule and spectrum must, therefore, be treated as a separate case, and much tedious computation is necessary before structural parameters can be determined. The best method of attack so far has been to consider the asymmetric top as falling somewhere between the oblate and prolate symmetric top; interpolation between the two sets of energy levels of the latter leads to a first approximation of the energy levels—and hence spectrum—of the asymmetric molecule. It suffices to say that arbitrary methods such as this have been quite successful, and much very precise structural data have been published.

In order to give an idea of the precision of such measurements, we collect in Table 2.4 some molecular data determined by microwave methods, including examples from diatomic and linear molecules, symmetric tops, and asymmetric tops.

2.5 TECHNIQUES AND INSTRUMENTATION

2.5.1 Outline

It is not proposed here to give more than a brief outline of the techniques of microwave spectroscopy since detailed accounts are available in some of the books listed in the bibliography. Microwave spectroscopy, of course, follows the usual pattern: source, monochromator, beam direction, sample, and detector. We shall discuss each in turn.

1. *The source and monochromator.* The usual source in this region is the klystron valve which, since it emits radiation of only a very narrow frequency range, is called 'monochromatic' and acts as its own monochromator. The actual emission frequency is variable electronically and hence a spectrum may be scanned over a limited range of frequencies using a single klystron.

   One slight disadvantage of this source is that the total energy radiated is very small—of the order of milliwatts only. However, since all this is concentrated into a narrow frequency band a sharply tuned detector can be sufficiently activated to produce a strong signal.

2. *Beam direction.* This is achieved by the use of 'waveguides'—hollow tubes of copper or silver, usually of rectangular cross-section—inside which the radiation is confined. The waveguides may be gently tapered or bent to allow focusing and directing of the radiation. Atmospheric absorption of the beam is considerable, so the system must be efficiently evacuated.

3. *Sample and sample space.* In almost all microwave studies so far the sample has been gaseous. However, pressures of 0·01 mmHg are sufficient to give a reasonable absorption spectrum, so many substances which are usually thought of as solid or liquid may be examined provided their vapour pressures are above this value. The sample is retained by very thin mica windows in a piece of evacuated waveguide.

4. *Detector.* It is possible to use an ordinary superheterodyne radio receiver as detector, provided this may be tuned to the appropriate high frequency; however, a simple crystal detector is found to be more sensitive and easier to use. This detects the radiation focused upon it by the waveguide, and the signal it gives is amplified electronically for display on an oscilloscope, or for permanent record on paper.

2.5.2 The Stark Effect

We cannot leave the subject of microwave spectroscopy without a brief description of the Stark effect and its applications. A more detailed discussion is to be found in the books by Kroto and by Townes and Schawlow mentioned in the bibliography.

Experimentally the Stark effect requires the placing of an electric field, either perpendicular or parallel to the direction of the radiation beam, across the sample. Practically it is simpler to have a perpendicular field. We shall consider three advantages of this field.

1. A molecule exhibiting a rotational spectrum must have an electric dipole moment, and so its rotational energy levels will be perturbed by the

application of an exterior field since interaction will occur. Put simply, the absorption lines of the spectrum will be shifted by an amount depending on the extent of the interaction, and thus depending on both E, the applied field, and μ, the dipole moment. For a linear molecule the shift is found to be:

$$\Delta v \propto (\mu E)^2 \qquad \text{(linear molecule)}$$

while for a symmetric top:

$$\Delta v \propto \mu E \qquad \text{(symmetric top)}$$

Thus we have immediately a very accurate method of determining dipole moments, simply by observation of the Stark shift. More important, the measurement is made on very dilute gas samples, so the dipole moment observed may be taken to be that of the actual molecule, uncomplicated by molecular interactions, solvent effects, etc. Some values determined in this way are included in Table 2.4.

2. The second valuable application of the Stark effect is in the assignment of observed spectral lines to particular J values. We have seen that, in the absence of marked departure from rigidity and good resolving power, the assignment of J values is not always obvious. The line of lowest frequency which we observe *may* happen to correspond with $J = 0$, or it may be that it is the first observable line of a series, either because earlier lines are intrinsically very weak or because of limitations in the apparatus used. However, we have seen that each line is $2J + 1$ degenerate because rotations can occur in $2J + 1$ orientations in space without violating quantum laws. In the absence of any orienting effect these transitions have precisely the same frequency, but a Stark field constitutes an orienting effect, and splits the degeneracy; thus multiplet structure is observed for all lines with $J > 0$. The number of components depends on J, and hence unambiguous assignments can be made.

3. The final application is purely an instrumental one, but is especially interesting in that it has its counterpart in other spectral regions. We have already referred to the concept of signal-to-noise ratio in Chapter 1; that part of the noise which arises from random fluctuations in the background radiation may be removed by modulating the beam by means of the Stark effect as explained below.

Imagine the application of a Stark field in a periodic manner such as the 'square-wave' variation of Fig. 2.13; while the field is switched on the signal is modified in the way described in 1 and 2 above. If we arrange the modulation frequency to be some 100 to 1000 Hz, and construct the amplifier so that it amplifies only the component of the signal which has the modulated frequency, stray radiation which has not been through the modulating field will be completely ignored. This results in a great improvement of the signal-to-noise ratio.

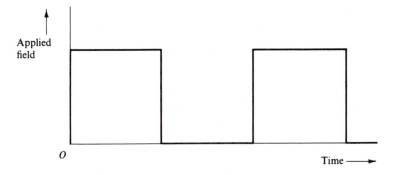

Figure 2.13 A 'square wave' potential as used for Stark modulation.

A further refinement is to arrange for both the modulated and unmodulated parts to be amplified separately and displayed on the same oscilloscope—the modulated part on the upper half, say, and the unmodulated on the lower. This much facilitates the measurement of the Stark splittings discussed in 1 and 2 above.

2.6 CHEMICAL ANALYSIS BY MICROWAVE SPECTROSCOPY

Improvements and simplifications in the techniques of microwave spectroscopy are now allowing it to move away from being purely a specialist research instrument towards becoming a technique for routine analysis. Even though effectively limited to gaseous samples, it has much to offer in this respect, since it is a highly sensitive (0·01 mmHg pressure is adequate) and specific analytical tool.

The microwave spectrum of a substance is very rich in lines since many rotational levels are populated at room temperatures, but since the lines are very sharp and their positions can be measured with great accuracy, observation of just a few of them is sufficient, after comparison with tabulated data, to establish the presence of a previously examined substance in a sample. And the technique is quantitative, since the intensity of a spectrum observed under given conditions is directly dependent on the amount of substance present. Thus mixtures can be readily analysed.

It is the *whole* molecule, by virtue of its moment(s) of inertia, which is examined by microwave spectroscopy. This means that the technique cannot detect the presence of particular molecular groupings in a sample, like —OH or —CH_3 (cf. the chapters on infra-red, raman, and magnetic resonance spectroscopy later), but it can readily distinguish the presence of isotopes in a sample, and it can even detect different conformational isomers, provided they have different moments of inertia.

One fascinating area where microwave analysis is being used is in the chemical examination of interstellar space. Electronic spectroscopy has long been able to detect the presence of various atoms, ions, and a few radicals (e.g., —OH) in the light of stars, but recently, use of microwaves has extended the analysis to the detection of simple stable molecules in space. Some 30 or so molecules have already been characterized in this way, the earliest among them (water, ammonia, and formaldehyde) giving new impetus to speculations regarding the origins of biological molecules and of life itself. Such observations concern the emission of microwaves by these molecules and, by comparing the relative intensities of various rotational transitions, particularly in the spectrum of ammonia, accurate estimates can be made of the temperature of interstellar material.

BIBLIOGRAPHY

Chantry, G. W.: *Modern Aspects of Microwave Spectroscopy*, Academic Press, 1979.

Gordy, W., and R. L. Cook: *Microwave Molecular Spectra* (*Techniques of Organic Chemistry*, vol. 9), John Wiley, 1970.

Gordy, W., W. V. Smith, and R. Trambarulo: *Microwave Spectroscopy*, John Wiley, 1966.

Ingram, D. J. E.: *Spectroscopy at Radio and Microwave Frequencies*, 2nd ed., Butterworth, 1967.

Kroto, H. W.: *Molecular Rotation Spectra*, Wiley Interscience, 1975.

Sugden, T. M., and N. C. Kenney: *Microwave Spectroscopy of Gases*, Van Nostrand, 1965.

Townes, C. H. and A. L. Schawlow: *Microwave Spectroscopy*, McGraw-Hill, 1955.

Wollrab, J. E.: *Rotational Spectra and Molecular Structure*, Academic Press, 1967.

PROBLEMS

(Useful constants: $h = 6 \cdot 626 \times 10^{-34}$ J s; $k = 1 \cdot 381 \times 10^{-23}$ J K^{-1}; $c = 2 \cdot 998 \times 10^8$ m s^{-1}; $8\pi^2 = 78 \cdot 956$; atomic masses: ^{1}H $= 1 \cdot 673 \times 10^{-27}$ kg; ^{2}D $= 3 \cdot 344 \times 10^{-27}$ kg; ^{19}F $= 31 \cdot 55 \times 10^{-27}$ kg; ^{35}Cl $= 58 \cdot 06 \times 10^{-27}$ kg; ^{37}Cl $= 61 \cdot 38 \times 10^{-27}$ kg; ^{79}Br $= 131 \cdot 03 \times 10^{-27}$ kg.)

2.1 The rotational spectrum of ^{79}Br^{19}F shows a series of equidistant lines spaced $0 \cdot 714\,33$ cm^{-1} apart. Calculate the rotational constant B, and hence the moment of inertia and bond length of the molecule. Determine the wavenumber of the $J = 9 \to J = 10$ transition, and find which transition gives rise to the most intense spectral line at room temperature (say 300 K).

2.2 Using your answers to Prob. 2.1, calculate the number of revolutions per second which the BrF molecule undergoes when in (*a*) the $J = 0$ state, (*b*) the $J = 1$ state, and (*c*) the $J = 10$ state.

Hint: Use $E = \frac{1}{2}I\omega^2$ in conjunction with Eqs (2.10) and (2.13), but remember that here ω is in *radians* per second.

2.3 The rotational constant for H^{35}Cl is observed to be $10 \cdot 5909$ cm^{-1}. What are the values of B for H^{37}Cl and for ^{2}D^{35}Cl?

2.4 Three consecutive lines in the rotational spectrum of H^{79}Br are observed at $84 \cdot 544$, $101 \cdot 355$ and $118 \cdot 112$ cm^{-1}. Assign the lines to their appropriate $J'' \to J'$ transitions, then deduce values for B and D, and hence evaluate the bond length and approximate vibrational frequency of the molecule.

2.5 Sketch a diagram similar to that of Fig. 2.7, using $B = 5$ cm^{-1} and a temperature of 1600 K.

Note: Find the maximum and calculate only two or three points on either side—don't attempt to carry out the calculation for every value of J.

2.6 The bond lengths of the linear molecule H—C≡N are given in Table 2.4. Calculate I and B for HCN and for DCN, using relative atomic masses of H = 1, D = 2, C = 12 and N = 14.

2.7 The diatomic molecule HCl has a B value of 10·593 cm^{-1} and a centrifugal distortion constant D of $5·3 \times 10^{-4}$ cm^{-1}. Estimate the vibrational frequency and force constant of the molecule. The observed vibrational frequency is 2991 cm^{-1}; explain the discrepancy.

THREE

INFRA-RED SPECTROSCOPY

We saw in the previous chapter how the elasticity of chemical bonds led to anomalous results in the rotational spectra of rapidly rotating molecules—the bonds stretched under centrifugal forces. In this chapter we consider another consequence of this elasticity—the fact that atoms in a molecule do not remain in fixed relative positions but vibrate about some mean position. We consider first the case of a diatomic molecule and the spectrum which arises if its only motion is vibration; then we shall deal with the more practical case of a diatomic molecule undergoing vibration and rotation simultaneously; finally we shall extend the discussion to more complex molecules.

3.1 THE VIBRATING DIATOMIC MOLECULE

3.1.1 The Energy of a Diatomic Molecule

When two atoms combine to form a stable covalent molecule (e.g., HCl gas) they may be said to do so because of some internal electronic rearrangement. This is not the place to discuss the detailed mechanisms of chemical bond formation; we may simply look on the phenomenon as a balancing of forces. On the one hand there is a repulsion between the positively charged nuclei of both atoms, and between the negative electron 'clouds'; on the other there is an attraction between the nucleus of one atom and the electrons of the other, and vice versa. The two atoms settle at a mean

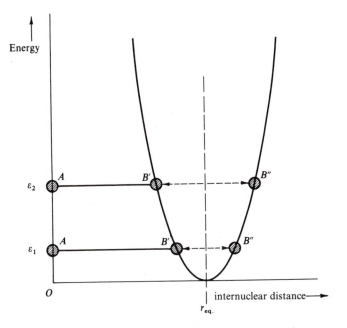

Figure 3.1 Parabolic curve of energy plotted against the extension or compression of a spring obeying Hooke's law.

internuclear distance such that these forces are just balanced and the energy of the whole system is at a minimum. Attempt to squeeze the atoms more closely together and the repulsive force rises rapidly; attempt to pull them further apart and we are resisted by the attractive force. In either case an attempt to distort the bond requires an input of energy and so we may plot energy against internuclear distance as in Fig. 3.1. At the minimum the internuclear distance is referred to as the equilbrium distance $r_{eq.}$, or more simply, as the bond length.

The compression and extension of a bond may be likened to the behaviour of a spring and we may extend the analogy by assuming that the bond, like a spring, obeys Hooke's law. We may then write

$$f = -k(r - r_{eq.}) \tag{3.1}$$

where f is the restoring force, k the force constant, and r the internuclear distance. In this case the energy curve is parabolic and has the form

$$E = \tfrac{1}{2}k(r - r_{eq.})^2 \tag{3.2}$$

This model of a vibrating diatomic molecule—the so-called simple harmonic oscillator model—while only an approximation, forms an excellent starting point for the discussion of vibrational spectra.

3.1.2 The Simple Harmonic Oscillator

In Fig. 3.1 we have plotted the energy according to Eq. (3.2). The zero of curve and equation is found at $r = r_{eq.}$, and any energy in excess of this, for example, ε_1, arises because of extension or compression of the bond. The figure shows that if one atom (A) is considered to be stationary on the $r = 0$ axis, the other will oscillate between B' and B''. If the energy is increased to ε_2 the oscillation will become more vigorous—that is to say, the degree of compression or extension will be greater—but the vibrational frequency will not change. An elastic bond, like a spring, has a certain vibration frequency dependent upon the mass of the system and the force constant, but independent of the amount of distortion. Classically it is easy to show that the oscillation frequency is:

$$\omega_{osc.} = \frac{1}{2\pi} \sqrt{\frac{k}{\mu}} \quad \text{Hz} \tag{3.3}$$

where μ is the reduced mass of the system (cf. Eq. (2.9)). To convert this frequency to wavenumbers, the unit most usually employed in vibrational spectroscopy, we must divide by the velocity of light, c, expressed in cm s^{-1} (cf. Sec. 1.1), obtaining:

$$\bar{\omega}_{osc.} = \frac{1}{2\pi c} \sqrt{\frac{k}{\mu}} \quad \text{cm}^{-1} \tag{3.4}$$

Vibrational energies, like all other molecular energies, are quantized, and the allowed vibrational energies for any particular system may be calculated from the Schrödinger equation. For the simple harmonic oscillator these turn out to be:

$$E_v = (v + \tfrac{1}{2}) h \omega_{osc.} \quad \text{joules} \qquad (v = 0, 1, 2, \ldots) \tag{3.5}$$

where v is called the *vibrational quantum number*. Converting to the spectroscopic units, cm^{-1}, we have:

$$\varepsilon_v = \frac{E_v}{hc} = (v + \tfrac{1}{2}) \bar{\omega}_{osc.} \quad \text{cm}^{-1} \tag{3.6}$$

as the only energies allowed to a simple harmonic vibrator. Some of these are shown in Fig. 3.2.

In particular we should notice that the *lowest* vibrational energy, obtained by putting $v = 0$ in Eq. (3.5) or (3.6), is

$$E_0 = \tfrac{1}{2} h \omega_{osc.} \quad \text{joules} \qquad [\omega_{osc.} \text{ in Hz}]$$

or

$$\varepsilon_0 = \tfrac{1}{2} \bar{\omega}_{osc.} \quad \text{cm}^{-1} \qquad [\bar{\omega}_{osc.} \text{ in cm}^{-1}] \tag{3.7}$$

The implication is that the diatomic molecule (and, indeed, *any* molecule) can never have zero vibrational energy; the atoms can never be completely

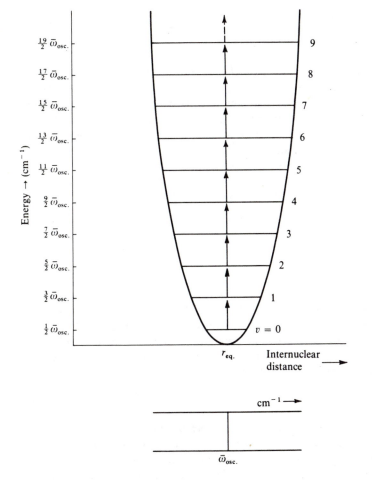

Figure 3.2 The allowed vibrational energy levels and transitions between them for a diatomic molecule undergoing simple harmonic motion.

at rest relative to each other. The quantity $\frac{1}{2}h\omega_{osc.}$ joules or $\frac{1}{2}\bar{\omega}_{osc.}$ cm^{-1} is known as the zero-point energy; it depends only on the classical vibration frequency and hence (Eq. (3.3) or (3.4)) on the strength of the chemical bond and the atomic masses.

The prediction of zero-point energy is the basic difference between the wave mechanical and classical approaches to molecular vibrations. Classical mechanics could find no objection to a molecule possessing no vibrational energy but wave mechanics insists that it must always vibrate to some extent; the latter conclusion has been amply borne out by experiment.

Further use of the Schrödinger equation leads to the simple *selection rule* for the harmonic oscillator undergoing vibrational changes:

$$\Delta v = \pm 1 \tag{3.8}$$

To this we must, of course, add the condition that vibrational energy changes will only give rise to an observable spectrum if the vibration can interact with radiation, i.e. (cf. Chapter 1), if the vibration involves a change in the dipole moment of the molecule. Thus vibrational spectra will be observable only in heteronuclear diatomic molecules since homonuclear molecules have no dipole moment.

Applying the selection rule we have immediately:

$$\varepsilon_{v+1 \to v} = (v + 1 + \tfrac{1}{2})\bar{\omega}_{\text{osc.}} - (v + \tfrac{1}{2})\bar{\omega}_{\text{osc.}}$$

$$= \bar{\omega}_{\text{osc.}} \quad \text{cm}^{-1} \tag{3.9a}$$

for emission and

$$\varepsilon_{v \to v+1} = \bar{\omega}_{\text{osc.}} \quad \text{cm}^{-1} \tag{3.9b}$$

for absorption, whatever the initial value of v.

Such a simple result is also obvious from Fig. 3.2—since the vibrational levels are equally spaced, transitions between any two neighbouring states will give rise to the same energy change. Further, since the difference between energy levels expressed in cm^{-1} gives directly the wavenumber of the spectral line absorbed or emitted

$$\bar{v}_{\text{spectroscopic}} = \varepsilon = \bar{\omega}_{\text{osc.}} \quad \text{cm}^{-1} \tag{3.10}$$

This, again, is obvious if one considers the mechanism of absorption or emission in classical terms. In absorption, for instance, the vibrating molecule will absorb energy only from radiation with which it can coherently interact (cf. Fig. 1.8) and this must be radiation of its own oscillation frequency.

3.1.3 The Anharmonic Oscillator

Real molecules do not obey exactly the laws of simple harmonic motion; real bonds, although elastic, are not so homogeneous as to obey Hooke's law. If the bond between atoms is stretched, for instance, there comes a point at which it will break—the molecule dissociates into atoms. Thus although for small compressions and extensions the bond may be taken as perfectly elastic, for larger amplitudes—say greater than 10 per cent of the bond length—a much more complicated behaviour must be assumed. Figure 3.3 shows, diagrammatically, the shape of the energy curve for a typical diatomic molecule, together with (dashed) the ideal, simple harmonic parabola.

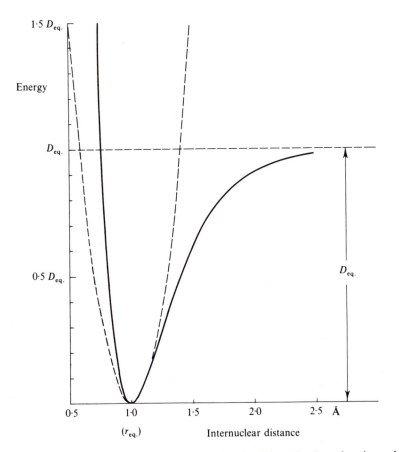

Figure 3.3 The Morse curve: the energy of a diatomic molecule undergoing anharmonic extensions and compressions.

A purely empirical expression which fits this curve to a good approximation was derived by P. M. Morse, and is called the Morse function:

$$E = D_{eq.}[1 - \exp\{a(r_{eq.} - r)\}]^2 \qquad (3.11)$$

where a is a constant for a particular molecule and $D_{eq.}$ is the dissociation energy.

When Eq. (3.11) is used instead of Eq. (3.2) in the Schrödinger equation, the pattern of the allowed vibrational energy levels is found to be:

$$\varepsilon_v = (v + \tfrac{1}{2})\bar{\omega}_e - (v + \tfrac{1}{2})^2\bar{\omega}_e x_e \quad \text{cm}^{-1} \qquad (v = 0, 1, 2, \ldots) \qquad (3.12)$$

where $\bar{\omega}_e$ is an oscillation frequency (expressed in wavenumbers) which we shall define more closely below, and x_e is the corresponding anharmonicity constant which, for bond stretching vibrations, is always small and positive

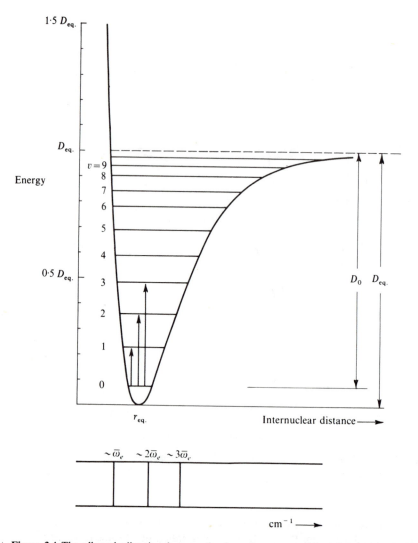

Figure 3.4 The allowed vibrational energy levels and some transitions between them for a diatomic molecule undergoing anharmonic oscillations.

($\approx + 0.01$), so that the vibrational levels crowd more closely together with increasing v. Some of these levels are sketched in Fig. 3.4.

It should be mentioned that Eq. (3.12), like (3.11), is an approximation only; more precise expressions for the energy levels require cubic, quartic, etc., terms in $(v + \frac{1}{2})$ with anharmonicity constants y_e, z_e, etc., rapidly diminishing in magnitude. These terms are important only at large values of v, and we shall ignore them.

If we rewrite Eq. (3.12), for the anharmonic oscillator, as:

$$\varepsilon_v = \bar{\omega}_e\{1 - x_e(v + \tfrac{1}{2})\}(v + \tfrac{1}{2}) \tag{3.13}$$

and compare with the energy levels of the *harmonic* oscillator (Eq. (3.6)), we see that we can write:

$$\bar{\omega}_{\text{osc.}} = \bar{\omega}_e\{1 - x_e(v + \tfrac{1}{2})\} \tag{3.14}$$

Thus the anharmonic oscillator behaves like the harmonic oscillator but with an oscillation frequency which decreases steadily with increasing v. If we now consider the hypothetical energy state obtained by putting $v = -\tfrac{1}{2}$ (at which, according to Eq. (3.13), $\varepsilon = 0$) the molecule would be at the equilibrium point with zero vibrational energy. Its oscillation frequency (in cm^{-1}) would be:

$$\bar{\omega}_{\text{osc.}} = \bar{\omega}_e$$

Thus we see that $\bar{\omega}_e$ may be defined as the (hypothetical) *equilibrium oscillation frequency* of the anharmonic system—the frequency for infinitely small vibrations about the equilibrium point. For any real state specified by a positive integral v the oscillation frequency will be given by Eq. (3.14). Thus in the ground state ($v = 0$) we would have:

$$\bar{\omega}_0 = \bar{\omega}_e(1 - \tfrac{1}{2}x_e) \quad \text{cm}^{-1}$$

and

$$\varepsilon_0 = \tfrac{1}{2}\bar{\omega}_e(1 - \tfrac{1}{2}x_e) \quad \text{cm}^{-1}$$

and we see that the zero point energy differs slightly from that for the harmonic oscillator (Eq. (3.7)).

The selection rules for the anharmonic oscillator are found to be:

$$\Delta v = \pm 1, \pm 2, \pm 3, \ldots$$

Thus they are the same as for the harmonic oscillator, with the additional possibility of larger jumps. These, however, are predicted by theory and observed in practice to be of rapidly diminishing probability and normally only the lines of $\Delta v = \pm 1, \pm 2$, and ± 3, at the most, have observable intensity. Further, the spacing between the vibrational levels is, as we shall shortly see, of order 10^3 cm^{-1} and, at room temperature, we may use the Boltzmann distribution (Eq. (1.12)) to show

$$\frac{N_{v=1}}{N_{v=0}} = \exp\left\{-\frac{6\cdot63 \times 10^{-34} \times 3 \times 10^{10} \times 10^3}{1\cdot38 \times 10^{-23} \times 300}\right\}$$

$$\approx \exp(-4\cdot8) \approx 0\cdot008.$$

In other words, the population of the $v = 1$ state is nearly $0\cdot01$ or some one per cent of the ground state population. Thus, to a very good approximation, we may ignore all transitions originating at $v = 1$ or more and restrict ourselves to the three transitions:

1. $v = 0 \rightarrow v = 1$, $\Delta v = +1$, with considerable intensity.

$$\Delta \varepsilon = \varepsilon_{v=1} - \varepsilon_{v=0}$$

$$= (1 + \tfrac{1}{2})\bar{\omega}_e - x_e(1 + \tfrac{1}{2})^2 \bar{\omega}_e - \{\tfrac{1}{2}\bar{\omega}_e - (\tfrac{1}{2})^2 x_e \bar{\omega}_e\}$$

$$= \bar{\omega}_e(1 - 2x_e) \quad \text{cm}^{-1} \qquad (3.15a)$$

2. $v = 0 \rightarrow v = 2$, $\Delta v = +2$, with small intensity.

$$\Delta \varepsilon = (2 + \tfrac{1}{2})\bar{\omega}_e - x_e(2 + \tfrac{1}{2})^2 \bar{\omega}_e - \{\tfrac{1}{2}\bar{\omega}_e - (\tfrac{1}{2})^2 x_e \bar{\omega}_e\}$$

$$= 2\bar{\omega}_e(1 - 3x_e) \quad \text{cm}^{-1} \qquad (3.15b)$$

3. $v = 0 \rightarrow v = 3$, $\Delta v = +3$, with normally negligible intensity.

$$\Delta \varepsilon = (3 + \tfrac{1}{2})\bar{\omega}_e - \{\tfrac{1}{2}\bar{\omega}_e - (\tfrac{1}{2})^2 x_e \bar{\omega}_e\}$$

$$= 3\bar{\omega}_e(1 - 4x_e) \quad \text{cm}^{-1} \qquad (3.15c)$$

These three transitions are shown in Fig. 3.4. To a good approximation, since $x_e \approx 0.01$, the three spectral lines lie very close to $\bar{\omega}_e$, $2\bar{\omega}_e$, and $3\bar{\omega}_e$. The line near $\bar{\omega}_e$ is called the *fundamental absorption*, while those near $2\bar{\omega}_e$ and $3\bar{\omega}_e$ are called the *first* and *second overtones*, respectively. The spectrum of HCl, for instance, shows a very intense absorption at 2886 cm^{-1}, a weaker one at 5668 cm^{-1}, and a very weak one at 8347 cm^{-1}. If we wish to find the equilibrium frequency of the molecule from these data, we must solve any two of the three equations (cf. Eqs. (3.15)):

$$\bar{\omega}_e(1 - 2x_e) = 2886$$

$$2\bar{\omega}_e(1 - 3x_e) = 5668$$

$$3\bar{\omega}_e(1 - 4x_e) = 8347 \text{ cm}^{-1}$$

and we find $\bar{\omega}_e = 2990 \text{ cm}^{-1}$, $x_e = 0.0174$. Thus we see that, whereas for the ideal harmonic oscillator the spectral absorption occurred *exactly* at the classical vibration frequency, for real, anharmonic molecules the observed fundamental absorption frequency and the equilibrium frequency may differ considerably.

The force constant of the bond in HCl may be calculated directly from Eq. (2.22) by inserting the value of $\bar{\omega}_e$:

$$k = 4\pi^2 \bar{\omega}_e^2 c^2 \mu \quad \text{N m}^{-1}$$

$$= 516 \text{ N m}^{-1}$$

when the fundamental constants and the reduced mass are inserted. These data, together with that for a few of the very many other diatomic molecules studied by infra-red techniques, are collected in Table 3.1.

Table 3.1 Some molecular data for diatomic molecules determined by infra-red spectroscopy

Molecule	Vibration (cm^{-1})	Anharmonicity constant, x_e	Force constant (N m^{-1})	Internuclear distance $r_{eq.}$ (nm)
HF	4138·5	0·0218	966	0·0927
HCl†	2990·6	0·0174	516	0·1274
HBr	2649·7	0·0171	412	0·1414
HI	2309·5	0·0172	314	0·1609
CO	2169·7	0·0061	1902	0·1131
NO	1904·0	0·0073	1595	0·1151
ICl†	384·2	0·0038	238	0·2321

† Data refers to the ^{35}Cl isotope.

Although we have ignored transitions from $v = 1$ to higher states, we should note that, if the temperature is raised or if the vibration has a particularly low frequency, the population of the $v = 1$ state may become appreciable. Thus at, say, 600 K (i.e., about 300°C) $N_{v=1}/N_{v=0}$ becomes exp $(-2·4)$ or about 0·09, and transitions from $v = 1$ to $v = 2$ will be some 10 per cent the intensity of those from $v = 0$ to $v = 1$. A similar increase in the excited state population would arise if the vibrational frequency were 500 cm^{-1} instead of 1000 cm^{-1}. We may calculate the wavenumber of this transition as:

4. $v = 1 \rightarrow v = 2$, $\Delta v = +1$, normally very weak,

$$\Delta\varepsilon = 2\tfrac{1}{2}\bar{\omega}_e - 6\tfrac{1}{4}x_e\bar{\omega}_e - \{1\tfrac{1}{2}\bar{\omega}_e - 2\tfrac{1}{4}x_e\bar{\omega}_e\}$$
$$= \bar{\omega}_e(1 - 4x_e) \quad \text{cm}^{-1} \tag{3.15d}$$

Thus, should this weak absorption arise, it will be found close to and at slightly *lower* wavenumber than the fundamental (since x_e is small and positive). Such weak absorptions are usually called *hot bands* since a high temperature is one condition for their occurrence. Their nature may be confirmed by raising the temperature of the sample when a true hot band will increase in intensity.

We turn now to consider a diatomic molecule undergoing simultaneous vibration and rotation.

3.2 THE DIATOMIC VIBRATING-ROTATOR

We saw in Chapter 2 that a typical diatomic molecule has rotational energy separations of 1–10 cm^{-1}, while in the preceding section we found that the vibrational energy separations of HCl were nearly 3000 cm^{-1}. Since the energies of the two motions are so different we may, as a first approximation, consider that a diatomic molecule can execute rotations and vibrations quite independently. This, which we shall call the Born–Oppenheimer approximation (although, cf. Eq. (6·1), this strictly includes electronic energies), is tantamount to assuming that the combined rotational–vibrational energy is simply the sum of the separate energies:

$$E_{\text{total}} = E_{\text{rot.}} + E_{\text{vib.}} \quad \text{(joules)}$$

$$\varepsilon_{\text{total}} = \varepsilon_{\text{rot.}} + \varepsilon_{\text{vib.}} \quad (\text{cm}^{-1}) \qquad (3.16)$$

We shall see later in what circumstances this approximation does not apply.

Taking the separate expressions for $\varepsilon_{\text{rot.}}$ and $\varepsilon_{\text{vib.}}$ from Eqs (2.26) and (3.12) respectively, we have:

$$\varepsilon_{J, v} = \varepsilon_J + \varepsilon_v$$
$$= BJ(J + 1) - DJ^2(J + 1)^2 + HJ^3(J + 1)^3 + \ldots$$
$$+ (v + \tfrac{1}{2})\bar{\omega}_e - x_e(v + \tfrac{1}{2})^2\bar{\omega}_e \quad \text{cm}^{-1} \qquad (3.17)$$

Initially, we shall ignore the small centrifugal distortion constants D, H, etc., and hence write

$$\varepsilon_{\text{total}} = \varepsilon_{J, v} = BJ(J + 1) + (v + \tfrac{1}{2})\bar{\omega}_e - x_e(v + \tfrac{1}{2})^2\bar{\omega}_e \qquad (3.18)$$

Note, however, that it is not logical to ignore D since this implies that we are treating the molecule as rigid, yet vibrating! The retention of D would have only a very minor effect on the spectrum.

The rotational levels are sketched in Fig. 3.5 for the two lowest vibrational levels, $v = 0$ and $v = 1$. There is, however, no attempt at scale in this diagram since the separation between neighbouring J values is, in fact, only some 1/1000 of that between the v values. Note that since the rotational constant B in Eq. (3.18) is taken to be the same for all J and v (a consequence of the Born–Oppenheimer assumption), the separation between two levels of given J is the same in the $v = 0$ and $v = 1$ states.

It may be shown that the selection rules for the combined motions are the same as those for each separately; therefore we have:

$$\Delta v = \pm 1, \pm 2, \text{ etc.} \qquad \Delta J = \pm 1 \qquad (3.19)$$

Strictly speaking we may also have $\Delta v = 0$, but this corresponds to the purely rotational transitions already dealt with in Chapter 2. Note carefully, however, that a *diatomic* molecule, except under very special and rare circumstances, may *not* have $\Delta J = 0$; in other words a vibrational change *must* be accompanied by a simultaneous rotational change.

In Fig. 3.6 we have drawn some of the relevant energy levels and transitions, designating rotational quantum numbers in the $v = 0$ state as J'' and in the $v = 1$ state as J'. The use of a single prime for the upper state

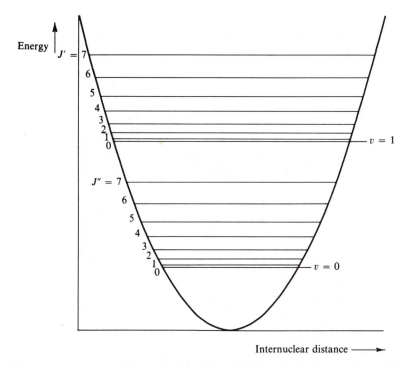

Energy

$J' = 7$
6
5
4
3
2
1
0

$v = 1$

$J'' = 7$
6
5
4
3
2
1
0

$v = 0$

Internuclear distance ⟶

Figure 3.5 The rotational energy levels for two different vibrational states of a diatomic molecule.

and a double for the lower state is conventional in all branches of spectroscopy.

Remember (and cf. Eq. (2.20)) that the rotational levels J'' are filled to varying degrees in any molecular population, so the transitions shown will occur with varying intensities. This is indicated schematically in the spectrum at the foot of Fig. 3.6.

An analytical expression for the spectrum may be obtained by applying the selection rules (Eq. (3.19)) to the energy levels (Eq. (3.18)). Considering only the $v = 0 \rightarrow v = 1$ transition we have in general:

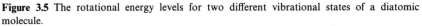

$$\Delta\varepsilon_{J,v} = \varepsilon_{J',v=1} - \varepsilon_{J'',v=0}$$
$$= BJ'(J'+1) + 1\tfrac{1}{2}\bar\omega_e - 2\tfrac{1}{4}x_e\bar\omega_e - \{BJ''(J''+1) + \tfrac{1}{2}\bar\omega_e - \tfrac{1}{4}x_e\bar\omega_e\}$$
$$= \bar\omega_o + B(J'-J'')(J'+J''+1) \quad \text{cm}^{-1}$$

where, for brevity, we write $\bar\omega_o$ for $\bar\omega_e(1-2x_e)$.

We should note that taking B to be identical in the upper and lower vibrational states is a direct consequence of the Born–Oppenheimer approximation—rotation is unaffected by vibrational changes.

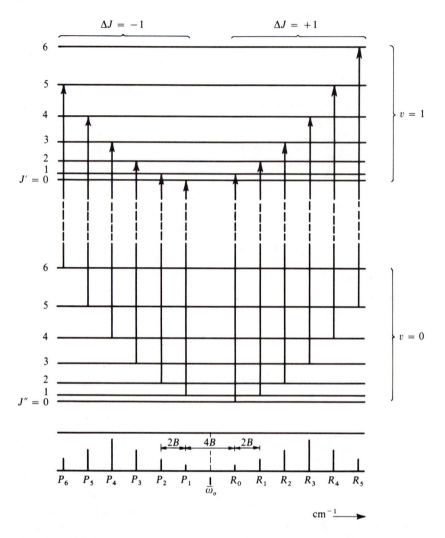

Figure 3.6 Some transitions between the rotational–vibrational energy levels of a diatomic molecule together with the spectrum arising from them.

Now we can have:

1. $\Delta J = +1$, i.e., $J' = J'' + 1$ or $J' - J'' = +1$; hence

$$\Delta\varepsilon_{J,\,v} = \bar{\omega}_o + 2B(J'' + 1) \text{ cm}^{-1} \qquad J'' = 0, 1, 2, \ldots \qquad (3.20a)$$

2. $\Delta J = -1$, i.e., $J'' = J' + 1$ or $J' - J'' = -1$; and

$$\Delta\varepsilon_{J,\,v} = \bar{\omega}_o - 2B(J' + 1) \text{ cm}^{-1} \qquad J' = 0, 1, 2, \ldots \qquad (3.20b)$$

These two expressions may conveniently be combined into:

$$\Delta \varepsilon_{J,\,v} = \bar{v}_{\text{spect.}} = \bar{\omega}_o + 2Bm \quad \text{cm}^{-1} \qquad m = \pm 1,\ \pm 2, \ldots \quad (3.20c)$$

where m, replacing $J'' + 1$ in Eq. (3.20a) and $J' + 1$ in Eq. (3.20b) has positive values for $\Delta J = +1$ and is negative if $\Delta J = +1$. Note particularly that m *cannot be zero* since this would imply values of J' or J'' to be -1. The frequency $\bar{\omega}_o$ is usually called the *band origin* or *band centre*.

Equation (3.20c), then, represents the combined vibration–rotation spectrum. Evidently it will consist of equally spaced lines (spacing $= 2B$) on each side of the band origin $\bar{\omega}_o$, but, since $m \neq 0$, the *line at $\bar{\omega}_o$ itself will not appear*. Lines to the low-frequency side of $\bar{\omega}_o$, corresponding to negative m (that is, $\Delta J = -1$) are referred to as the *P branch*, while those to the high-frequency side (m positive, $\Delta J = +1$) are called the *R branch*. This apparently arbitrary notation may become clearer if we state here that later, in other contexts, we shall be concerned with ΔJ values of 0 and ± 2, in addition to ± 1 considered here; the labelling of line series is then quite consistent:

Lines arising from $\Delta J =$	-2	-1	0	$+1$	$+2$	
called:	O	P	Q	R	S	branch

The P and R notation, with the *lower J* (J'') value as a suffix, is illustrated on the diagrammatic spectrum of Fig. 3.6. This is the conventional notation for such spectra.

It is readily shown that the inclusion of the centrifugal distortion constant D leads to the following expression for the spectrum:

$$\Delta \varepsilon = \bar{v}_{\text{spect.}} = \bar{\omega}_o + 2Bm - 4Dm^3 \quad \text{cm}^{-1} \qquad (m = \pm 1,\ \pm 2,\ \pm 3, \ldots)$$

$$(3.21)$$

But we have seen in Chapter 2 that B is some 10 cm^{-1} or less, while D is only some 0·01 per cent of B. Since a good infra-red spectrometer has a resolving power of about 0·5 cm^{-1} it is obvious that D is negligible to a very high degree of accuracy.

The anharmonicity factor, on the other hand, is not negligible. It affects not only the position of the band origin (since $\bar{\omega}_o = \bar{\omega}_e(1 - 2x_e)$), but, by extending the selection rules to include $\Delta v = \pm 2,\ \pm 3$, etc., also allows the appearance of overtone bands having identical rotational structure. This is illustrated in Fig. 3.7(a), where the fundamental absorption and first overtone of carbon monoxide are shown. From the band centres we can calculate, as shown in Sec. 1.3, the equilibrium frequency $\bar{\omega}_e$ and the anharmonicity constant x_e.

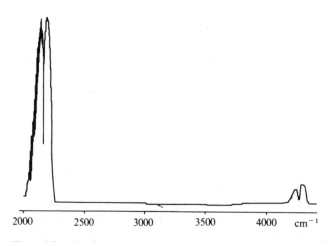

Figure 3.7(*a*) The fundamental absorption (centred at about 2143 cm^{-1}) and the first overtone (centred at about 4260 cm^{-1}) of carbon monoxide. The fine structure of the *P* branch in the fundamental is partially resolved. (Gas pressure 650 mmHg in a 10 cm cell.)

3.3 THE VIBRATION–ROTATION SPECTRUM OF CARBON MONOXIDE

In Fig. 3.7(*b*) we show the fundamental vibration–rotation band of carbon monoxide under high resolution, with some lines in the *P* and *R* branches numbered according to their *J″* values. Table 3.2 gives the observed wavenumbers of the first five lines in each branch. We shall discuss shortly the slight decrease in separation between the rotational lines as the wavenumber increases; this decrease is apparent from the table and from a close inspection of the 'wings' of the spectrum.

From the table we see that the band centre is at about 2143 cm^{-1} while the average line separation near the centre is 3·83 cm^{-1}. This immediately gives:

$$2B = 3·83 \text{ cm}^{-1} \qquad B = 1·915 \text{ cm}^{-1}$$

Table 3.2 Part of the infra-red spectrum of carbon monoxide

Line	$\bar{\nu}$	Separation $\Delta\bar{\nu}$	Line	$\bar{\nu}$	Separation $\Delta\bar{\nu}$
$P_{(1)}$	2139·43		$R_{(0)}$	2147·08	
		3·88			3·78
$P_{(2)}$	2135·55		$R_{(1)}$	2150·86	
		3·92			3·73
$P_{(3)}$	2131·63		$R_{(2)}$	2154·59	
		3·95			3·72
$P_{(4)}$	2127·68		$R_{(3)}$	2158·31	
		3·98			3·66
$P_{(5)}$	2123·70		$R_{(4)}$	2161·97	

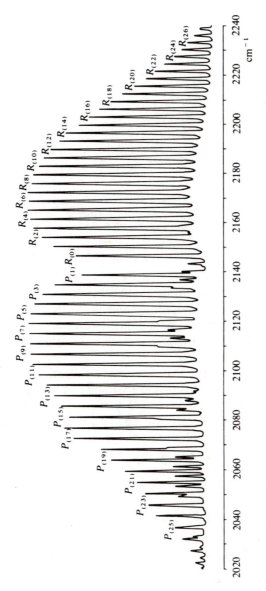

Figure 3.7(b) The centre of the fundamental band of carbon monoxide under higher resolution than in in (*a*). (Gas pressure 100 mmHg in a 10 cm cell.) The lines are labelled according to their *J″* values. The *P* branch is complicated by the presence of a band centred at about 2100 cm^{-1} due to the one per cent of ^{13}CO in the sample; some of the rotational lines from this band appear between *P* branch lines, others are overlapped by a *P* branch line and give it an enhanced intensity (e.g. lines $P_{(16)}$, $P_{(17)}$, $P_{(23)}$, and $P_{(24)}$).

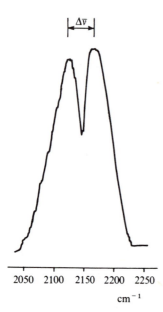

2050   2100   2150   2200   2250

cm^{-1}

Figure 3.8 The fundamental band of Fig. 3.7(*b*) under very low resolution. All rotational fine structure has been lost and a typical PR contour is seen.

This is in satisfactory agreement with the value $B = 1.921\,18\ cm^{-1}$ derived by microwave studies (cf. Sec. 2.3.1) and we could, therefore, have obtained quite good values for the rotational constant and hence the moment of inertia and bond length from infra-red data alone. Historically, of course, the infra-red values came first, the more precise microwave values following much later.

It is worth noting at this point that approximate rotational data is obtainable from spectra even if the separate rotational lines are not re-solved. Thus Fig. 3.8 shows the spectrum of carbon monoxide under much poorer resolution, when the rotational fine structure is blurred out to an envelope. Now we saw in Eq. (2.21) that the maximum population of levels, and hence maximum intensity of transition, occurs at a J value of $\sqrt{kT/2Bhc} - \frac{1}{2}$. Remembering that $m = J + 1$ we substitute in Eq. (3.20c) $m = \pm\sqrt{kT/2Bhc} + \frac{1}{2}$ and obtain:

$$\bar{v}_{max.\ intensity} = \bar{\omega}_0 \pm 2B(\sqrt{kT/2Bhc} + \tfrac{1}{2})$$

where the $+$ and $-$ signs refer to the R and P branches, respectively. The *separation* between the two maxima, $\Delta\bar{v}$, is then:

$$\Delta\bar{v} = 4B(\sqrt{kT/2Bhc} + \tfrac{1}{2}) = \sqrt{8kTB/hc} + 2B$$

or, since B is small compared with $\Delta\bar{v}$, we can write

$$\Delta\bar{v} \approx \sqrt{8kTB/hc}, \qquad B \approx hc(\Delta\bar{v})^2/8kT \quad cm^{-1}$$

where c is expressed in cm s^{-1}. In the case of carbon monoxide the separation is about 55 cm^{-1} (Fig. 3.8), while the temperature at which the spectrum was obtained was about 300 K. We are led, then, to a B value of about 1·8 cm^{-1} which is in fair agreement with the earlier values, but much less precise.

From Table 3.2 we see that the band origin, at the midpoint of $P_{(1)}$ and $R_{(0)}$, is at 2143·26 cm^{-1}. This, then, is the fundamental vibration frequency of carbon monoxide, if anharmonicity is ignored. The latter can be taken into account, however, since the first overtone is found to have its origin at 4260·04 cm^{-1}. We have:

$$\bar{\omega}_e(1 - 2x_e) = \bar{\omega}_o = 2143·26$$

$$2\bar{\omega}_e(1 - 3x_e) = 4260·04$$

from which $\omega_e = 2169·74$ cm^{-1}, $x_e = 0·0061$.

3.4 BREAKDOWN OF THE BORN–OPPENHEIMER APPROXIMATION: THE INTERACTION OF ROTATIONS AND VIBRATIONS

So far we have assumed that vibration and rotation can proceed quite independently of each other. A molecule vibrates some 10^3 times during the course of a single rotation, however, so it is evident that the bond length (and hence the moment of inertia and B constant) also changes continually during the rotation. If the vibration is simple harmonic the mean bond length will be the same as the equilibrium bond length and it will not vary with vibrational energy; this is seen in Fig. 3.1. However, the rotational constant B depends on $1/r^2$ and, as shown by an example in Sec. 2.3.4, the average value of this quantity is not the same as $1/r_{eq}^2$ where r_{eq} is the equilibrium length. Further an increase in the vibrational energy is accompanied by an increase in the vibrational amplitude and hence the value of B will depend on the v quantum number.

In the case of anharmonic vibrations the situation is rather more complex. Now an increase in vibrational energy will lead to an increase in the average bond length—this is perhaps most evident from Fig. 3.4. The rotational constant then varies even more with vibrational energy.

In general, it is plain that, since $r_{av.}$ *increases* with the vibrational energy, B is *smaller* in the upper vibrational state than in the lower. In fact an equation of the form:

$$B_v = B_e - \alpha(v + \tfrac{1}{2}) \tag{3.22}$$

gives, to a high degree of approximation, the value of B_v, the rotational constant in vibrational level v, in terms of the equilibrium value B_e and α, a small positive constant for each molecule.

Here we restrict our discussion to the fundamental vibrational change, i.e., the change $v = 0 \rightarrow v = 1$, and we may take the respective B values as B_0 and B_1 with $B_0 > B_1$. For this transition:

$$\Delta\varepsilon = \varepsilon_{J', v=1} - \varepsilon_{J'', v=0}$$

$$= \bar{\omega}_o + B_1 J'(J' + 1) - B_0 J''(J'' + 1) \text{ cm}^{-1}$$

where, as before, $\bar{\omega}_o = \bar{\omega}_e(1 - 2x_e)$.

We then have the two cases:

1. $$\Delta J = +1 \qquad J' = J'' + 1$$

$$\Delta\varepsilon = \bar{v}_R = \bar{\omega}_o + (B_1 + B_0)(J'' + 1) + (B_1 - B_0)(J'' + 1)^2 \text{ cm}^{-1}$$

$$(J'' = 0, 1, 2, \ldots) \quad (3.23a)$$

and

2. $$\Delta J = -1 \qquad J'' = J' + 1$$

$$\Delta\varepsilon = \bar{v}_P = \bar{\omega}_o - (B_1 + B_0)(J' + 1) + (B_1 - B_0)(J' + 1)^2 \text{ cm}^{-1}$$

$$(J' = 0, 1, 2, \ldots) \quad (3.23b)$$

where we have written $\bar{v}_P$ and $\bar{v}_R$ to represent the wavenumbers of the P and R branch lines respectively. These two equations can be combined into the expression:

$$\bar{v}_{P,R} = \bar{\omega}_o + (B_1 + B_0)m + (B_1 - B_0)m^2 \quad \text{cm}^{-1} \qquad (m = \pm 1, \pm 2, \ldots)$$

$$(3.23c)$$

where positive m values refer to the R branch and negative to P.

We see that ignoring vibration–rotation interaction involves setting $B_1 = B_0$, when Eq. (3.23c) immediately simplifies to (3.20c). Since $B_1 < B_0$ the last term of (3.23c) is always negative, irrespective of the sign of m, and the effect on the spectrum of a diatomic molecule is to crowd the rotational lines more closely together with increasing m on the R branch side, while the P branch lines become more widely spaced as (negative) m increases. Normally B_1 and B_0 differ only slightly and the effect is marked only for high m values. This is exactly the situation shown in the spectrum of carbon monoxide, Fig. 3.7(b).

In Table 3.3 some of the data for carbon monoxide are tabulated, together with the positions of lines calculated from the equation:

$$\bar{v}_{\text{spect.}} = 2143 \cdot 28 + 3 \cdot 813m - 0 \cdot 0175m^2 \quad \text{cm}^{-1}.$$

From this we see that, for this molecule:

$$B_1 = 1 \cdot 898 \text{ cm}^{-1} \qquad B_0 = 1 \cdot 915 \text{ cm}^{-1}$$

Table 3.3 Observed and calculated wavenumbers of some lines in the spectrum of carbon monoxide

m	J''	$\bar{v}_{obs.}$	$\bar{v}_{calc.}$ †
30	29	2241·64	2241·91
25	24	2227·63	2227·65
20	19	2212·62	2212·54
15	14	2196·66	2196·53
10	9	2179·77	2179·66
5	4	2161·97	2161·90
0	—	(Band centre)	2143·28
− 5	5	2123·70	2123·78
− 10	10	2103·27	2103·40
− 15	15	2082·01	2082·15
− 20	20	2059·91	2060·02
− 25	25	2037·03	2037·02
− 30	30	2013·35	2013·14

† Values calculated from: $\bar{v} = 2143·28 + 3·813m - 0·0175m^2$.

and hence, using Eq. (3.22), we have:

$$\alpha = 0·018 \qquad B_e = 1·924 \text{ cm}^{-1}$$

Further, we can calculate the equilibrium bond length and the bond lengths in the $v = 0$ and $v = 1$ states (cf. pp. 46–47) to be:

$$r_{eq.} = 0·1130 \text{ nm} \qquad r_0 = 0·1133 \text{ nm} \qquad r_1 = 0·1136 \text{ nm}$$

3.5 THE VIBRATIONS OF POLYATOMIC MOLECULES

In this section and the next, just as in the corresponding one dealing with the pure rotational spectra of polyatomic molecules, we shall find that although there is an increase in the complexity, only slight and quite logical extensions to the simple theory are adequate to give us an understanding of the spectra. We shall need to discuss:

1. The number of fundamental vibrations and their symmetry
2. The possibility of overtone and combination bands
3. The influence of rotation on the spectra.

3.5.1 Fundamental Vibrations and their Symmetry

Consider a molecule containing N atoms: we can refer to the position of each atom by specifying three coordinates (e.g., the x, y, and z cartesian coordinates). Thus the total number of coordinate values is $3N$ and we say the molecule has $3N$ *degrees of freedom* since each coordinate value may be specified quite independently of the others. However, once all $3N$ coordinates have been fixed, the bond distances and bond angles of the molecule are also fixed and no further arbitrary specifications can be made.

Now the molecule is free to move in three-dimensional space, as a whole, without change of shape. We can refer to such movement by noting the position of its centre of gravity at any instant—to do this requires a statement of three coordinate values. This translational movement uses three of the $3N$ degrees of freedom leaving $3N - 3$. In general, also, the rotation of a non-linear molecule can be resolved into components about three perpendicular axes (cf. Sec. 1.1). Specification of these axes also requires three degrees of freedom, and the molecule is left with $3N - 6$ degrees of freedom. The only other motion allowed to it is internal vibration, so we know immediately that a non-linear N-atomic molecule can have $3N - 6$ different internal vibrations:

$$\text{Non-linear:} \quad 3N - 6 \text{ fundamental vibrations} \qquad (3.24a)$$

If, on the other hand, the molecule is linear, we saw in Chapter 2 that there is no rotation about the bond axis; hence only two degrees of rotational freedom are required, leaving $3N - 5$ degrees of vibrational freedom—one more than in the case of a non-linear molecule:

$$\text{Linear:} \quad 3N - 5 \text{ fundamental vibrations} \qquad (3.24b)$$

In both cases, since an N-atomic molecule has $N - 1$ bonds (for acyclic molecules) between its atoms, $N - 1$ of the vibrations are bond-stretching motions, the other $2N - 5$ (non-linear) or $2N - 4$ (linear) are bending motions.

Let us look briefly at examples of these rules. First, we see that for a diatomic molecule (perforce linear) such as we have already considered in this chapter: $N = 2$, $3N - 5 = 1$ and thus there can be only one fundamental vibration. Note, however, that the $3N - 5$ rule says nothing about the presence, absence, or intensity of overtone vibrations—these are governed solely by anharmonicity.

Next, consider water, H_2O. This (Fig. 3.9) is non-linear and triatomic. Also in the figure are the $3N - 6 = 3$ allowed vibrational modes, the arrows attached to each atom showing the direction of its motion during half of the vibration. Each motion is described as stretching or bending depending on the nature of the change in molecular shape.

These three vibrational motions are also referred to as the *normal modes of vibration* (or *normal vibrations*) of the molecule; in general a normal vibration is defined as a molecular motion in which all the atoms move in phase and with the same frequency.

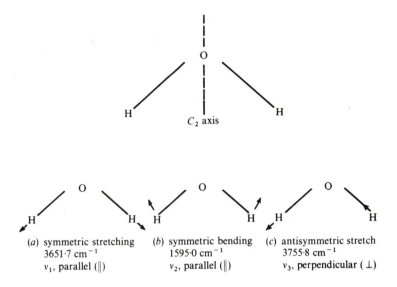

(a) symmetric stretching
3651·7 cm^{-1}
v_1, parallel ($\parallel$)

(b) symmetric bending
1595·0 cm^{-1}
v_2, parallel ($\parallel$)

(c) antisymmetric stretch
3755·8 cm^{-1}
v_3, perpendicular ($\perp$)

Figure 3.9 The symmetry of the water molecule and its three fundamental vibrations. The motion of the oxygen atom, which must occur to keep the centre of gravity of the molecule stationary, is here ignored.

Further each motion of Fig. 3.9 is labelled either symmetric or antisymmetric. It is not necessary here to go far into the matter of general molecular symmetry since other excellent texts already exist for the interested student, but we can see quite readily that the water molecule contains some elements of symmetry. In particular consider the dashed line at the top of Fig. 3.9 which bisects the HOH angle; if we rotate the molecule about this axis by 180° its final appearance is identical with the initial one. This axis is thus referred to as a C_2 axis since twice in every complete revolution the molecule presents an identical aspect to an observer. This particular molecule has only the one rotational symmetry axis, and it is conventional to refer the molecular vibrations to this axis. Thus consider the first vibration, Fig. 3.9(a). If we rotate the *vibrating* molecule by 180° the vibration is quite unchanged in character—we call this a symmetric vibration. The bending vibration, v_2, is also symmetric. Rotation of the stretching motion of Fig. 3.9(c) about the C_2 axis, however, produces a vibration which is in antiphase with the original and so this motion is described as the antisymmetric stretching mode.

In order to be infra-red active, as we have seen, there must be a dipole change during the vibration and this change may take place either along the line of the symmetry axis (parallel to it, or $\parallel$) or at right angles to the line (perpendicular, $\perp$). Figure 3.10 shows the nature of the dipole changes for the three vibrations of water, and justifies the labels parallel or perpen-

Vibrational Mode	Distorted Molecule	Normal Molecule	Distorted Molecule

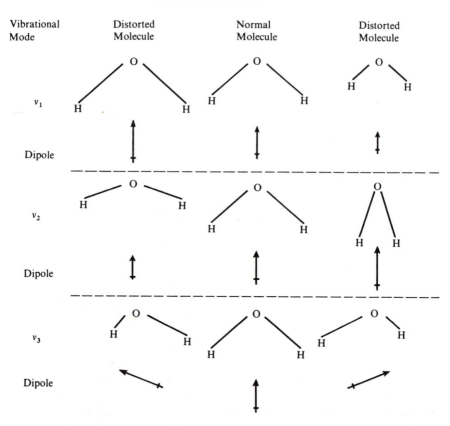

Figure 3.10 The change in the electric dipole moment produced by each vibration of the water molecule. This is seen to occur either along (∥) or across (⊥), the symmetry axis. The amplitudes are greatly exaggerated for clarity.

dicular attached to them in Fig. 3.9. We shall see later that the distinction is important when considering the influence of *rotation* on the spectrum.

Finally the vibrations are labelled in Fig. 3.9 as v_1, v_2, and v_3. By convention it is usual to label vibrations in decreasing frequency within their symmetry type. Thus the symmetric vibrations of H_2O are labelled v_1 for the highest fully symmetric frequency (3651·7 cm^{-1}), and v_2 for the next highest (1595·0 cm^{-1}); the antisymmetric vibration at 3755·8 cm^{-1} is then labelled v_3.

Our final example is of the linear triatomic molecule CO_2, for which the normal vibrations are shown in Fig. 3.11. For this molecule there are two different sets of symmetry axes. There is an infinite number of twofold axes (C_2) passing through the carbon atom at right angles to the bond direction, and there is an ∞-fold axis (C_∞) passing through the bond axis

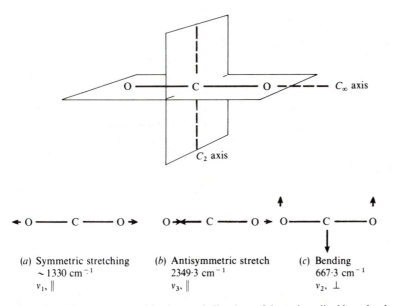

(a) Symmetric stretching
~ 1330 cm⁻¹
v_1, ‖

(b) Antisymmetric stretch
2349·3 cm⁻¹
v_3, ‖

(c) Bending
667·3 cm⁻¹
v_2, ⊥

Figure 3.11 The symmetry and fundamental vibrations of the carbon dioxide molecule.

itself (this is referred to as ∞-fold since rotation of the molecule about the bond axis through *any* angle gives an identical aspect). The names symmetric stretch and antisymmetric stretch are self-evident, but it should be noted that the symmetric stretch produces no change in the dipole moment (which remains zero) so that this vibration is not infra-red active; the vibration frequency may be obtained in other ways, however, which we shall discuss in the next chapter.

For linear triatomic molecules, $3N - 5 = 4$, and we would expect four vibrational modes instead of the three shown in Fig. 3.11. However, consideration shows that v_2 in fact consists of *two* vibrations—one in the plane of the paper as drawn, and the other in which the oxygen atoms move simultaneously into and out of the plane. The two sorts of motion are, of course, identical in all respects except direction and are termed *degenerate*; they must, nevertheless, be considered as separate motions, and it is always in the degeneracy of a bending mode that the extra vibration of a linear molecule over a non-linear one is to be found.

It might be thought that v_2 of H_2O (Fig. 3.9(b)) could occur by the hydrogens moving simultaneously in and out of the plane of the paper. Such a motion is not a vibration, however, but a *rotation*. As the molecule approaches linearity this rotation degenerates into a vibration, and the molecule loses one degree of rotational freedom in exchange for one of vibration.

3.5.2 Overtone and Combination Frequencies

If one were able to observe the molecules of H_2O or CO_2 directly their overall vibrations would appear extremely complex; in particular, each atom would not follow tidily any one of the separate paths depicted in Figs 3.9 or 3.11, but its motion would essentially be a superposition of all such paths, since every possible vibration is always excited, at least to the extent of its zero point energy. However, such superposition could be resolved into its components if, for instance, we could examine the molecules under stroboscopic light flashing at each fundamental frequency in turn. This is, so to speak, the essence of infra-red spectroscopy—instead of flashing we have the radiation frequency, and the 'examination' is a sensing of dipole alteration. Thus, as we would expect, the infra-red spectrum of a complex molecule consists essentially of an absorption band at each of the $3N - 6$ (non-linear) or $3N - 5$ (linear) fundamental frequencies.

This is, of course, an over-simplification, in which two approximations are implicit: (1) that each vibration is simple harmonic, (2) that each vibration is quite independent and unaffected by the others. We shall consider (2) in more detail later; for the moment we can accept it as a good working approximation.

When the restriction to simple harmonic motion is lifted we have again, as in the case of the diatomic molecule (Sec. 3.1.3), the possibility of first, second, etc., overtones occurring at frequencies near $2v_1, 3v_1, \ldots, 2v_2, 3v_2, \ldots, 2v_3, \ldots$, etc., where each v_i is a fundamental mode. The intensities fall off rapidly. However, in addition, the selection rules now permit *combination bands* and *difference bands*. The former arise simply from the addition of two or more fundamental frequencies or overtones. Such combinations as $v_1 + v_2$, $2v_1 + v_2$, $v_1 + v_2 + v_3$, etc., become allowed, although their intensities are normally very small. Similarly the difference bands, for example, $v_1 - v_2$, $2v_1 - v_2$, $v_1 + v_2 - v_3$, have small intensities but are often to be found in a complex spectrum.

The intensities of overtone or combination bands may sometimes be considerably enhanced by a resonance phenomenon. It may happen that two vibrational modes in a particular molecule have frequencies very close to each other—they are described as *accidentally* degenerate. Note that we are not here referring to identical vibrations, such as the two identical v_2's of CO_2 (Fig. 3.11), but rather to the possibility of two quite different modes having similar energies. Normally the fundamental modes are quite different from each other and accidental degeneracy is found most often between a fundamental and some overtone or combination. A simple example is to be found in CO_2 where v_1, described as at about 1330 cm^{-1}, is very close to that of $2v_2 = 1334 \text{ cm}^{-1}$. (As mentioned earlier, these bands are not observable in the infra-red, but both may be seen in the Raman spectrum discussed in the next chapter; the principles of resonance apply equally to both techniques.) Quantum mechanics shows that two such bands may interfere with each other in such a way that the higher is raised in frequency, the lower depressed—and in fact the Raman spectrum shows two bands, one at 1285 cm^{-1}, the other at 1385 cm^{-1}. Their mean is plainly at about 1330 cm^{-1}.

Note, however, that one of these bands arises from a fundamental mode (v_1), the other from the overtone $2v_2$, and we would normally expect the former to be much more intense than the latter. In fact, they are found to be of about the same intensity—the overtone has gained intensity at the expense of the fundamental. This is an extreme case—normally the overtone takes only a small part of the intensity from the fundamental. The situation is often likened to that of two pendulums connected to a common bar—when the pendulums have quite different frequencies they oscillate independently; when their frequencies are similar they can readily exchange energy, one with the other, and an oscillation given to one is transferred to and fro between them. They are said to resonate. Similarly two close molecular vibrational frequencies resonate and exchange energy—the phenomenon being known as *Fermi resonance* when a fundamental resonates with an overtone. In the spectrum of a complex molecule exhibiting many fundamentals and overtones, there is a good chance of accidental degeneracy, and Fermi resonance, occurring. However, it should be mentioned that not all such degeneracies lead to resonance. It is necessary, also, to consider the molecular symmetry and the type of degenerate vibrations; we shall not, however, pursue the topic further here.

3.6 THE INFLUENCE OF ROTATION ON THE SPECTRA OF POLYATOMIC MOLECULES

In Sec. 3.2 we found that the selection rule for the simultaneous rotation and vibration of a diatomic molecule was

$$\Delta v = \pm 1, \pm 2, \pm 3, \dots \qquad \Delta J = \pm 1 \qquad \Delta J \neq 0$$

and that this gave rise to a spectrum consisting of approximately equally spaced line series on each side of a central minimum designated as the band centre.

Earlier in the present section we showed that the vibrations of complex molecules could be subdivided into those causing a dipole change either (1) parallel or (2) perpendicular to the major axis of rotational symmetry. The purpose of this distinction, and the reason for repeating it here, is that the selection rules for the *rotational* transitions of complex molecules depend, rather surprisingly, on the type of *vibration*, ∥ or ⊥, which the molecule is undergoing. Less surprisingly, the selection rules and the energies depend on the *shape* of the molecule also. We shall deal first with the linear molecule as the simplest, and then say a few words about the other types of molecule.

3.6.1 Linear Molecules

Parallel vibrations The selection rule for these is identical with that for diatomic molecules, i.e.,

$$\Delta J = \pm 1 \qquad \Delta v = \pm 1 \qquad\qquad \text{for simple harmonic motion} \qquad (3.25a)$$

$$\Delta J = \pm 1 \qquad \Delta v = \pm 1, \pm 2, \pm 3, \ldots \quad \text{for anharmonic motion} \qquad (3.25b)$$

(This is, in fact, as expected, since a diatomic molecule is linear and can undergo only parallel vibrations.) The spectra will thus be similar in appearance, consisting of P and R branches with lines about equally spaced on each side, no line occurring at the band centre. Now, however, the moment of inertia may be considerably larger, the B value correspondingly smaller, and the P or R line spacing will be less. Figure 3.12 shows part of the spectrum of HCN, a linear molecule whose structure is H—C≡N. The band concerned is the symmetric stretching frequency at about 3310 cm^{-1}, (corresponding to the v_1 mode of CO_2 in Fig. 3.11), and the spacing is observed to be about 2·8–3·0 cm^{-1} near the band centre. This is to be compared, for example, with the spacing of about 4·0 cm^{-1} in the case of CO.

For still larger molecules the value of B may be so small that separate lines can no longer be resolved in the P and R branches. In this case the situation is exactly analogous to that shown previously in Fig. 3.8 and the same remarks apply as to the possibility of deriving a rough value of B from the separation between the maxima of the P and R envelopes. We shall shortly see that a non-linear molecule cannot give rise to this type of band

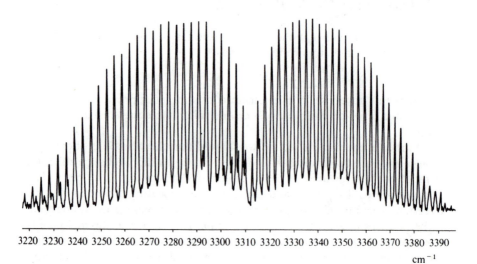

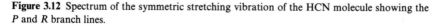

3220 3230 3240 3250 3260 3270 3280 3290 3300 3310 3320 3330 3340 3350 3360 3370 3380 3390

cm^{-1}

Figure 3.12 Spectrum of the symmetric stretching vibration of the HCN molecule showing the P and R branch lines.

shape, so its observation somewhere within a spectrum is sufficient proof that a linear, or nearly linear, molecule is being studied.

Perpendicular vibrations   For these the selection rule is found to be:

$$\Delta v = \pm 1 \qquad \Delta J = 0, \pm 1 \qquad \text{for simple harmonic motion} \qquad (3.26)$$

which implies that now, for the first time, a vibrational change can take place with *no simultaneous rotational transition*. The result is illustrated in Fig. 3.13, which shows the same energy levels and transitions as Fig. 3.6

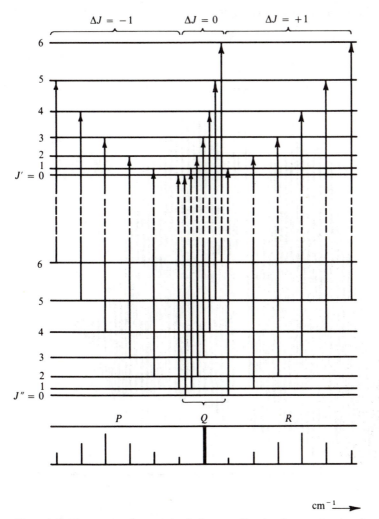

Figure 3.13 The rotational energy levels for two vibrational states showing the effect on the spectrum of transitions for which $\Delta J = 0$.

with the addition of $\Delta J = 0$ transitions. If the oscillation is taken as simple harmonic the energy levels are identical with those of Eq. (3.18) and the P and R branch lines are given, as before, by Eqs (3.20) or (3.21). Transitions with $\Delta J = 0$, however, correspond to a Q *branch* whose lines may be derived from the equations:

$$\Delta\varepsilon = \varepsilon_{J,v+1} - \varepsilon_{J,v}$$
$$= 1\tfrac{1}{2}\bar{\omega}_e - 2\tfrac{1}{4}x_e\bar{\omega}_e + BJ(J+1) - \{\tfrac{1}{2}\bar{\omega}_e - \tfrac{1}{4}x_e\bar{\omega}_e + BJ(J+1)\}$$
$$= \bar{\omega}_o \quad \text{cm}^{-1} \qquad \text{for all } J \tag{3.27}$$

Thus the Q branch consists of lines superimposed upon each other at

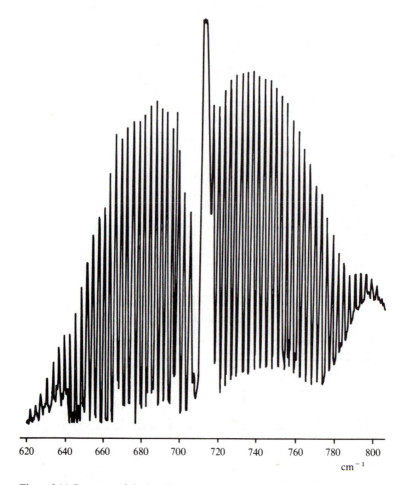

620   640   660   680   700   720   740   760   780   800

$$\text{cm}^{-1}$$

Figure 3.14 Spectrum of the bending mode of the HCN molecule showing the PQR structure. The broad absorption centred at 800 cm^{-1} is due to an impurity.

the band centre $\bar{\omega}_o$, one contribution arising for each of the populated J values. The resultant line is usually very intense.

If we take into account the fact that the B values differ slightly in the upper and lower vibrational states (cf. Sec. 3.4), we would write instead:

$$\Delta\varepsilon = \varepsilon_{J,v+1} - \varepsilon_{J,v}$$
$$= 1\tfrac{1}{2}\bar{\omega}_e - 2\tfrac{1}{4}x_e\bar{\omega}_e + B'J(J+1) - \{\tfrac{1}{2}\bar{\omega}_e - \tfrac{1}{4}x_e\bar{\omega}_e + B''J(J+1)\}$$
$$= \bar{\omega}_o + J(J+1)(B' - B''). \tag{3.28}$$

Further, if $B' < B''$, we see that the Q branch line would become split into a series of lines on the *low*-frequency side of $\bar{\omega}_o$ (since $B' - B''$ is negative). Normally, however, $B' - B''$ is so small that the lines cannot be resolved, and the Q branch appears as a somewhat broad absorption centred around $\bar{\omega}_o$. This is illustrated in Fig. 3.14, which is a spectrum of the bending mode of HCN (corresponding to v_2 of CO_2 in Fig. 3.11). Finally, if the rotational fine structure is unresolved, this type of band has the distinctive contour shown in Fig. 3.15.

It should be remembered (see Chapter 2) that polyatomic molecules with zero dipole moment do not give rise to pure rotation spectra in the microwave region (for example, CO_2, $HC{\equiv}CH$, CH_4). Such molecules do, however, show vibrational spectra in the infra-red region (or Raman, cf. Chapter 4) and, if these spectra exhibit resolved fine structure, the moment of inertia of the molecule can be obtained.

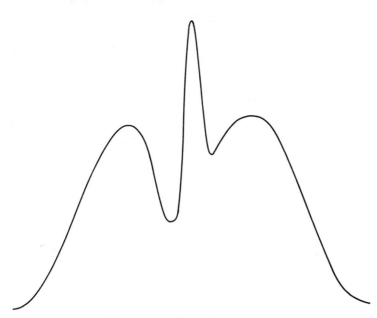

Figure 3.15 The contour of a PQR band under low resolution.

3.6.2 The Influence of Nuclear Spin

It is necessary here to say a brief word about the spectrum of carbon dioxide and other linear molecules possessing a centre of symmetry. A centre of symmetry means that identical atoms are symmetrically disposed with respect to the centre of gravity of the molecule. Thus, plainly both CO_2 [O=C=O] and acetylene [H—C≡C—H] possess a centre of symmetry, while HCN, or N_2O [N≡N=O] do not.

The reader may have noticed that, although we used CO_2 as an example of a vibrating molecule in Fig. 3.11, we did not use it to illustrate real spectra in the subsequent discussion. This is because the centre of symmetry has an effect on the intensity of alternate lines in the P and R branches. The effect is due to the existence of nuclear spin (cf. Chapter 7) and is an additional factor determining the populations of rotational levels. In the case of CO_2 every alternate rotational level is completely unoccupied and so alternate lines in the P and R branches have zero intensity. This leads to a line-spacing of 4B instead of the usual 2B discussed above. That the spacing is indeed 4B (and not 2B with an unexpectedly large value of B) can be shown in several ways, perhaps the most convincing of which is to examine the spectrum of the isotopic molecule ^{18}O—C—^{16}O. Here there is no longer a centre of symmetry, nuclear spin does not now affect the spectrum and the line-spacing is found to be just half that for 'normal' CO_2.

In the case of acetylene, alternate levels have populations which differ by a factor of 3:1 (this, due to nuclear spin alone, is superimposed on the normal thermal distribution and degeneracy) so that the P and R branch lines show a strong, weak, strong, weak, . . . alternation in intensity, as shown in Fig. 3.16.

3.6.3 Symmetric Top Molecules

Following the Born–Oppenheimer approximation we can take the vibrational–rotational energy levels for this type of molecule to be the sum of the vibrational levels:

$$\varepsilon_{\text{vib.}} = (v + \tfrac{1}{2})\bar{\omega}_e - (v + \tfrac{1}{2})^2 x_e \bar{\omega}_e \quad \text{cm}^{-1} \qquad [v = 0, 1, 2, 3, \ldots]$$

and the rotational levels (cf. Eq. (2.38)),

$$\varepsilon_{\text{rot.}} = BJ(J + 1) + (A - B)K^2 \quad \text{cm}^{-1} \qquad [J = 0, 1, 2, \ldots ;$$

$$K = J, (J - 1), (J - 2), \ldots, -J]$$

thus

$$\varepsilon_{J,v} = \varepsilon_{\text{vib.}} + \varepsilon_{\text{rot.}} = (v + \tfrac{1}{2})\bar{\omega}_e - (v + \tfrac{1}{2})^2 x_e \bar{\omega}_e$$

$$+ BJ(J + 1) + (A - B)K^2 \quad \text{cm}^{-1} \quad (3.29)$$

This equation assumes, of course, that centrifugal distortion is negligible.

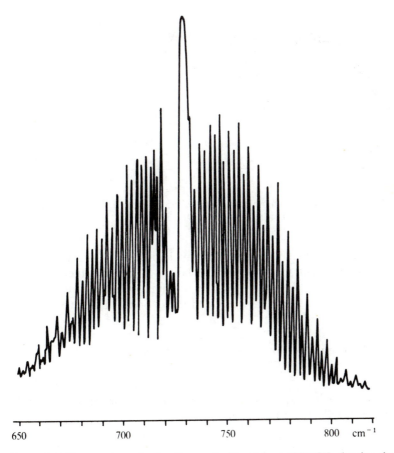

650   700   750   800   cm⁻¹

Figure 3.16 The spectrum of a bending mode of acetylene, HC≡CH, showing the strong, weak, strong, weak, ... intensity alternation in the rotational fine structure due to the nuclear spin of the hydrogen atoms.

Again it is necessary to divide the vibrations into those which change the dipole (1) parallel and (2) perpendicular to the main symmetry axis— which is nearly always the axis about which the 'top' rotates. The rotational selection rules differ for the two types.

Parallel vibrations   Here the selection rule is:

$$\Delta v = \pm 1 \qquad \Delta J = 0, \pm 1 \qquad \Delta K = 0 \qquad (3.30)$$

Since here $\Delta K = 0$, terms in K will be identical in the upper and lower state and so the spectral frequencies will be independent of K. Thus the situation will be identical to that discussed for the *perpendicular* vibrations of a linear molecule. The spectrum will contain P, Q, and R branches with a P, R line spacing of $2B$ (which is unlikely to be resolved) and a strong central Q

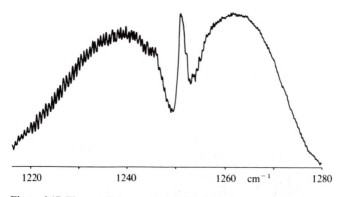

1220　　　　　　1240　　　　　　1260　cm^{-1}　　　1280

Figure 3.17 The parallel stretching vibration, centred at 1251 cm^{-1}, of the symmetric top molecule methyl iodide, CH_3I, showing the typical PQR contour.

branch. Such a spectrum, a $\parallel$ band of methyl iodide, CH_3I, is shown in Fig. 3.17. The intensity of the Q branch (relative to lines in the P and R branches) varies with the ratio I_A/I_B; in the limit, when $I_A \to 0$, the symmetric top becomes a linear molecule and the Q branch has zero intensity, as discussed earlier.

Perpendicular vibrations For these the selection rule is:

$$\Delta v = \pm 1 \qquad \Delta J = 0, \pm 1 \qquad \Delta K = \pm 1 \qquad (3.31)$$

Each of the following expressions is readily derivable for the spectral lines, taking the energy levels of Eq. (3.29).

(1) $\Delta J = +1, \Delta K = \pm 1$ (R branch lines):

$$\Delta \varepsilon = \bar{v}_{\text{spect.}} = \bar{\omega}_o + 2B(J+1) + (A-B)(1 \pm 2K) \quad \text{cm}^{-1} \qquad (3.32a)$$

(2) $\Delta J = -1, \Delta K = \pm 1$ (P branch lines):

$$\bar{v}_{\text{spect.}} = \bar{\omega}_o - 2B(J+1) + (A-B)(1 \pm 2K) \quad \text{cm}^{-1} \qquad (3.32b)$$

(3) $\Delta J = 0, \Delta K = \pm 1$ (Q branch lines):

$$\bar{v}_{\text{spect.}} = \bar{\omega}_o + (A-B)(1 \pm 2K) \quad \text{cm}^{-1} \qquad (3.32c)$$

We see, then, that this type of vibration gives rise to many sets of P and R branch lines since for each J value there are many allowed values of K ($K = J, J-1, \ldots, -J$). The wings of the spectrum will thus be quite complicated and will not normally be resolvable into separate lines. The Q branch is also complex, since it too will consist of a series of lines on both

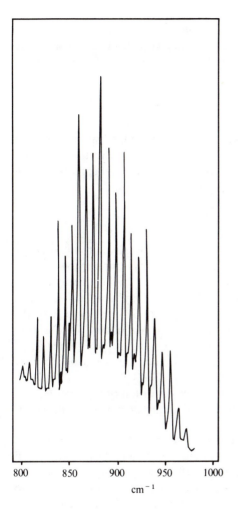

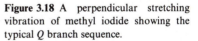

Figure 3.18 A perpendicular stretching vibration of methyl iodide showing the typical Q branch sequence.

sides of $\bar{\omega}_o$ separated by $2(A - B)$. This latter term may not be small (and is equal to zero only for *spherical top* molecules which have all their moments of inertia equal). For $A \gg B$ (for example, CH_3I) the Q branch lines will be well separated and will appear as a series of maxima above the P, R envelope. This spectrum is shown in Fig. 3.18.

It will be noted in this figure that the lines have a distinct periodical variation in intensity—strong, weak, weak, strong, weak, weak, ... This behaviour reminds us of CO_2 and C_2H_2, discussed earlier, in which the presence or absence of nuclear spin altered the relative populations of the rotational levels. In that case, where the molecule had a twofold axis of symmetry, the periodicity also was two—strong, weak, strong, weak, ... It is not surprising, therefore, that the threefold periodicity, strong, weak,

weak, strong, ... seen in CH_3I, arises because of its threefold axis of symmetry to rotations about the C—I axis. The appearance of such a spectrum confirms immediately that we are dealing with a molecule containing an XY_3 grouping.

Other Polyatomic Molecules

We shall not go further with the discussion of their detailed spectra here—it suffices to state that the complexity increases, naturally, with the molecular complexity. An excellent treatment is to be found in Herzberg's book, but the subject is not for the beginner in spectroscopy.

Summary

We have seen that the infra-red spectrum of even a simple diatomic molecule may contain a great many lines, while that of a polyatom may be extraordinarily complex, even though some of the details of fine structure are blurred by insufficient resolving power. Although in favourable cases much information may be obtained about bond lengths and angles or at least the general shape of a molecule, in others even the assignment of observed bands to particular molecular vibrations is not trivial. Assignments are based mainly on experience with related molecules, on the band contour (from which the type of vibration, $\parallel$ or $\perp$, can usually be deduced), and on the use of Raman spectra (see Chapter 4). Consideration of the symmetry of the molecule is also important because this determines which vibrations are likely to be infra-red active.

Fortunately the usefulness of infra-red spectroscopy extends far beyond the measurement of precise vibrational frequencies and molecular structural features. In the next section we discuss briefly the application of infra-red techniques to chemical analysis—a branch of the subject where it is by no means essential always to be able to assign observed bands precisely.

3.7 ANALYSIS BY INFRA-RED TECHNIQUES

Because of the $3N - 6$ and $3N - 5$ rules it is evident that a complex molecule is likely to have an infra-red spectrum exhibiting a large number of normal vibrations. Each normal mode involves some displacement of all, or nearly all, the atoms in the molecule, but while in some of the modes all atoms may undergo approximately the same displacement, in others the displacements of a small group of atoms may be much more vigorous than those of the remainder. Thus we may divide the normal modes into two classes: the *skeletal vibrations*, which involve all the atoms to much the same extent, and the *characteristic group vibrations*, which involve only a small portion of the molecule, the remainder being more or less stationary. We deal with these classes separately.

Skeletal frequencies usually fall in the range 1400–700 cm^{-1} and arise from linear or branched chain structures in the molecule. Thus such groups as

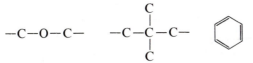

etc., each give rise to several skeletal modes of vibration and hence several absorption bands in the infra-red. It is seldom possible to assign particular bands to specific vibrational modes, but the whole complex of bands observed is highly typical of the molecular structure under examination. Further, changing a substituent (on the chain, or in the ring) usually results in a marked change in the pattern of the absorption bands. Thus these bands are often referred to as the 'fingerprint' bands, because a molecule or structural moiety may often be recognized merely from the appearance of this part of the spectrum. An excellent example of this is shown in Fig. 3.19(*a*) which compares the infra-red spectra of natural and synthetic thymidine. The remarkably exact correlation between the spectra shows that the synthetic product cannot differ in the slightest degree from the natural substance.

Group frequencies, on the other hand, are usually almost independent of the structure of the molecule as a whole and, with a few exceptions, fall in the regions well above and well below that of the skeletal modes; Table 3.4 collects some of the data, of which a much more complete selection is to be found in the books by Bellamy and Nakamoto mentioned in the bibliography at the end of this chapter. We see that the vibrations of light atoms in terminal groups (for example, —CH$_3$, —OH, —C≡N, >C=O, etc.) are of high frequency, while those of heavy atoms (—C—Cl, —C—Br, metal–metal, etc.) are low in frequency. Their frequencies, and consequently their spectra, are highly characteristic of the group, and can be used for analysis. For example, the —CH$_3$ group gives rise to a symmetric C—H stretching absorption invariably falling between 2850 and 2890 cm^{-1}, an asymmetric stretching frequency at 2940–2980 cm^{-1}, a symmetric deformation (i.e., the

opening and closing of the $\overset{\displaystyle |}{\underset{\diagup\, |\, \diagdown}{\underset{\text{H\ \ H\ \ H}}{C}}}$ 'umbrella') at about 1375 cm^{-1}, and

an asymmetric deformation at about 1470 cm^{-1}. Again, the >C=O group shows a very sharp and intense absorption between 1600 and 1750 cm^{-1}, depending largely on the other substituents of the group. An example of the application of group frequency data is shown in Fig. 3.19(*b*); this is the spectrum of thioacetic acid—acetic acid in which one oxygen atom has been

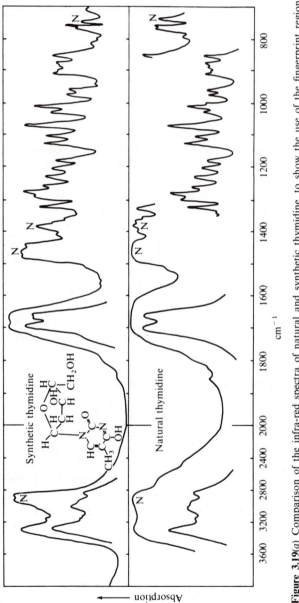

Figure 3.19(a) Comparison of the infra-red spectra of natural and synthetic thymidine, to show the use of the fingerprint region. (N = absorption from liquid paraffin (nujol) in which the solid thymidine is suspended.) *(These spectra and that of Fig. 3.19(b) are reproduced by kind permission of Professor N. Sheppard of the University of East Anglia, Norwich.)*

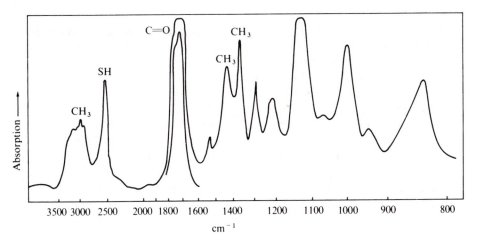

Figure 3.19(*b*) The spectrum of thiocetic acid, CH₃CO.SH.

Table 3.4 Characteristic stretching frequencies of some molecular groups

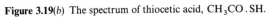

Group	Approximate frequency (cm⁻¹)	Group	Approximate frequency (cm⁻¹)
—OH	3600	>C=O	1750–1600
—NH₂	3400	>C=C<	1650
≡CH	3300	>C=N~	1600
$\bigcirc$—H	3060	≫C—C< ⎫ ≫C—N< ⎬ 1200–1000 ≫C—O~ ⎭	
=CH₂	3030	>C=S	1100
—CH₃	2970 (asym. stretch) 2870 (sym. stretch) 1460 (asym. deform.) 1375 (sym. deform.)	≫C—F	1050
		≫C—Cl	725
—CH₂—	2930 (asym. stretch) 2860 (sym. stretch) 1470 (deformation)	≫C—Br	650
		≫C—I	550
—SH	2580		
—C≡N	2250		
—C≡C—	2220		

replaced by sulphur; the question might be asked: is the molecule $CH_3CO.SH$, or $CH_3CS.OH$? The infra-red spectrum gives a very clear answer. It shows a very sharp absorption at about 1730 cm^{-1}, and one at about 2600 cm^{-1}, and these are consistent with the presence of $>C=O$ and $-SH$ groups, respectively (cf. Table 3.4). Also there is little or no absorption at 1100 cm^{-1} (apart from the general background caused by the skeletal vibrations), thus indicating the absence of $>C=S$.

The idea of group vibrations also covers the motions of isolated features of a molecule which have frequencies not too near those of the skeletal vibrations. Thus isolated multiple bonds (for example, $>C=C<$ or $-C≡C-$) have frequencies which are highly characteristic. When, however, two such groups which, in isolation, have comparable frequencies, occur together in a molecule, resonance occurs and the group frequencies may be shifted considerably from the expected value. Thus the isolated

carbonyl in a ketone $\left(\begin{array}{c} R \\ \diagdown \\ \diagup \\ R \end{array} C=O \right)$ and the $>C=C<$ double bond, have

group frequencies of 1715 and 1650 cm^{-1} respectively; however, when the

grouping $>C=C-\overset{|}{\underset{|}{C}}=O$ occurs, their separate frequencies are shifted to

1675 and about 1600 cm^{-1} respectively and the intensity of the $>C=C<$ absorption increases to become comparable with that of the inherently strong $>C=O$ band (cf. Fermi resonance, p. 97). Closer coupling of the two groups, as in the ketene radical, $>C=C=O$, gives rise to absorptions at about 2100 and 1100 cm^{-1}, which are very far removed from the 'characteristic' frequencies of the separate groups.

Shifts in group frequencies can arise in other ways too, particularly as the result of interactions between different molecules. Thus the $-OH$ stretching frequency of alcohols is very dependent on the degree of hydrogen bonding, which lengthens and weakens the $-OH$ bond, and hence lowers its vibrational frequency. If the hydrogen bond is formed between the $-OH$ and, say, a carbonyl group, then the latter frequency is also lowered, although to a less extent than the $-OH$, since hydrogen bonding weakens the $>C=O$ linkage also. However, shifts in group frequency position caused by resonance or intermolecular effects are in themselves highly characteristic and very useful for diagnostic purposes.

In a similar way a change of physical state may cause a shift in the frequency of a vibration, particularly if the molecule is rather polar. In general the more condensed phase gives a lower frequency: $v_{gas} > v_{liquid} \approx v_{solution} > v_{solid}$. Thus in the relatively polar molecule HCl there is a shift of some 100 cm^{-1} in passing from vapour to liquid and a further decrease of 20 cm^{-1} on solidification. Non-polar CO_2, on the other hand, shows negligible shifts in its symmetric vibrations (Fig. 3.11(a) and (b)) but a lowering of some 60 cm^{-1} in v_3 on solidification.

Examination of Table 3.4 shows that there are logical trends in group frequencies, since Eq. (3.4):

$$\bar{\omega} = \frac{1}{2\pi c} \sqrt{\frac{k}{\mu}} \quad \text{cm}^{-1}$$

is approximately obeyed. Thus we see that increasing the mass of the atom undergoing oscillation within the group (i.e., increasing μ) tends to decrease the frequency—cf. the series CH, CF, CCl, CBr, or the values for $>C=O$ and $>C=S$. Also, increasing the strength of the bond, and hence increasing the force constant k, tends to increase the frequency, e.g., the series $-C-X$, $-C=X$, $-C\equiv X$, where X is C, N, or (in the first two fragments) O.

We should at this point consider very briefly the intensities of infra-red bands. We have seen that an infra-red spectrum only appears if the vibration produces a change in the permanent electric dipole of the molecule. It is reasonable to suppose, then, that the more polar a bond, the more intense will be the infra-red spectrum arising from vibrations of that bond. This is generally borne out in practice. Thus the intensities of the $>C=O$, $>C=N-$, and $>C=C<$ bands decrease in that order, as do those of the $-OH$, $>NH$, and $\geqslant CH$ bands. For this reason, too, the vibrations of ionic crystal lattices often give rise to very strong absorptions. We shall see in the next chapter that the reverse is true in Raman spectroscopy—there the less polar (and hence usually more *polarizable*) bonds give the most intense spectral lines.

In summary, then, experience coupled with comparison spectra of known compounds enables one to deduce a considerable amount of structural information from an infra-red spectrum. It should perhaps be mentioned that the *complete* interpretation of the spectrum of a complex molecule can be a very difficult or impossible task. One is usually content to assign the strongest bands and to be able to explain some of the weaker ones as overtones or combinations.

3.8 TECHNIQUES AND INSTRUMENTATION

3.8.1 Outline

We first deal briefly with each component of the spectrometer as it is usually assembled for infra-red work.

1. *Source.* The source is always some form of filament which is maintained at red- or white-heat by an electric current. Two common sources are the Nernst filament, consisting of a spindle of rare-earth oxides about 1 inch long and 0·1 inch in diameter, and the 'globar' filament, a rod of carborundum, somewhat thicker and longer than the Nernst. The Nernst

requires to be pre-heated before it will conduct electricity, but once red-heat is reached the temperature is maintained by the current.

2. *Optical path and monochromator.* The beam is guided and focused by mirrors silvered on their surfaces. Normally a focus is produced at the point where the sample is to be placed. Ordinary lenses and mirrors are not suitable as glass absorbs strongly over most of the frequencies used. Any windows which are essential (e.g., to contain a sample, or to protect the detector) must be made of mineral salts transparent to infra-red radiation (NaCl and KBr are much used) which have been highly polished in order to reduce scattering to a minimum.

Similarly, the monochromator in some instruments is made of a rock salt or potassium bromide prism, which is rotatable to produce the required frequency in a manner similar to that of Fig. 1.11 (except, of course, that the lens is replaced by a condensing mirror). Modern instruments, however, use a rotatable grating instead of a prism, since this gives much better resolving power.

3. *Detector.* Two main types are in common use, one sensing the heating effect of the radiation, the other depending on photoconductivity. In both the greater the effect (temperature or conductivity rise) at a given frequency, the greater the transmittance (and the less the absorbance) of the sample at that frequency.

An example of the temperature method is to be found in the Golay cell which is pneumatic in operation. The radiation falls on to a very small cell containing air, and temperature changes are measured in terms of pressure changes within the cell which can be recorded directly as 'transmittance'. Alternative examples of this type of detector are small, sensitive thermocouples or bolometers.

The phenomenon of conductivity in substances is thought to arise as a consequence of the movement of loosely held electrons through the lattice; insulators, on the other hand, have no such loosely bound electrons. Semiconductors are essentially midway between these materials, having no loosely bound electrons in the normal state, but having 'conduction bands' or raised electron energy levels into which electrons may be readily excited by the absorption of energy from an outside source. Photoconductors are a particular class of semiconductor in which the energy required comes from incident radiation, and some materials, such as lead sulphide, have been found sufficiently sensitive to infra-red radiation (although only above some 3500 cm^{-1}) that they make excellent detectors. The conductivity of the material can be measured continuously by a type of Wheatstone bridge network and, when plotted against frequency, this gives directly the transmittance of the sample.

4. *Sample.* For reasons just stated, the sample is held between plates of polished mineral salt rather than glass. Pure liquids are studied in thicknesses of about 0·01 mm, while solutions are usually 0·1–10 mm thick,

depending on the dilution. Gas samples at pressures of up to 1 atmosphere or greater are usually contained in glass cells either 5 or 10 cm long, closed at their ends with rock salt windows. Special long-path cells, in which the radiation is repeatedly reflected up and down the cell, may be used for gases at low pressure, perhaps less than 100 mmHg.

Solid samples are more difficult to examine because the particles reflect and scatter the incident radiation and transmittance is always low. If the solid cannot be dissolved in a suitable solvent, it is best examined by grinding it very finely in paraffin oil (nujol) and thus forming a suspension, or 'mull'. This can then be held between salt plates in the same way as a pure liquid or solvent. Provided the refractive indices of liquid and solid phase are not very different, scattering will be slight.

Another technique for handling solids is to grind them very finely with potassium bromide. Under very high pressure this material will flow slightly, and the mixture can usually be pressed into a transparent disk. This may then be placed directly in the infra-red beam in a suitable holder. Although superficially attractive the method is not generally recommended because of the difficulty in obtaining really reproducible results.

A further technique which is often used to study otherwise intractible samples is known as attenuated total reflectance (atr) spectroscopy. Consider a trapezoidal block of a transparent material, as in Fig. 3.20(a). If the chamfer angle is properly chosen, radiation shone into one end will strike the flat surfaces at less than the critical angle and so will undergo total internal reflection to emerge, only slightly diminished in intensity, at the far end, as in Fig. 3.20(b). Now, although the internal reflection is conventionally called 'total', in fact the radiation beam penetrates slightly beyond the surface of the block during each reflection. If sample material is pressed closely to the outside of the block (Fig. 3.20(c)) the beam will travel a small distance through the sample at each reflection and so, on emerging at the far end of the block, it will 'carry' the absorption spectrum of the sample— the internal reflection is *attenuated*, or diminished by sample absorption, hence the name of this type of spectroscopy. The amount of penetration into the sample depends on the wavelength of the radiation and the angle of incidence, but it is of the order of 10^{-4} to 10^{-3} cm for infra-red waves. During its passage through the block it may undergo some 10–20 reflections, so the total path length through the sample is 10^{-3}–10^{-2} cm, which is a short, but often perfectly adequate, path length for the production of a reasonable spectrum.

The block must be of material which is infra-red transparent, and must have a refractive index higher than that of the sample, otherwise internal reflection will not occur. Suitable materials are silver chloride, thallium halides, or germanium, and the block is typically some 5 cm long, 2 cm

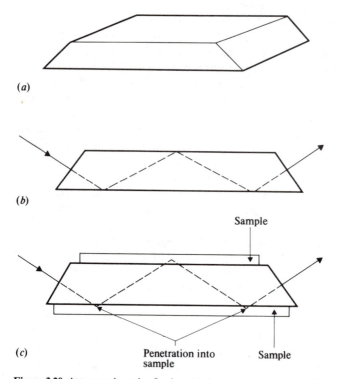

Figure 3.20 Attenuated total reflection. (*a*) the transparent block, (*b*) internal reflection in the block, and (*c*) penetration into the sample pressed against the block.

wide, and 0·5 cm thick. The sample material can be in any form (except gaseous, for which a path length of 10^{-2} cm is far too short) provided it can be kept in very close contact with the block, but the technique is usually reserved for samples difficult to study by ordinary means. Thus, it is virtually impossible to study fibrous material by transmission—the rough surface scatters all the radiation falling on it; but if the fibres are clamped firmly to the outside of an atr block, quite acceptable spectra result. And it is an excellent method of studying surface coatings, since it is only the surface of the sample which is penetrated by the radiation. Further, since the depth of penetration can, to some extent, be changed by varying the angle of incidence of the beam, the change in composition of a surface with depth can be studied. Thus one can measure the degree of oxidation of a polymer surface, or the diffusion of materials into a surface.

3.8.2 Double- and Single-Beam Operation

Figure 3.21 shows the spectrum of the atmosphere between 4000 and 400 cm^{-1} taken with a path length of some 2 m—this is not abnormally long

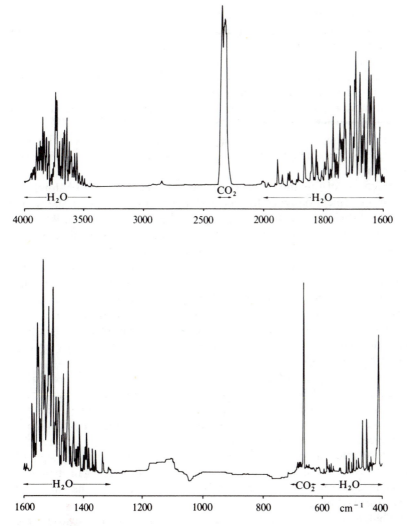

Figure 3.21 The infra-red spectrum of atmospheric water vapour and carbon dioxide.

for the beam paths in a spectrometer. It is evident that, although H_2O and CO_2 occur in air only in small percentages, their absorbance over much of the spectrum is considerable. Not only would this absorbance have to be subtracted from the spectrum of any sample run under comparable conditions but, since the percentage of water vapour in the atmosphere is variable, such a 'background' spectrum as Fig. 3.21 would have to be run afresh for each sample.

If the regions of these absorbances are not to be denied us in spectroscopic studies, some action must be taken either to remove the H_2O and

CO_2 from the air, or to remove the effects of their spectra. It is possible to remove these gases either by complete evacuation of the spectrometer, or by sweeping them out with a current of dry nitrogen, or dry CO_2-free air. The first is not easy since a modern spectrometer may have a volume of some 0.3 m^3 and there will be a great many places in its container where leaks may occur. Nor is it ever completely effective, since water vapour proves to be remarkably tenacious and weeks of hard evacuation may be necessary before all the water vapour is desorbed from the surfaces inside the spectrometer. For this reason, also, sweeping with a dry inert gas is not very effective. However, these methods do, quite rapidly, reduce the interference considerably.

The *effects* of this interference can be removed much more simply by using an instrument designed for *double-beam* operation. In this, the source radiation is divided into two by means of the mirrors M_1 and M_2 (Fig. 3.22). One beam is brought to a focus at the sample space, while the other follows an exactly equivalent path and is referred to as the reference beam. The two beams meet at the sector mirror M_3, which is sketched in plan view in Fig. 3.21(*b*). As this mirror rotates it alternately reflects the reference beam, or allows the sample beam through the spaces, into the monochromator. Thus the detector 'sees' the sample beam and reference beam alternately. Both beams have travelled the same distance through the atmosphere and thus both are reduced in energy to the same extent by absorption by CO_2 and H_2O.

If a sample, capable of absorbing energy from the beam at the particular frequency passed by the monochromator, is now placed in the sample beam, the detector will receive a signal alternating in intensity, since the sample beam carries less energy than the reference beam. It is a simple matter, electronically, to amplify this alternating signal and to arrange that a calibrated attenuator is driven into the reference beam until the signal is reduced to zero, i.e., until sample and reference beams are again balanced. The distance moved by the attenuator is a direct measure of the amount of energy absorbed by the sample.

By balancing sample and reference beams in this way, the absorption of atmospheric CO_2 and H_2O do not appear in the infra-red spectra since both beams are reduced in energy to the same extent. The double-beam spectrometer has other advantages, however.

Firstly, it is much simpler to amplify the alternating signal produced than the d.c. signal resulting from a single-beam detector.

Secondly, the sector mirror acts as a modulator (cf. p. 68) since it interrupts the beam periodically and, by amplifying only that component of the signal having the sector mirror frequency (usually 10–100 rotations per second), a great improvement in the signal-to-noise ratio results.

Thirdly, when examining the spectra of solutions, one can put a cell containing the appropriate quantity of pure solvent into the reference beam,

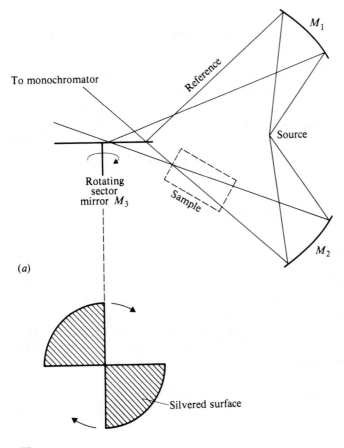

(a)

(b)

Figure 3.22 (a) Schematic diagram of a double-beam infra-red spectrometer. (b) A plan view of the rotating sector mirror M_3.

thus eliminating the solvent spectrum from the final trace. On a single-beam instrument the solvent spectrum must be taken separately and 'subtracted' from the solution spectrum in order to arrive at the spectrum of the substance of interest.

It should be pointed out, however, that a double-beam instrument is never *completely* effective in removing traces of water vapour or CO_2 from the spectra. No matter how carefully the instrument is assembled small differences occur in the beam paths and a small residual spectrum results. This can usually be removed, however, by sweeping with dry, inert gas as well as using the double-beam principle.

A further, more serious disadvantage which is not always appreciated by users of spectrometers is that the double-beam instrument only removes the spectral trace of CO_2 and H_2O; the very strong absorption of energy by these gases still remains in both beams. This means that at some parts of the spectrum the actual amount of energy reaching the detector may be extremely small. Under these conditions, unless the spectrometer is very carefully operated, the spectral trace of a substance may be quite false. Fortunately, regions of very high atmospheric absorption are few and narrow but they should be borne in mind when examining infra-red spectra. This disadvantage can only be removed by sweeping out or evacuating the spectrometer. Similar, but more pronounced, effects occur in regions of strong solvent absorbance when a compensating cell is put in the reference beam.

3.8.3 Fourier Transform Spectroscopy

Infra-red spectroscopy extends outside the limits we have discussed so far in this chapter, and in particular a good deal of useful molecular information is contained in spectra below 400 cm^{-1}, that is, the far infra-red region, from about 400 cm^{-1} to 20 cm^{-1} or 10 cm^{-1}. Because sources are weak and detectors insensitive, this region is known as 'energy-limited' and difficulty is experienced in obtaining good signal-to-noise ratios by conventional means. The advent of Fourier transform spectroscopy, already introduced in Sec. 1.8, has made the far infra-red much more accessible, and has considerably speeded and improved spectroscopy in the infra-red region in in general.

In this region Fourier transform (FT) methods are used in absorption. The apparatus derives from the classical attempt by Michelson to measure

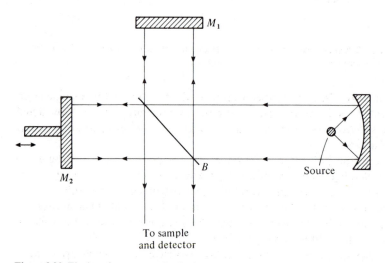

Figure 3.23 The interferometer unit of a Fourier transform spectrometer.

the 'ether wind' by determining the velocity of light in two perpendicular directions. A parallel beam of radiation is directed from the source to the interferometer, consisting of a beam splitter B and two mirrors M_1 and M_2 (Fig. 3.23). The beam splitter is a plate of suitably transparent material (e.g., potassium bromide) coated so as to reflect just 50 per cent of the radiation falling on it. Thus half the radiation goes to M_1, and half to M_2, returns from both these mirrors along the same path, and is then recombined to a single beam at the beam splitter (clearly half the total radiation is sent back to the source, but this is immaterial).

Now it is well known (and the essence of the Michelson experiment) that if *monochromatic* radiation is emitted by the source, the recombined beam leaving B shows constructive or destructive interference, depending on the relative path lengths B to M_1 and B to M_2. Thus if the path lengths are identical or differ by an integral number of wavelengths, constructive interference gives a bright beam leaving B, whereas if the difference is a half-integral number of wavelengths, the beams cancel at B. As the mirror M_2 is moved smoothly towards or away from B, therefore, a detector sees radiation alternating in intensity. It is fairly easy to imagine that if the source emits *two* separate monochromatic frequencies, v_1 and v_2, then the interference pattern (beat pattern) of v_1 and v_2 would overlay the interference caused by M_1 and M_2; the detector would see a more complicated intensity fluctuation as M_2 is moved, but computing the Fourier transform of the resultant signal is a very rapid way of obtaining the original frequencies and intensities emitted by the source. Taking the process further, even 'white' radiation emitted by the source produces an interference pattern which can be transformed back to the original frequency distribution.

Clearly then, if the recombined beam from such a source is directed through a *sample* before reaching the detector, sample absorptions will show up as gaps in the frequency distribution which, after transformation, yields a normal absorption spectrum. The production of a spectrum, then, may be thought of as follows: mirror M_2 is moved smoothly over a period of time (e.g., one second) through about 1 cm distance, while the detector signal—the interferogram—is collected into a multi-channel computer (it may be, for instance, that the detector signal is monitored every thousandth of a second during the mirror traverse, and each piece of information put serially into one of 1000 different storage points in the computer); the computer then carries out Fourier transformation on the stored data, and replaces the 'proper' spectrum piecemeal into the same 1000 locations, ready for plotting out on to paper.

The great advantage of FT spectroscopy is its speed. Since the whole spectrum is contained in the interferogram, which is recorded in the computer within one second, this is effectively the scan time. Even adding the computing and plotting time of, say, 15 seconds, the overall time to obtain a spectrum is very short compared with the 10 minutes or so required by conventional methods to obtain similar resolution. Essentially, to achieve a given resolution each element, or observation point, of the spectrum must be examined for a given time; in the conventional method each element is examined *consecutively*, whereas in the interferometer all the elements are examined *simultaneously* and later sorted out by rapid computation; the advantage of the latter is obvious.

Other advantages are also offered by the FT technique; briefly these are:

1. In a conventional instrument the radiation is invariably brought to a focus on a slit, and it is essentially the image of the slit which is seen by the detector; a very fine slit gives good resolving power since only a narrow spread of frequencies falls on the detector at any one moment, but the total amount of energy passing through the instrument is severely limited, requiring high-gain and hence 'noisy' amplifiers. In FT work parallel beams are used throughout, and there is no need to bring the radiation to a focus except for convenience at the sample and at the detector—no slit is required and *all* the source energy passes through the instrument; consequently amplifiers are less critical and the resolving power is governed solely by the mirror traverse and computer capacity. It is for this reason that FT instruments were first developed for use in the energy-limited far infra-red region.
2. The resolving power of an FT instrument is constant over the entire spectrum; in a grating or prism instrument the resolving power depends on the angle which that component makes with the radiation beam, and hence varies with frequency—in particular it is usually especially poor at the ends of the spectrum.
3. The presence of a computer for FT work means that other jobs can be carried out automatically; thus several scans of the same sample may be added into the computer store in order to improve the signal-to-noise ratio; the spectrum output may be modified to suit the user by removing solvent peaks, correcting baseline drift, expanding parts of the spectrum, offsetting window absorptions, etc. It is even possible for calculation of peak intensities or sample concentrations to be made on a routine basis.

3.8.4 The Carbon Dioxide Laser

Currently the only laser which seems to have some potential for use in routine infra-red spectroscopy is that made from a mixture of carbon dioxide and nitrogen, the so-called CO_2 laser. The primary excitation step in this is the excitation of nitrogen molecules into their first excited vibrational level, which, as Fig. 3.24 shows, is at about 2360 cm^{-1} above the ground state. This happens to be very close in energy to ν_3, the asymmetric stretching vibration of CO_2 at 2350 cm^{-1}, and so resonance excitation of CO_2 to

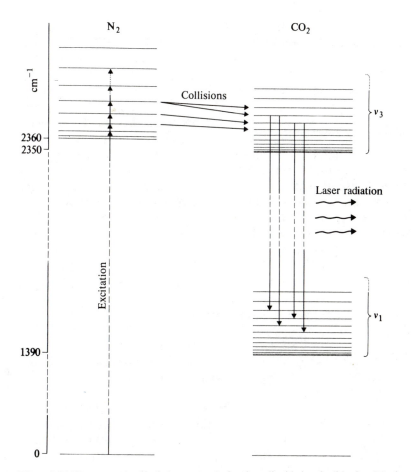

Figure 3.24 The energy levels of nitrogen and of carbon dioxide involved in the CO_2 laser.

this level occurs by collision with excited N_2. Now both N_2 and CO_2 have rotational states associated with these vibrational energies, which are also indicated (schematically and not to scale) in Fig. 3.24. This means that many different rotational levels of N_2 will be populated in the initial excitation, and consequently many rotational levels of the v_3 vibration of CO_2 will be collisionally activated. In fact, levels up to some 200 cm^{-1} above the lowest state of v_3 are effectively populated in this way.

The excited CO_2 can, and of course does, decay spontaneously and directly to the ground state, liberating energy as heat. For this reason some helium is incorporated into the gas mixture, to help in transferring the heat to the walls of the containing tube, and the laser is normally operated in the pulsed mode, although pulses can be as high as several per second. In addition, however, stimulated emission is possible to v_1, the symmetric stretching mode of CO_2 at about 1390 cm^{-1}, and this mode also has associated rotational levels. The system, then, is basically a four-level laser (cf. Sec. 1.10), although the many additional rotational levels also play their part.

When acting as a laser, any activated molecules in v_3 can drop to any *allowed* levels of v_1, that is, those levels to which transitions are allowed under the $\Delta J = \pm 1$ selection rule. This means that a large number (perhaps 80–100) of discrete laser frequencies are emitted over the range 900–1100 cm^{-1}—the separation of the v_1 and v_3 vibrations of CO_2. The spacing between the emission lines is about 2 cm^{-1}, the spacing of the rotational transitions. Now although this cannot be regarded as a 'continuous' source of radiation, and although its frequency range is very limited compared with the useful infra-red spectrum (covering 3000–400 cm^{-1} or lower), nonetheless it has some potential as an infra-red source, as we illustrate in the following paragraphs.

Most organic materials, especially when in solution, have fairly broad infra-red absorptions, with a bandwidth of some 5–20 cm^{-1}; in this respect a source with discrete lines spaced at only 2 cm^{-1} is essentially continuous, in the sense that there is little loss of spectral information between the lines. Equally, many such molecules do have useful diagnostic absorptions in the 900–1100 cm^{-1} region. So spectra obtained with a CO_2 laser, although perhaps looking slightly different from those obtained with a normal source, may well have useful 'fingerprinting' possibilities.

Further the laser radiation is very intense—in fact, since its main industrial use is in cutting and welding materials, an unmodified CO_2 laser tends to melt the spectrometer. Once its intensity has been reduced so that it is 'only' some 10^3 to 10^5 times stronger than a normal infra-red source, however, it proves ideally suited to applications such as monitoring air pollution. Here one is looking for parts-per-million quantities of pollutants, and the best method is to shine the infra-red beam through a very long path of the material. Clearly an intense beam that does not diverge over a large distance is exactly what is required.

Finally, quantitative measurements of materials are most accurately obtained by using monochromatic radiation, and in this respect one of the emission lines from the CO_2 laser, selected by the use of a normal dispersion grating, is ideal, and potentially very useful for continuous monitoring of production materials.

BIBLIOGRAPHY

Alpert, N. A., W. E. Keiser, and H. A. Szymanski: *Theory and Practice of Infra-Red Spectroscopy*, Plenum Press, 1970.

Barnes, A. J., and W. J. Orville-Thomas (eds): *Vibrational Spectroscopy—Modern Trends*, Elsevier, 1977.

Bellamy, L. J.: *The Infra-Red Spectra of Complex Molecules*, Chapman and Hall, 1975.

Bellamy, L. J.: *Advances in Infra-Red Group Frequencies*, Methuen, 1968.

Ferraro, J. R., and L. J. Basile: *Fourier Transform Infra-Red Spectroscopy: Applications to Chemical Systems*, Academic Press; vol. 1, 1978; vol. 2, 1979.

Finch, A., F. N. Dickson, and F. F. Bentley: *Chemical Applications of Far Infra-Red Spectroscopy*, Academic Press, 1970.

Herzberg, G.: *Molecular Structure and Molecular Spectra:* vol. 1, *Spectra of Diatomic Molecules*, 2nd ed., Van Nostrand, 1950.

Herzberg, G.: *Molecular Spectra and Molecular Structure:* vol. 2, *Infra-Red and Raman Spectra of Polyatomic Molecules*, Van Nostrand, 1945.

Meloan, C. E.: *Elementary Infra-Red Spectroscopy*, Macmillan, 1963.

Nakamoto, K.: *Infra-Red Spectra of Inorganic and Co-ordination Compounds*, 3rd ed., John Wiley, 1978.

Ross, S. D.: *Inorganic Infra-Red and Raman Spectra*, McGraw-Hill, 1972.

Sherwood, P. M. A.: *Vibrational Spectroscopy of Solids*, Cambridge University Press, 1972.

Siesler, H. W., and K. Holland-Moritz: *Infra-Red and Raman Spectroscopy of Polymers*, Marcel Dekker Inc., 1980.

Smith, A. L.: *Applied Infra-Red Spectroscopy*, Wiley, 1979.

Steele, D.: *The Interpretation of Vibrational Spectra*, Chapman and Hall, 1971.

Williams, D. H., and I. Fleming: *Spectroscopic Methods in Organic Chemistry*, 3rd ed., McGraw-Hill, 1980.

PROBLEMS

(Useful constants: $h = 6 \cdot 626 \times 10^{-34}$ J s; $c = 2 \cdot 998 \times 10^{8}$ m s^{-1}; $N = 6 \cdot 023 \times 10^{23}$ mol^{-1}; $k = 1 \cdot 381 \times 10^{-23}$ J K^{-1}; $4\pi^2 = 39 \cdot 478$; 1 cm$^{-1} \equiv 11 \cdot 958$ J mol^{-1}; atomic masses: ^{14}N $= 23 \cdot 25 \times 10^{-27}$ kg; ^{16}O $= 26 \cdot 56 \times 10^{-27}$ kg.)

3.1 The fundamental and first overtone transitions of ^{14}N^{16}O are centred at 1876·06 cm^{-1} and 3724·20 cm^{-1} respectively. Evaluate the equilibrium vibration frequency, the anharmonicity, the exact zero point energy, and the force constant of the molecule. Assuming that in Eq. (3.12) v is a continuous variable, use calculus to determine the maximum value of ε_v, and hence calculate a value for the dissociation energy of NO. Criticize this method.

3.2 The vibrational wavenumbers of the following molecules in their $v = 0$ states are: HCl: 2885 cm^{-1}; DCl: 1990 cm^{-1}; D$_2$: 2990 cm^{-1}; and HD: 3627 cm^{-1}. Calculate the energy change, in kJ mol^{-1} of the reaction

$$\text{HCl} + \text{D}_2 \rightarrow \text{DCl} + \text{HD}$$

and determine whether energy is liberated or absorbed.

Hint: Consider the zero point energies of the four molecules concerned.

3.3 The equilibrium vibration frequency of the iodine molecule I$_2$ is 215 cm^{-1}, and the anharmonicity constant x is 0·003; what, at 300 K, is the intensity of the 'hot band' ($v = 1 \rightarrow v = 2$ transition) relative to that of the fundamental ($v = 0 \rightarrow v = 1$)?

3.4 An infra-red spectrum of OCS is obtained in which the rotational fine structure is not resolved. Using data from Table 2.2, calculate the separation between the P and R branch maxima at $T = 300$ K.

3.5 How many normal modes of vibration are possible for (a) HBr, (b) OCS (linear), (c) SO$_2$ (bent), and (d) C$_6$H$_6$?

3.6 Estimate, using data from Table 3.4, the vibrational wave-number of (a) $-$ OD,

(b) $\overset{\diagdown}{\underset{\diagup}{\text{C}}}$$-S-$. (Relative atomic masses are: H = 1, D = 2, C = 12, O = 16, S = 32.)

FOUR

RAMAN SPECTROSCOPY

4.1 INTRODUCTION

When a beam of light is passed through a transparent substance, a small amount of the radiation energy is scattered, the scattering persisting even if all dust particles or other extraneous matter are rigorously excluded from the substance. If monochromatic radiation, or radiation of a very narrow frequency band, is used the scattered energy will consist almost entirely of radiation of the incident frequency (the so-called *Rayleigh scattering*) but, in addition, certain discrete frequencies above and below that of the incident beam will be scattered; it is this which is referred to as *Raman scattering*.

4.1.1 Quantum Theory of Raman Effect

The occurrence of Raman scattering may be most easily understood in terms of the quantum theory of radiation. This treats radiation of frequency v as consisting of a stream of particles (called photons) having energy hv where h is Planck's constant. Photons can be imagined to undergo collisions with molecules and, if the collision is perfectly elastic, they will be deflected unchanged. A detector placed to collect energy at right angles to an incident beam will thus receive photons of energy hv, that is, radiation of frequency v.

However, it may happen that energy is exchanged between photon and molecule during the collision: such collisions are called 'inelastic'. The molecule can gain or lose amounts of energy only in accordance with the quantal laws; i.e., its energy change, ΔE joules, must be the difference in energy between two of its allowed states. That is to say, ΔE must represent a change in the vibrational and/or rotational energy of the molecule. If the molecule *gains* energy ΔE, the photon will be scattered with energy $hv - \Delta E$ and the equivalent radiation will have a frequency $v - \Delta E/h$. Conversely if the molecule *loses* energy ΔE, the scattered frequency will be $v + \Delta E/h$.

Radiation scattered with a frequency lower than that of the incident beam is referred to as Stokes' radiation, while that at higher frequency is called anti-Stokes'. Since the former is accompanied by an *increase* in molecular energy (which can always occur, subject to certain selection rules) while the latter involves a *decrease* (which can only occur if the molecule is originally in an excited vibrational or rotational state), Stokes' radiation is generally more intense than anti-Stokes'. Overall, however, the total radiation scattered at any but the incident frequency is extremely small, and sensitive apparatus is needed for its study.

4.1.2 Classical Theory of the Raman Effect; Molecular Polarizability

The classical theory of the Raman effect, while not wholly adequate, is worth consideration since it leads to an understanding of a concept basic to this form of spectroscopy—the polarizability of a molecule. When a molecule is put into a static electric field it suffers some distortion, the positively charged nuclei being attracted towards the negative pole of the field, the electrons to the positive pole. This separation of charge centres causes an *induced electric dipole moment* to be set up in the molecule and the molecule is said to be *polarized*. The size of the induced dipole, μ, depends both on the magnitude of the applied field, E, and on the ease with which the molecule can be distorted. We may write

$$\mu = \alpha E \qquad (4.1)$$

where α is the *polarizability* of the molecule.

Consider first a diatomic molecule, such as H_2, shown at Fig. 4.1(*a*). The polarizability is *anisotropic*, i.e., the electrons forming the bond are more easily displaced by an electric field applied along the bond axis than one across this direction: this may be confirmed experimentally, for example by a study of the absolute intensity of lines in the Raman spectrum, when it is found that the induced dipole moment for a given field applied along the axis is approximately twice as large as that induced by the same field applied across the axis; fields in other directions induce intermediate dipole moments. We can represent the polarizability in various directions most conveniently by drawing a *polarizability ellipsoid*, as in Fig. 4.1(*b*),

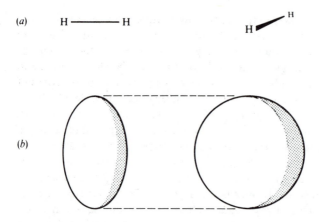

Figure 4.1 The hydrogen molecule and its polarizability ellipsoid seen from two directions at right angles.

where the ellipsoid is a three-dimensional surface whose distance from the electrical centre of the molecule (here the centre of gravity also) is proportional to $1/\sqrt{\alpha_i}$, where α_i is the polarizability along the line joining point i on the ellipsoid with the electrical centre. Thus where the polarizability is greatest the axis of the ellipsoid is least, and vice versa. (This representation is chosen because of an analogy with the momentum of a body—the momental ellipsoid is defined similarly using $1/\sqrt{I_i}$, where I_i is the moment of inertia about an axis i.)

Since the polarizability of a diatomic molecule is the same for all directions at right angles to the bond axis, the ellipsoid has a circular cross-section in this direction; thus it is shaped rather like a tangerine. All diatomic molecules, for example CO, HCl, and linear polyatomic molecules, for example CO_2, HC≡CH, etc., having ellipsoids of the same general shape, differing only in the relative sizes of their major and minor axes.

When a sample of such molecules is subjected to a beam of radiation of frequency v the electric field experienced by each molecule varies according to the equation (cf. Eq. (1.1)):

$$E = E_0 \sin 2\pi v t \tag{4.2}$$

and thus the induced dipole also undergoes oscillations of frequency v:

$$\mu = \alpha E = \alpha E_0 \sin 2\pi v t \tag{4.3}$$

Such an oscillating dipole emits radiation of its own oscillation frequency and we have immediately in Eq. (4.3) the classical explanation of Rayleigh scattering.

If, in addition, the molecule undergoes some internal motion, such as vibration or rotation, which *changes the polarizability* periodically, then the oscillating dipole will have superimposed upon it the vibrational or rotational oscillation. Consider, for example, a vibration of frequency $v_{\text{vib.}}$ which changes the polarizability: we can write

$$\alpha = \alpha_0 + \beta \sin 2\pi v_{\text{vib.}} t \qquad (4.4)$$

where α_0 is the equilibrium polarizability and β represents the rate of change of polarizability with the vibration. Then we have:

$$\mu = \alpha E = (\alpha_0 + \beta \sin 2\pi v_{\text{vib.}} t)E_0 \sin 2\pi v t$$

or, expanding and using the trigonometric relation,

$$\sin A \sin B = \tfrac{1}{2}\{\cos (A - B) - \cos (A + B)\}$$

we have

$$\mu = \alpha_0 E_0 \sin 2\pi v t + \tfrac{1}{2}\beta E_0 \{\cos 2\pi(v - v_{\text{vib.}})t - \cos 2\pi(v + v_{\text{vib.}})t\} \quad (4.5)$$

and thus the oscillating dipole has frequency components $v \pm v_{\text{vib.}}$ as well as the exciting frequency v.

It should be carefully noted, however, that if the vibration does not alter the polarizability of the molecule (and we shall later give examples of such vibrations) then $\beta = 0$ and the dipole oscillates only at the frequency of the incident radiation; the same is true of a rotation. Thus we have the general rule:

> In order to be Raman active a molecular rotation or vibration must cause some change in a component of the molecular polarizability. A change in polarizability is, of course, reflected by a change in either the *magnitude* or the *direction* of the polarizability ellipsoid.

(This rule should be contrasted with that for infra-red and microwave activity, which is that the molecular motion must produce a change in the electric dipole of the molecule.)

Let us now consider briefly the shapes of the polarizability ellipsoids of more complicated molecules, taking first the bent triatomic molecule H_2O shown at Fig. 4.2(*a*). By analogy with the discussion for H_2 given above, we might expect the polarizability surface to be composed of *two* similar ellipsoids, one for each bond. While this may be correct in minute detail, we must remember that the oscillating electric field which we wish to apply for Raman spectroscopy is usually that of radiation in the visible or ultra-violet region, i.e., having a wavelength of some 1 μm–10 nm (cf. Fig. 1.4); molecular bonds, on the other hand, have dimensions of only some 0·1 nm, so we cannot expect our radiation to probe the finer details of bond polarizability—even the hardest of X-rays can scarcely do that. Instead the radiation can only sense the average polarizability in various directions through the molecule, and the polarizability ellipsoid, it may be shown, is

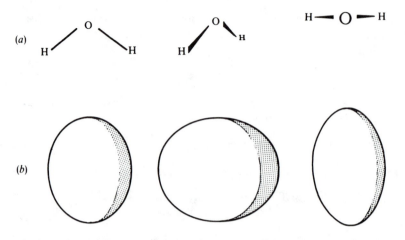

Figure 4.2 The water molecule and its polarizability ellipsoid seen along the three coordinate axes.

always a true ellipsoid—i.e., a surface having *all* sections elliptical (or possibly circular). In the particular case of H_2O the polarizability is found to be different along all three of the major axes of the molecule (which lie along the line in the molecular plane bisecting the HOH angle, at right angles to this in the plane, and perpendicularly to the plane), and so all three of the ellipsoidal axes are also different; the ellipsoid is sketched in various orientations in Fig. 4.2(*b*). Other such molecules, for example, H_2S or SO_2, have similarly shaped ellipsoids but with different dimensions.

Symmetric top molecules, because of their axial symmetry, have polarizability ellipsoids rather similar to those of linear molecules, i.e., with a circular cross-section at right angles to their axis of symmetry. It should be stressed, however, that sections in other planes are truly *elliptical*. For a molecule such as chloroform, $CHCl_3$ (Fig. 4.3(*a*)), where the chlorine atoms are bulky, the usual tendency is to draw the polarizability surface as egg-shaped, fatter at the chlorine-containing end. This is not correct; the polarizability ellipsoid for chloroform is shown at Fig. 4.3(*b*) where it will be seen that, since the polarizability is greater across the symmetry axis, the *minor* axis of the ellipsoid lies in this direction. Similar molecules are, for example, CH_3Cl and NH_3, etc. (although the latter fortuitously has a virtually spherical 'ellipsoid').

Finally, spherical top molecules, such as CH_4, CCl_4, SiH_4, etc., have spherical polarizability surfaces, since they are completely isotropic as far as incident radiation is concerned.

We are now in a position to discuss in detail the Raman spectra of various types of molecule. Since we shall be dealing with rotational and vibrational changes it is evident that expressions for the energy levels and

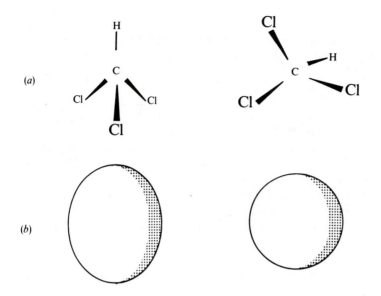

Figure 4.3 The chloroform molecule, $CHCl_3$, and its polarizability ellipsoid from across and along the symmetry axis.

for many of the allowed transitions will be identical with those already discussed in the previous two chapters. For clarity we shall repeat any such expressions but not rederive them, being content to give a cross-reference to where their derivation may be found.

4.2 PURE ROTATIONAL RAMAN SPECTRA

4.2.1 Linear Molecules

The rotational energy levels of linear molecules have already been stated (cf. Eq. (2.24)):

$$\varepsilon_J = BJ(J + 1) - DJ^2(J + 1)^2 \text{ cm}^{-1} \qquad (J = 0, 1, 2, \ldots)$$

but, in Raman spectroscopy, the precision of the measurements does not normally warrant the retention of the term involving D, the centrifugal distortion constant. Thus we take the simpler expression:

$$\varepsilon_J = BJ(J + 1) \text{ cm}^{-1} \qquad (J = 0, 1, 2, \ldots) \tag{4.6}$$

to represent the energy levels.

Transitions between these levels follow the formal selection rule:

$$\Delta J = 0, \text{ or } \pm 2 \text{ only} \tag{4.7}$$

which is to be contrasted with the corresponding selection rule for microwave spectroscopy, $\Delta J = \pm 1$, given in Eq. (2.17). The fact that in Raman work the rotational quantum number changes by two units rather than one is connected with the symmetry of the polarizability ellipsoid. For a linear molecule, such as is depicted in Fig. 4.1, it is evident that during end-over-end rotation the ellipsoid presents the same appearance to an observer *twice* in every complete rotation. It is equally clear that rotation about the bond axis produces no change in polarizability, hence, as in infra-red and microwave spectroscopy, we need concern ourselves only with end-over-end rotations.

If, following the usual practice, we define ΔJ as $(J_{\text{upper state}} - J_{\text{lower state}})$ then we can ignore the selection rule $\Delta J = -2$ since, for a pure rotational change, the upper state quantum number must necessarily be greater than that in the lower state. Further the 'transition' $\Delta J = 0$ is trivial since this represents no change in the molecular energy and hence Rayleigh scattering only.

Combining, then, $\Delta J = +2$ with the energy levels of Eq. (4.6) we have:

$$\Delta \varepsilon = \varepsilon_{J' = J+2} - \varepsilon_{J'' = J}$$

$$= B(4J + 6) \text{ cm}^{-1} \tag{4.8}$$

Since $\Delta J = +2$, we may label these lines S branch lines (cf. Sec. 3.2) and write

$$\Delta \varepsilon_S = B(4J + 6) \text{ cm}^{-1} \qquad (J = 0, 1, 2, \ldots) \tag{4.9}$$

where J is the rotational quantum number in the *lower* state.

Thus if the molecule gains rotational energy from the photon during collision we have a series of S branch lines to the low wavenumber side of the exciting line (Stokes' lines), while if the molecule loses energy to the photon the S branch lines appear on the high wavenumber side (anti-Stokes' lines). The wavenumbers of the corresponding spectral lines are given by:

$$\bar{\nu}_S = \bar{\nu}_{\text{ex.}} \pm \Delta \varepsilon_S = \bar{\nu}_{\text{ex.}} \pm B(4J + 6) \text{ cm}^{-1} \tag{4.10}$$

where the plus sign refers to anti-Stokes' lines, the minus to Stokes', and $\bar{\nu}_{\text{ex.}}$ is the wavenumber of the exciting radiation.

The allowed transitions and the Raman spectrum arising are shown schematically in Fig. 4.4. Each transition is labelled according to its lower J value and the relative intensities of the lines are indicated assuming that the population of the various energy levels varies according to Eq. (2.21) and Fig. 2.7. In particular it should be noted here that Stokes' and anti-Stokes' lines have comparable intensity because many rotational levels are populated and hence downward transitions are approximately as likely as upward ones.

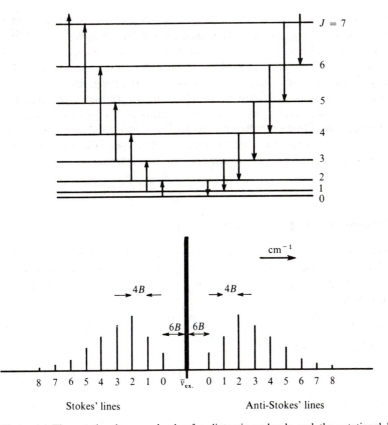

Figure 4.4 The rotational energy levels of a diatomic molecule and the rotational Raman spectrum arising from transitions between them. Special lines are numbered according to their lower J values.

When the value $J = 0$ is inserted into Eq. (4.10) it is seen immediately that the separation of the first line from the exciting line is $6B$ cm^{-1}, while the separation between successive lines is $4B$ cm^{-1}. For diatomic and light triatomic molecules the rotational Raman spectrum will normally be resolved and we can immediately obtain a value of B, and hence the moment of inertia and bond lengths for such molecules. If we recall that homonuclear diatomic molecules (for example, O_2, H_2) give no infra-red or microwave spectra since they possess no dipole moment, whereas they *do* give a rotational Raman spectra, we see that the Raman technique yields structural data unobtainable from the techniques previously discussed. It is thus complementary to microwave and infra-red studies, not merely confirmatory.

It should be mentioned that, if the molecule has a centre of symmetry (as, for example, do H_2, O_2, CO_2), then the effects of nuclear spin will be observed in the Raman as in the infra-red. Thus for O_2 and CO_2 (since the spin of oxygen is zero) every alternate rotational level is absent; for example, in the case of O_2, every level with *even* J values is missing, and thus every transition labelled $J = 0, 2, 4, \ldots$ in Fig. 4.4 is also completely missing from the spectrum. In the case of H_2, and other molecules composed of nuclei with non-zero spin, the spectral lines show an alternation of intensity.

Linear molecules with more than three heavy atoms have large moments of inertia and their rotational fine structure is often unresolved in the Raman spectrum. Direct structural information is not, therefore, obtainable, but we shall see shortly that, taken in conjunction with the infra-red spectrum, the Raman can still yield much very useful information.

4.2.2 Symmetric Top Molecules

The polarizability ellipsoid for a typical symmetric top molecule, for example, $CHCl_3$, was shown in Fig. 4.3(b). Plainly rotation about the top axis produces no change in the polarizability, but end-over-end rotations will produce such a change.

From Eq. (2.38) we have the energy levels:

$$\varepsilon_{J, K} = BJ(J + 1) + (A - B)K^2 \quad \text{cm}^{-1}$$

$$(J = 0, 1, 2, \ldots; K = \pm J, \pm (J - 1), \ldots) \quad (4.11)$$

The selection rules for Raman spectra are:

$$\Delta K = 0$$

$$\Delta J = 0, \pm 1, \pm 2 \qquad \text{(except for } K = 0 \text{ states}$$
$$\text{when } \Delta J = \pm 2 \text{ only)} \qquad (4.12)$$

K, it will be remembered, is the rotational quantum number for axial rotation, so the selection rule $\Delta K = 0$ implies that changes in the angular momentum about the top axis will not give rise to a Raman spectrum—such rotations are, as mentioned previously, Raman inactive. The restriction of ΔJ to ± 2 for $K = 0$ states means effectively that ΔJ cannot be ± 1 for transitions involving the ground state ($J = 0$) since $K = \pm J$, $\pm (J - 1), \ldots, 0$. Thus for all J values other than zero, K also may be different from zero and $\Delta J = \pm 1$ transitions are allowed.

Restricting ourselves, as before, to positive ΔJ we have the two cases:

1. $\Delta J = +1$ (R branch lines)

$$\Delta \varepsilon_R = \varepsilon_{J' = J+1} - \varepsilon_{J'' = J}$$

$$= 2B(J + 1) \, \text{cm}^{-1} \qquad (J = 1, 2, 3, \ldots \text{ (but } J \neq 0)) \qquad (4.13a)$$

2. $\Delta J = +2$ (S branch lines)

$$\Delta\varepsilon_S = \varepsilon_{J'=J+2} - \varepsilon_{J''=J}$$
$$= B(4J + 6) \text{ cm}^{-1} \quad (J = 0, 1, 2, \ldots). \tag{4.13b}$$

Thus we shall have two series of lines in the Raman spectrum:

$$\left.\begin{array}{ll}
\bar{\nu}_R = \bar{\nu}_{\text{ex.}} \pm \Delta\varepsilon_R = \bar{\nu}_{\text{ex.}} \pm 2B(J+1) \text{ cm}^{-1} & (J = 1, 2, \ldots) \\
\bar{\nu}_S = \bar{\nu}_{\text{ex.}} \pm \Delta\varepsilon_S = \bar{\nu}_{\text{ex.}} \pm B(4J+6) \text{ cm}^{-1} & (J = 0, 1, 2, \ldots)
\end{array}\right\} \tag{4.14}$$

These series are sketched separately in Fig. 4.5(a) and (b), where each line is labelled with its corresponding *lower* J value. In the R branch, lines appear

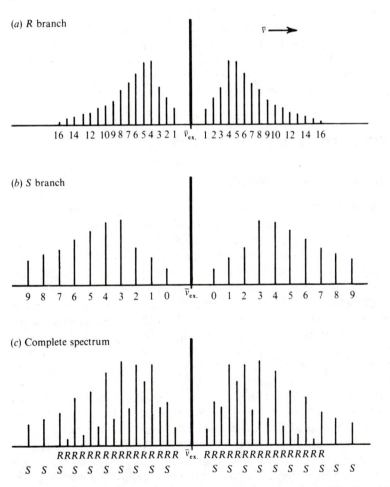

Figure 4.5 Rotational Raman spectrum of a symmetric top molecule. The R and S branch lines are shown separately in (a) and (b) respectively, with the total spectrum in (c).

at $4B$, $6B$, $8B$, $10B$, ... cm^{-1} from the exciting line, while the S branch series occurs at $6B$, $10B$, $14B$, ... cm^{-1}. The complete spectrum, shown at Fig. 4.5(c) illustrates how every alternate R line is overlapped by an S line. Thus a marked intensity alternation is to be expected which, it should be noted, is not connected with nuclear spin statistics.

4.2.3 Spherical Top Molecules; Asymmetric Top Molecules

Examples of spherical top molecules are those with tetrahedral symmetry such as methane, CH_4, or silane, SiH_4. The polarizability ellipsoid for such molecules is a spherical surface and it is evident that rotation of this ellipsoid will produce no change in polarizability. Therefore the pure rotations of spherical top molecules are completely inactive in the Raman.

Normally *all* rotations of asymmetric top molecules, on the other hand, are Raman active. Their Raman spectra are thus quite complicated and will not be dealt with in detail here; it suffices to say that, as in the microwave region, the spectra may often be interpreted by considering the molecule as intermediate between the oblate and prolate types of symmetric top.

4.3 VIBRATIONAL RAMAN SPECTRA

4.3.1 Raman Activity of Vibrations

If a molecule has little or no symmetry it is a very straightforward matter to decide whether its vibrational modes will be Raman active or inactive: in fact, it is usually correct to assume that *all* its modes are Raman active. But when the molecule has considerable symmetry it is not always easy to make the decision, since it is sometimes not clear, without detailed consideration, whether or not the polarizability changes during the vibration.

We consider first the simple asymmetric top molecule H_2O whose polarizability ellipsoid was shown in Fig. 4.2. In Fig. 4.6 we illustrate at (a), (b), and (c) respectively the three fundamental modes v_1, v_2, and v_3, sketching for each mode the equilibrium configuration in the centre with the extreme positions to right and left. The approximate shapes of the corresponding polarizability ellipsoids are also shown.

During the symmetric stretch, in (a), the molecule as a whole increases and decreases in size; now when a bond is stretched, the electrons forming it are less firmly held by the nuclei and so the bond becomes more polarizable. Thus the polarizability ellipsoid of H_2O may be expected to decrease in size while the bonds stretch, and to increase while they compress, but to maintain an approximately constant shape. On the other hand while undergoing the bending motion, in (b), it is the shape of the ellipsoid which changes most; thus if we imagine vibrations of very large amplitude, at one

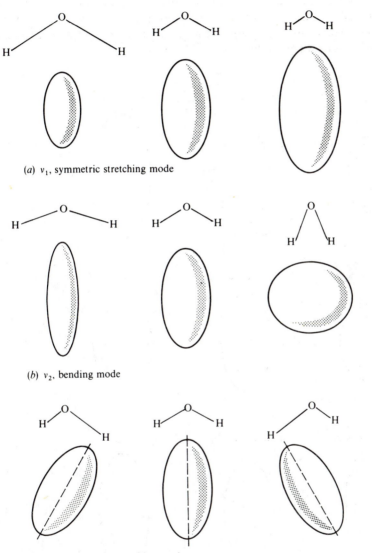

(a) v_1, symmetric stretching mode

(b) v_2, bending mode

(c) v_3, asymmetric stretching mode

Figure 4.6 The change in size, shape, or direction of the polarizability ellipsoid of the water molecule during each of its three vibrational modes. The centre column shows the equilibrium position of the molecule while to right and left are the extremes of each vibration.

extreme (on the left) the molecule approaches the linear configuration with a horizontal axis, while at the other extreme (on the right) it approximates to a diatomic molecule (if the two H atoms are almost coincidental) with a vertical axis. Finally in (c) we have the asymmetric stretching motion, v_3, where both the size and shape remain approximately constant, but the direction of the major axis (shown dashed) changes markedly. Thus all three vibrations involve obvious changes in at least one aspect of the polarizability ellipsoid, and all are Raman *active*.

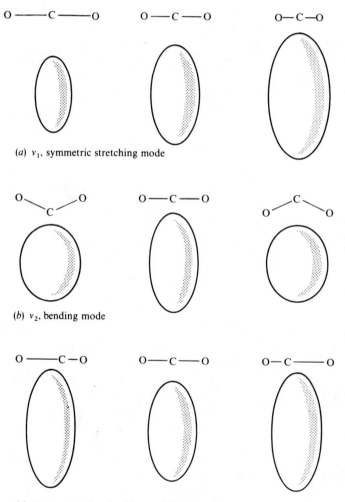

Figure 4.7 The shape of the polarizability ellipsoid of the carbon dioxide molecule during its vibrations.

Now consider the linear triatomic molecule CO_2, whose three fundamental vibrational modes have been shown in Fig. 3.11; in Fig. 4.7 we illustrate the extreme and equilibrium configurations of the molecule and their approximate polarizability ellipsoids. The question of the Raman behaviour of the symmetric stretching mode, v_1, is easily decided—during the motion the molecule changes size, and so there is a corresponding fluctuation in the size of the ellipsoid; the motion is thus Raman *active*. It might be thought that the v_2 and v_3 vibrations are also Raman active, because the molecule changes shape during each vibration and hence, presumably, so does the ellipsoid; however, both these modes are observed to be Raman *inactive*. We must, then, consider this example rather more carefully.

To do this it is usual to discuss the change of polarizability with some *displacement coordinate*, normally given the symbol ξ; thus for a stretching motion, ξ is a measure of the extension (positive ξ) or compression (negative ξ) of the bond under consideration; while for a bending mode, ξ measures the displacement of the bond angle from its equilibrium value, positive and negative ξ referring to opposite displacement directions.

Consider, as an example, the v_1 stretch of carbon dioxide sketched in Fig. 4.7(a). If the equilibrium value of the polarizability is α_0 (centre picture) then, when the bonds stretch (ξ positive), α increases (remember that the extent of the ellipsoid measures the *reciprocal* of α), while when the bonds contract (negative ξ) α decreases. Thus we can sketch the variation of α with ξ as in Fig. 4.8(a). The details of the curve are not important since we are concerned only with *small* displacements; it is plain that near the equilibrium position ($\xi = 0$) the curve has a distinct slope, that is $d\alpha/d\xi \neq 0$ at $\xi = 0$. Thus for small displacements the motion produces a change in polarizability and is therefore Raman *active*.

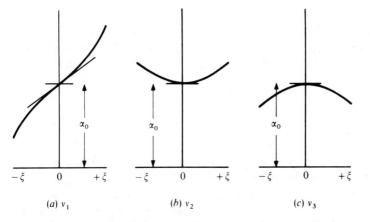

Figure 4.8 The variation of the polarizability, α, with the displacement coordinate, ξ, during the three vibrational modes of the carbon dioxide molecule.

If we now consider the situation for v_2, the bending motion of Fig. 4.7(b), we can count a downwards displacement of the oxygen atoms as negative ξ and an upwards displacement as positive. Although it is not clear from the diagrams whether the motion causes an increase or a decrease in polarizability (actually it is an *increase*) it *is* plain that the change is exactly the same for both positive and negative ξ. Thus we can plot α against ξ as in Fig. 4.8(b) with, as before, $\alpha = \alpha_0$ at $\xi = 0$. Now for small displacements we evidently have $d\alpha/d\xi = 0$ and hence for *small displacements* there is effectively no change in the polarizability and the motion is Raman *inactive*.

Exactly the same argument applies to the asymmetric stretch, v_3, shown at Fig. 4.7(c). Here the polarizability decreases equally for positive and negative ξ, so the plot of polarizability against ξ has the appearance of Fig. 4.8(c). Again $d\alpha/d\xi = 0$ for small displacements and the motion is Raman *inactive*.

We could have followed the same reasoning for the three vibrations of water discussed previously. In each case we would have discovered that the α versus ξ curve has the general shape of Fig. 4.8(a) or its mirror image; in other words, in each case $d\alpha/d\xi \neq 0$ and the motion is Raman *active*. In general, however, the slopes of the three curves would be different at $\xi = 0$, that is, $d\alpha/d\xi$ would have different values. Since we have seen that the Raman spectrum is forbidden for $d\alpha/d\xi = 0$ but allowed for $d\alpha/d\xi \neq 0$ we can imagine that the 'degree of allowedness' varies with $d\alpha/d\xi$. Thus if the polarizability curve has a large slope at $\xi = 0$ the Raman line will be strong; if the slope is small it will be weak; and if zero, not allowed at all. From this stems the following very useful general rule:

symmetric vibrations give rise to intense Raman lines; non-symmetric ones are usually weak and sometimes unobservable.

In particular, a bending motion usually yields only a very weak Raman line; for example the v_2 motion of H_2O (Fig. 3.6(b)), although allowed in the Raman, has not been observed, nor has v_3, for which $d\alpha/d\xi$ is also small.

4.3.2 Rule of Mutual Exclusion

A further extremely important general rule has been established whose operation may be exemplified by carbon dioxide. We can summarize our conclusions about the Raman and infra-red activities of the fundamental vibrations of this molecule in Table 4.1, and we see that, for this molecule, no vibration is simultaneously active in both Raman and infra-red. The corresponding general rule is:

Table 4.1 Raman and infra-red activities of carbon dioxide

Mode of vibration of CO_2	Raman	Infra-red
v_1: symmetric stretch	Active	Inactive
v_2: bend	Inactive	Active
v_3: asymmetric stretch	Inactive	Active

Rule of mutual exclusion. If a molecule has a *centre of symmetry* then Raman active vibrations are infra-red inactive, and vice versa. If there is no centre of symmetry then some (but not necessarily all) vibrations may be both Raman and infra-red active.

The converse of this rule is also true, i.e., the observance of Raman and infra-red spectra showing no common lines implies that the molecule has a centre of symmetry; but here caution is necessary since, as we have already seen, a vibration may be Raman active but too weak to be observed. However, if some vibrations are observed to give coincident Raman and infra-red absorptions it is certain that the molecule has no centre of symmetry. Thus extremely valuable structural information is obtainable by comparison of the Raman and infra-red spectra of a substance; we shall show examples of this in Sec. 4.5.

4.3.3 Overtone and Combination Vibrations

Without detailed consideration of the symmetry of a molecule and of its various modes of vibration, it is no easy matter to predict the activity, either in Raman or infra-red, of its overtone and combination modes. The nature of the problem can be seen by considering v_1 and v_2 of carbon dioxide; the former is Raman active only, the latter infra-red active. What, then, of the activity of $v_1 + v_2$? In fact it is only infra-red active, but this is not at all obvious merely from considering the dipole or polarizability changes during the motions. Again, when discussing Fermi resonance (Sec. 3.5.2) we chose as an example the resonance of v_1 and $2v_2$ of carbon dioxide in the *Raman* effect. Thus $2v_2$ is Raman active although the fundamental v_2 is only infra-red active.

We shall not attempt here to discuss this matter further, being content to leave the reader with a warning that the activity or inactivity of a fundamental in a particular type of spectroscopy does not necessarily imply corresponding behaviour of its overtones or combinations, particularly if the molecule has considerable symmetry. A more detailed discussion is to be found in Herzberg's book *Infra-red and Raman Spectra* and others mentioned in the bibliography.

4.3.4 Vibrational Raman Spectra

The structure of vibrational Raman spectra is easily discussed. For every vibrational mode we can write an expression of the form:

$$\varepsilon = \bar{\omega}_e(v + \tfrac{1}{2}) - \bar{\omega}_e x_e(v + \tfrac{1}{2})^2 \text{ cm}^{-1} \qquad (v = 0, 1, 2, \ldots) \qquad (4.15)$$

where, as before (cf. Eq. (3.12)), $\bar{\omega}_e$ is the equilibrium vibrational frequency expressed in wavenumbers and x_e is the anharmonicity constant. Such an expression is perfectly general, whatever the shape of the molecule or the nature of the vibration. Quite general, too, is the selection rule:

$$\Delta v = 0, \pm 1, \pm 2, \ldots \qquad (4.16)$$

which is the same for Raman as for infra-red spectroscopy, the probability of $\Delta v = \pm 2, \pm 3, \ldots$ decreasing rapidly.

Particularizing, now, to Raman active modes, we can apply the selection rule (4.16) to the energy level expression (4.15) and obtain the transition energies (cf. Eq. (3.15)):

$$
\begin{aligned}
v = 0 \rightarrow v = 1: \Delta\varepsilon_{\text{fundamental}} &= \bar{\omega}_e(1 - 2x_e) \text{ cm}^{-1} \\
v = 0 \rightarrow v = 2: \Delta\varepsilon_{\text{overtone}} &= 2\bar{\omega}_e(1 - 3x_e) \text{ cm}^{-1} \\
v = 1 \rightarrow v = 2: \Delta\varepsilon_{\text{hot}} &= \bar{\omega}_e(1 - 4x_e) \text{ cm}^{-1} \quad \text{etc.}
\end{aligned}
\qquad (4.17)
$$

Since the Raman scattered light is, in any case, of low intensity we can ignore completely all the weaker effects such as overtones and 'hot' bands, and restrict our discussion merely to the fundamentals. This is not to say that active overtones and hot bands cannot be observed, but they add little to the discussion here.

We would expect Raman lines to appear at distances from the exciting line corresponding to each active fundamental vibration. In other words we can write:

$$\bar{v}_{\text{fundamental}} = \bar{v}_{\text{ex.}} \pm \Delta\varepsilon_{\text{fundamental}} \text{ cm}^{-1} \qquad (4.18)$$

where the minus sign represents the Stokes' lines (i.e., for which the molecule has gained energy at the expense of the radiation) and the plus sign refers to the anti-Stokes' lines. The latter are often too weak to be observed, since as we saw earlier (cf. p. 79) very few of the molecules exist in the $v = 1$ state at normal temperatures.

The vibrational Raman spectrum of a molecule is, then, basically simple. It will show a series of reasonably intense lines to the low-frequency side of the exciting line with a much weaker, mirror-image series on the high-frequency side. The separation of each line from the centre of the exciting line gives immediately the Raman active fundamental vibration frequencies of the molecule.

As an example we illustrate the Raman spectrum of chloroform, $CHCl_3$, a symmetric top molecule (Fig. 4.9(a)). The exciting line in this case is the

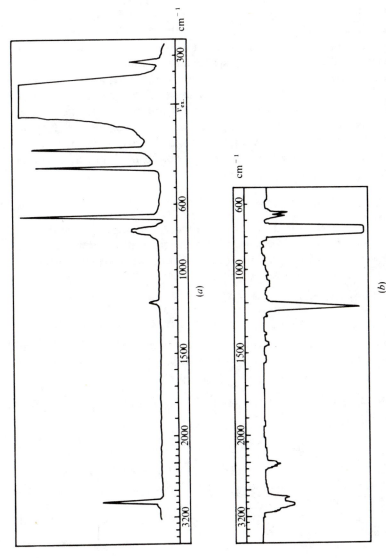

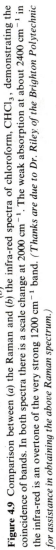

Figure 4.9 Comparison between (*a*) the Raman and (*b*) the infra-red spectra of chloroform, $CHCl_3$, demonstrating the coincidence of bands. In both spectra there is a scale change at 2000 cm^{-1}. The weak absorption at about 2400 cm^{-1} in the infra-red is an overtone of the very strong 1200 cm^{-1} band. (*Thanks are due to Dr. Riley of the Brighton Polytechnic for assistance in obtaining the above Raman spectrum.*)

very intense mercury line at 4358·3 Å, and a wavenumber scale is drawn from this line as zero. Raman lines appear at 262, 366, 668, 761, 1216, and 3019 cm^{-1} on the low-frequency (Stokes') side of the exciting line while the line at 262 cm^{-1} on the high-frequency (anti-Stokes') side is included for a comparison of its intensity.

For comparison also we show at Fig. 4.9(b) the *infra-red* spectrum of the same molecule. The range of the instrument used precluded measurements below 600 cm^{-1}, but we see clearly that strong (and hence fundamental) lines appear in the spectrum at wavenumbers corresponding very precisely with those of lines in the Raman spectrum.

For this molecule, containing five atoms, nine fundamental vibrations (that is, $3N - 6$) are to be expected. The molecule has considerable symmetry, however, and three of these vibrations are doubly degenerate (see Herzberg: *Infra-Red and Raman Spectra* for details) leaving six different fundamental absorptions; we see that these are all active in both the infrared and Raman. The immediate conclusion, not at all surprisingly, is that the molecule has no centre of symmetry.

4.3.5 Rotational Fine Structure

We need not consider in detail the rotational fine structure of Raman spectra in general, if only because such fine structure is rarely resolved, except in the case of diatomic molecules. For the latter we can write the vibration–rotation energy levels (cf. Eq. (3.18)) as:

$$\varepsilon_{J,v} = \bar{\omega}_e(v + \tfrac{1}{2}) - \bar{\omega}_e x_e(v + \tfrac{1}{2})^2 + BJ(J + 1) \text{ cm}^{-1}$$

$$(v = 0, 1, 2, \ldots, J = 0, 1, 2, \ldots) \quad (4.19)$$

where we write $\bar{v}_o$ for $\bar{\omega}_e(1 - 2x_e)$ and use the subscripts O, Q, and S to refer to the O branch lines ($\Delta J = -2$), Q branch lines ($\Delta J = 0$), and S branch lines ($\Delta J = +2$), respectively.

$$\Delta J = 0: \quad \Delta\varepsilon_Q = \bar{v}_o \quad \text{cm}^{-1} \qquad \text{(for all } J) \qquad (4.20)$$

$$\Delta J = +2: \quad \Delta\varepsilon_S = \bar{v}_o + B(4J + 6) \qquad (J = 0, 1, 2, \ldots)$$
$$\Delta J = -2: \quad \Delta\varepsilon_O = \bar{v}_o - B(4J + 6) \qquad (J = 2, 3, 4, \ldots) \Bigg\} \quad (4.21)$$

where we write $\bar{v}_o$ for $\bar{\omega}_e(1 - 2x_e)$ and use the subscripts O, Q, and S to refer to the O branch lines ($\Delta J = -2$), Q branch lines ($\Delta J = 0$), and S branch lines ($\Delta J = +2$), respectively.

Stokes' lines (i.e., lines to *low* frequency of the exciting radiation) will occur at wavenumbers given by:

$$\bar{v}_Q = \bar{v}_{\text{ex.}} - \Delta\varepsilon_Q = \bar{v}_{\text{ex.}} - \bar{v}_o \quad \text{cm}^{-1} \qquad \text{(for all } J)$$

$$\bar{v}_O = \bar{v}_{\text{ex.}} - \Delta\varepsilon_O = \bar{v}_{\text{ex.}} - \bar{v}_o + B(4J + 6) \text{ cm}^{-1} \qquad (J = 2, 3, 4, \ldots)$$

$$\bar{v}_S = \bar{v}_{\text{ex.}} - \Delta\varepsilon_S = \bar{v}_{\text{ex.}} - \bar{v}_o - B(4J + 6) \text{ cm}^{-1} \qquad (J = 0, 1, 2, \ldots).$$

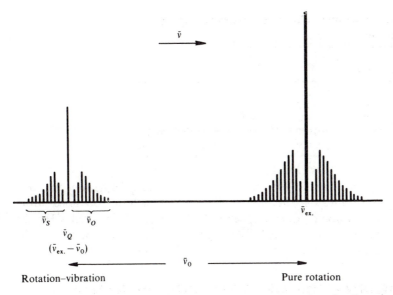

Figure 4.10 The pure rotation and the rotation–vibration spectrum of a diatomic molecule having a fundamental vibration frequency of $\bar{v}_0$ cm^{-1}. Stokes' lines only are shown.

The spectrum arising is sketched in Fig. 4.10 where, for completeness, the pure rotation lines in the immediate vicinity of the exciting line are also shown. The presence of the strong Q branch in the Raman spectrum is to be noted and compared with the P and R branches only which occur for a diatomic molecule in the infra-red (cf. the spectrum of carbon monoxide in Fig. 3.7). The analysis of the O and S branches in the Raman spectrum to give a value for B and hence for the moment of inertia and bond length is straightforward.

Much weaker anti-Stokes' lines will occur at the same distance from, but to high frequency of, the exciting line.

The resolution of Raman spectra is not sufficient to warrant the inclusion of finer details such as centrifugal distortion or the breakdown of the Born–Oppenheimer approximation which were discussed in Chapter 3 for the corresponding infra-red spectra.

For larger molecules we can, in fact, ignore the rotational fine structure altogether since it is not resolved. Even the O and S (or O, P, R, and S) band contours are seldom observed since they are very weak compared with the Q branch. Thus, while the infra-red spectrum of chloroform, shown in Fig. 4.9(b) shows distinct PR or PQR structure on some bands, the Raman spectrum, with the possible exception of the band at 760 cm^{-1}, shows only the strong Q branches. While some information is denied us in Raman spectra because of this, it does represent a considerable simplification of the overall appearance of such spectra.

Table 4.2 Some molecular data determined by Raman spectroscopy

Molecule	Bond length (nm)	Vibration (cm^{-1})
H_2	$0.074\,13 \pm 0.000\,01$	4395.2
N_2	$0.109\,76 \pm 0.000\,01$	2359.6
F_2	$0.141\,8 \pm 0.000\,1$	802.1
CS_2	$0.155\,3 \pm 0.000\,5$	656.6 (symmetrical stretch)
CH_4	$0.109\,4 \pm 0.000\,1$	2914.2 (symmetrical stretch)

In Table 4.2 we collect together some of the information on bond lengths and vibration frequencies which have been obtained from vibrational–rotational Raman spectra. In the case of CS_2 and CH_4 the symmetrical stretching modes only are given since the wavenumbers of the other modes are determined from infra-red techniques.

4.4 POLARIZATION OF LIGHT AND THE RAMAN EFFECT

4.4.1 The Nature of Polarized Light

It is well known that when a beam of light is passed through a Nicol prism or a piece of crystal filter (e.g., polaroid) the only light passing has its electric (or magnetic) vector confined to a particular plane; it is *plane polarized light*. Although superficially this light is indistinguishable from ordinary (or unpolarized) light, it has a very important property which can be demonstrated by using a second Nicol prism or crystal filter. When previously polarized light falls on the second polarizing device (now called the 'analyser') it will be passed with undiminished intensity only if the polarizing axes of the two prisms or crystal sheets are parallel to each other. At any other orientation of these axes the intensity passed will decrease until, when the axes are perpendicular, no light at all passes through the analyser. Thus the analyser serves both to detect polarized light and to determine its plane of polarization.

If the light incident upon the analyser is only partially polarized—i.e., if the majority, but not all, of the rays have their electric vectors parallel to a given plane—then the light will not be *completely* extinguished at any orientation of the analyser; its intensity will merely go through a minimum when the analyser is perpendicular to the plane of maximum polarization. We could, then, measure the degree of polarization in terms of the intensity of light transmitted parallel and perpendicular to this plane; it is more convenient, however, to measure the *degree of depolarization*, ρ, as:

$$\rho = I_{\perp}/I_{\parallel} \tag{4.22}$$

where $I_{\parallel}$ is the maximum and $I_{\perp}$ the minimum intensity passed by the analyser. Thus for completely plane-polarized light $I_{\perp} = 0$ and hence the degree of depolarization is zero also; for completely unpolarized (i.e., ordinary) light, $I_{\perp} = I_{\parallel}$ and $\rho = 1$. For intermediate degrees of polarization ρ lies between 0 and 1.

The relevance of this to Raman spectroscopy is that lines in some Raman spectra are found to be plane-polarized to different extents even though the exciting radiation is completely depolarized. The reason for this is most easily seen if we consider the vibrations of spherical top molecules.

4.4.2 Vibrations of Spherical Top Molecules

The tetrahedral molecule methane, CH_4, is a good example of a spherical top and we can see, from Fig. 4.11, that its polarizability ellipsoid is spherical. During the vibration known as the symmetric stretch all four C—H distances increase and diminish in phase so that the polarizability ellipsoid contracts and expands but *remains spherical*. For this reason the motion is often referred to as the 'breathing frequency'; it is plainly Raman active.

Let us now consider a beam of unpolarized radiation falling on this molecule, and let us designate the direction of this exciting radiation as the z axis. Since all diameters of a sphere are equal the molecule is equally polarizable in all directions; hence the induced dipole in the molecule will lie along the direction of greatest electric vector in the exciting radiation, i.e., perpendicular to the direction of propagation. Thus the induced dipole will lie in the xy plane whatever the plane of the incident radiation. This behaviour is illustrated in Fig. 4.12, where we show an incident beam with its electric vector in (a) the vertical (zy) plane and (b) some other plane making an angle α with the horizontal. In both cases the induced dipole is

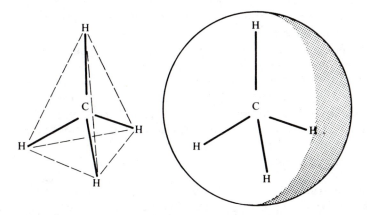

Figure 4.11 The tetrahedral structure of methane, CH_4, and the spherical polarizability ellipsoid of the molecule.

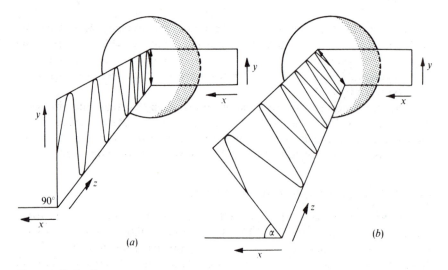

Figure 4.12 To illustrate the plane polarization of Raman scattering from the symmetric vibration ('breathing vibration') of a spherical top molecule.

in the xy plane. A non-polarized incident beam will contain components having all values α.

To an observer studying the scattered radiation at right angles to the incident beam, i.e., along the x axis, the oscillating dipole emitting the radiation is confined to the xy plane—the radiation is plane-polarized. When the molecule undergoes the breathing vibration, the polarizability ellipsoid remains spherical and the dipole change remains in the xy plane. Thus for this vibration the Raman line will be completely plane-polarized, and $\rho = 0$, quite irrespective of the nature of the exciting radiation.

We will now consider a less symmetric vibration of this molecule, for example, the asymmetric stretching mode (Fig. 4.13(a)) where one C—H bond stretches while the other three contract, and vice versa. During this vibration the polarizability surface loses its spherical symmetry and becomes ellipsoid at the extremes of the vibration. One such extreme is shown in Fig. 4.13(b). Now when exciting radiation interacts with the molecule the induced dipole moment will be greatest along the direction of easiest polarizability, i.e., along one of the *minor* axes of the ellipsoid. In a sample of molecules these axes will be oriented in random directions to the incident radiation, so now, because of the lack of spherical symmetry in the vibration, the induced dipole will be randomly oriented and the observed Raman line will be depolarized.

Thus we have immediately a method of assigning some observed Raman lines to their appropriate molecular vibrations—in the case of methane the totally symmetric vibration gives rise to a completely polarized Raman line whereas the non-symmetric vibrations give depolarized lines.

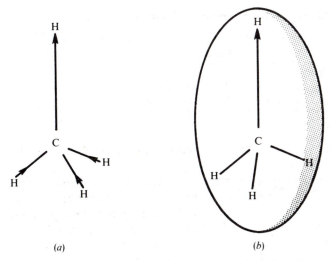

(a)　　　　　　　　　　　　　　(b)

Figure 4.13 An asymmetrical stretching vibration of a spherical top molecule together with the polarizability ellipsoid resulting at the extreme of the motion.

The degree of polarization of spectral lines can be readily estimated by noting how the intensity of each line varies when a piece of polaroid or other analyser is put into the scattered radiation first with its polarizing axis parallel to the xy plane (where z is defined by the direction of the incident beam) and secondly perpendicular to this plane.

4.4.3 Extension to Other Types of Molecule

Precise calculation, rather than the somewhat pictorial argument used above, shows that Raman scattering may be to some extent polarized when emitted by molecules with less symmetry than the tetrahedral ones. In general it can be stated that a symmetric vibration gives rise to a polarized or partially polarized Raman line while a non-symmetric vibration gives a depolarized line. Theoretically, if the degree of depolarization, ρ, is less than or equal to $\frac{6}{7}$, then the vibration concerned is symmetric and the Raman line is described as 'polarized', while if $\rho > \frac{6}{7}$ the line is 'depolarized' and the vibration non-symmetric. If we can speak loosely of molecules with increasing symmetry—e.g., linear molecules are less symmetric than symmetric tops which, in turn, are less symmetric than spherical tops—then the higher the molecular symmetry the smaller will be the degree of depolarization of the Raman line for a particular type of vibration.

We can see the usefulness of polarization measurements by considering a simple example. The molecule nitrous oxide has the formula N_2O. Knowing nothing about the structure of this molecule we might turn for help to its infra-red and Raman spectra. The strongest lines in these spectra

Table 4.3 Infra-red and Raman spectra of nitrous oxide

$\bar{v}$ (cm^{-1})	Infra-red	Raman
589	Strong; *PQR* contour	—
1285	Very strong; *PR* contour	Very strong; polarized
2224	Very strong; *PR* contour	Strong; depolarized

are collected in Table 4.3 together with their band contours (infra-red) and state of polarization (Raman).

The data tells us immediately that the molecule has no centre of symmetry (Raman and infra-red lines occur at the same wavenumber) and so the structure is *not* N—O—N. The fact that some infra-red bands have *PR* contour indicates that the molecule is linear, however, so we are led to the conclusion that the structure is N—N—O. Such a molecule should have $3N - 5 = 4$ fundamental modes but two of these (the bending modes) will be degenerate; all three different fundamental frequencies should be both infra-red and Raman active but we note that the perpendicular infra-red band (plainly to be associated with the bending mode) does not appear in the Raman. This accords with expectations—bending modes are often weak and even unobservable in the Raman.

We are left with the assignment of the 1285 and 2224 cm^{-1} bands to the symmetric and asymmetric stretching modes. Both infra-red bands have the same *PR* (parallel) contour, but we note that only the 1285 cm^{-1} is Raman polarized. This, then, we assign to the symmetric mode, leaving the 2224 cm^{-1} band as the asymmetric.

The analysis would not normally rest there. The overtone and combination bands would also be studied to ensure that their activities and contours are in agreement with the molecular model proposed; the fine structure of the infra-red bands also supports the structure; and finally isotopic substitution leads to changes in vibrational frequencies in excellent agreement with the model and assignments.

In this rather simple case polarization data were hardly essential to the analysis, but certainly useful. In more complicated molecules it can give very valuable information indeed.

4.5 STRUCTURE DETERMINATION FROM RAMAN AND INFRA-RED SPECTROSCOPY

In this section we shall discuss some examples of the combined use of Raman and infra-red spectroscopy to determine the shape of some simple molecules. The discussion must necessarily be limited and the molecules considered (CO_2, N_2O, SO_2; NO_3^-, ClO_3^-, and ClF_3) have been chosen to

Table 4.4 Infra-red and Raman bands of sulphur dioxide

Wavenumber	Infra-red contours	Raman
519	‖ type band	Polarized
1151	‖ type band	Polarized
1361	⊥ type band	Depolarized

illustrate the principles used; extension to other molecular types should be obvious.

Dealing first with the triatomic AB_2 molecules, the questions to be decided are whether each molecule is linear or not and, if linear, whether it is symmetrical $(B—A—B)$ or asymmetrical $(B—B—A)$. In the case of carbon dioxide and nitrous oxide, both molecules give rise to some infra-red bands with PR contours; they must, therefore, be linear. The mutual exclusion rule (cf. Sec. 4.3.2) shows that CO_2 has a centre of symmetry $(O—C—O)$ while N_2O has not $(N—N—O)$, since only the latter has bands common to both its infra-red and Raman spectra. Thus the structures of these molecules are completely determined.

The infra-red and Raman absorptions of SO_2 are collected in Table 4.4. We see immediately that the molecule has no centre of symmetry, since all three fundamentals are both Raman and infra-red active. In the infra-red all three bands show very complicated rotational fine structure, and it is evident that the molecule is non-linear—no band shows the simple PR structure of, say, carbon dioxide. The molecule has, then, the bent shape.

The AB_3 type molecules require rather more discussion. In general we would expect $3N - 6 = 6$ fundamental vibrations for these four-atomic molecules. However, if the molecular shape has some symmetry this number will be reduced by degeneracy. In particular, for the symmetric planar and symmetric pyramidal shapes, one stretching mode and one angle deformation mode are each doubly degenerate and so only four different fundamental frequencies should be observed. These are sketched in Table 4.5 where their various activities and band contours or polarizations are also collected. Both molecular shapes are in fact symmetric tops with the main (threefold) axis passing through atom A perpendicular to the B_3 plane. It is with respect to this axis that the vibrations can be described as ‖ or ⊥. The symmetric modes of vibration are parallel and Raman polarized while the asymmetric are perpendicular and depolarized. All the vibrations of the pyramidal molecule change both the dipole moment and the polarizability, hence all are both infra-red and Raman active. The symmetric stretching mode (v_1) of the planar molecule, however, leaves the dipole moment unchanged (it remains zero throughout) and so is infra-red inactive, while the

Table 4.5 Activities of vibrations of planar and pyramidal AB_3 molecules

Symmetric planar	Activity (R = Raman, I = infra-red)	Vibration	Pyramidal	Activity (R = Raman, I = infra-red)
	R: active (pol.) strong I: inactive	v_1 symmetric stretch		R: active (pol.) strong I: active ∥
 (⊕ = upwards ⊖ = downwards)	R: inactive I: active ∥	v_2 out-of-plane symmetric deformation		R: active (pol.) strong I: active ∥
	R: active (depol.) weak I: active ⊥	v_3 asymmetric stretch		R: active (depol.) weak I: active ⊥
	R: active (depol.) weak I: active ⊥	v_4 asymmetric deformation		R: active (depol.) weak I: active ⊥

symmetric bending mode does not change the polarizability (cf. the discussion of the bending mode of CO_2 in Sec. 4.3.1) and so v_2 is Raman inactive for planar AB_3.

The overall pattern of the spectra, then, should be as follows:

Planar AB_3: 1 vibration Raman active only (v_1)
              1 infra-red active only (v_2)
              2 vibrations both Raman and infra-red active (v_3, v_4).
Pyramidal AB_3: All four vibrations both Raman and infra-red active.
Non-symmetric AB_3: Possibly more than four different fundamental
              frequencies.

With this pattern in mind we can consider the spectra of NO_3^- and ClO_3^- ions. The spectroscopic data are summarized in Table 4.6. Without considering any assignment of the various absorption bands to particular vibrations, we can see immediately that the nitrate ion fits the expected pattern for a planar system, while the chlorate ion is pyramidal. Detailed assignments follow by comparison with Table 4.5. Thus for the nitrate ion,

Table 4.6 Infra-red and Raman spectra of NO_3^- and ClO_3^-

Nitrate ion (NO_3)			Chlorate ion (ClO_3^-)		
Raman (cm^{-1})	Infra-red (cm^{-1})	Assignment	Raman (cm^{-1})	Infra-red (cm^{-1})	Assignment
690	680 $\perp$	v_4	450 (depol.)	434 $\perp$	v_4
—	830 $\parallel$	v_2	610 (pol.)	624 $\parallel$	v_2
1049	—	v_1	940 (depol.)	950 $\perp$	v_3
1355	1350 $\perp$	v_3	982 (pol.)	994 $\parallel$	v_1

the band which is Raman active only is obviously v_1 while that which appears only in the infra-red is v_2. If we make the very reasonable assumption that stretching frequencies are larger than bending, then the assignment of v_3 and v_4 is self-evident. This same assumption, coupled with polarization and band contour data, gives the assignment shown in the table for the chlorate ion.

Finally we consider the spectroscopic data for ClF_3. This is found to have no less than *six* strong (and hence fundamental) infra-red absorptions, some of which also occur in the Raman. We know immediately, then, that the molecule is neither symmetric planar nor pyramidal. A complete analysis is not possible from the Raman and infra-red spectra alone, but the use of microwave spectroscopy shows that the molecule is T-shaped with bond angles of nearly 90°.

4.6 TECHNIQUES AND INSTRUMENTATION

Raman spectroscopy is essentially emission spectroscopy, and the bulk of the instrumentation is simply a typical visible-region spectrometer; the distinguishing characteristic of Raman work, of course, is the exciting source. The advent of accessible and relatively inexpensive laser sources during the past few years has caused a minor revolution in Raman techniques, by largely displacing the traditional mercury discharge lamp as an exciting source. Formerly the process of obtaining a good Raman spectrum of anything but the most straightforward samples involved as much art as science, required 10–20 ml of sample, and was often a very time-consuming operation; now Raman spectra of virtually all samples can be run on a completely routine basis using one millilitre or less of sample and taking a few minutes only.

In fact the laser is almost ideal as a Raman source; it gives a very narrow, highly monochromatic beam of radiation, which may be focused very finely into a small sample, and which packs a relatively large power—

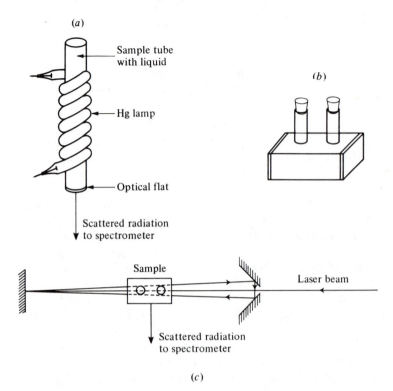

Figure 4.14 (*a*) Excitation of a sample with a spiral mercury lamp. (*b*) Sample container for excitation by a laser. (*c*) Plan view of multi-pass operation using a laser beam.

from several milliwatts to a few watts, depending on the type of laser—into its small frequency spread.

Figure 4.14 compares the two excitation techniques. In part (*a*) the mercury discharge lamp is shown in the form of a spiral round the sample; this form of excitation pumps a good deal of energy into the sample, but it is spread over the many emission lines of the mercury vapour—notably 435·8 and 253·6 nm—and it suffers from three main disadvantages. Firstly, the extended source allows a good deal of the exciting radiation to be scattered directly into the spectrometer where it masks all the Raman lines less than some 100 cm^{-1} from the exciting frequency; secondly the sample tube needs to be about 20–30 cm long and 1–2 cm in diameter, so that a considerable quantity of sample is required; thirdly, the relatively high frequency of the mercury radiation often causes the sample to fluoresce (cf. Sec. 6.3.3), and the resulting fluorescence spectrum swamps the very weak Raman spectrum. Nonetheless the mercury lamp is often still preferred for gaseous samples, where a large sample tube can have a very carefully tailored mercury lamp built round it, and thus collect the exciting radiation with maximum efficiency.

Figure 4.14(c) illustrates how the laser beam is directed through a sample and, by means of mirrors, caused to undergo multiple passes; only three such passes are shown but, by careful alignment of the mirrors up to 10 passes may be achieved, thus enhancing the Raman signal. The standard sample container is a 'box' of quartz, as shown in Fig. 4.14(b), about 2 cm long and 0.5 cm^2 in cross-section, provided with filling ports; its capacity is thus 1 ml. The quantity of sample required can be reduced considerably, however, by using a thin capillary tube, filled with sample and sealed at one end, with the laser beam directed along its length; in this case the multiple traverse facility cannot be used, but nonetheless a few microlitres of liquid give a spectrum of about 60 per cent of the intensity of that from a 1 ml sample with multiple traverse. Solids, in powder or transparent block form, are equally suitable for laser excitation. Further, lasers typically operate at frequencies lower than that of the mercury lamp (e.g., the He–Ne laser at 632.8 nm and the argon laser at 514.5 and 488.0 nm), and are thus very much less likely to cause the sample to fluoresce; also if one particular laser does produce fluorescence, it is a simple matter technically (although perhaps not financially) to switch to another laser and run a successful Raman spectrum. Finally, the Rayleigh scattering is much diminished with the very confined laser beam, and it is routine to detect Raman scattering to within some 20 cm^{-1} of the exciting line.

Radiation scattered from the sample is directed, via mirrors, into a spectrometer operating in the visible region; the monochromator is either a quartz prism or a grating, and the radiation is detected, in the case of a laser instrument, by a photoelectric detector with its output fed to an amplifier and pen recorder. Instruments using a mercury discharge lamp as exciting source may also use a photomultiplier and pen recorder, but they have, in addition, provision for the insertion of a photographic plate as detector, since the very weakest of Raman lines may be observed by allowing the exposure to continue for many hours, or even several days.

BIBLIOGRAPHY

Anderson, A.: *The Raman Effect*: vol. 1 *Principles*, Marcel Dekker Inc., 1971.

Anderson, A.: *The Raman Effect*: vol. 2 *Applications*, Marcel Dekker Inc., 1973.

Barnes, A. J., and W. J. Orville-Thomas (eds): *Vibrational Spectroscopy—Modern Trends*, Elsevier, 1977.

Gilson, T. R., and P. J. Hendra: *Laser Raman Spectroscopy*, Wiley, 1970.

Herzberg, G.: *Molecular Structure and Molecular Spectra*: vol. 1, *Spectra of Diatomic Molecules*, 2nd ed., Van Nostrand, 1950.

Herzberg, G.: *Molecular Spectra and Molecular Structure*: vol. 2, *Infra-Red and Raman Spectra of Polyatomic Molecules*, Van Nostrand, 1945.

Loader, J.: *Basic Laser Raman Spectroscopy*, Heyden & Son Ltd., 1970.

Ross, S. D.: *Inorganic Infra-Red and Raman Spectra*, McGraw-Hill, 1972.
Siesler, H. W., and K. Holland-Moritz: *Infra-Red and Raman Spectroscopy of Polymers*, Marcel Dekker Inc., 1980.
Tobin, M. C.: *Laser Raman Spectroscopy*, Wiley-Interscience, 1971.
Williams, D. H., and I. Fleming: *Spectroscopic Methods in Organic Chemistry*, 3rd ed., McGraw-Hill, 1980.

PROBLEMS

(Useful constants: $h = 6.626 \times 10^{-34}$ J s; $c = 2.998 \times 10^8$ m s^{-1}; $8\pi^2 = 78.957$; mass of H atom $= 1.673 \times 10^{-27}$ kg.)

4.1 With which type of spectroscopy would one observe the pure rotation spectrum of H_2? If the bond length of H_2 is 0.074 17 nm, what would be the spacing of the lines in the spectrum?

4.2 The spin of the hydrogen nucleus is $\frac{1}{2}$; does this make any difference to your answer to Prob. 4.1?

4.3 A molecule A_2B_2 has infra-red absorptions and Raman spectral lines as in the following table:

cm^{-1}	Infra-red	Raman
3374	—	Strong
3287	Very strong; PR contour	—
1973	—	Very strong
729	Very strong; PQR contour	—
612	—	Weak

Deduce what you can about the structure of the molecule and assign the observed vibrations to particular molecular modes as far as possible.

Hint: Use data from Table 3.4 to help with your answer.

4.4 A molecule AB_2 has the following infra-red and Raman spectra:

cm^{-1}	Infra-red	Raman
3756	Very strong: perpendicular	—
3652	Strong; parallel	Strong; polarized
1595	Very strong; parallel	—

The rotational fine structure of the infra-red bands is complex and does not show simple PR or PQR characteristics. Comment on the molecular structure, and assign the observed lines to specific molecular vibrations as far as possible.

4.5 Both N_2O and NO_2 exhibit three different fundamental vibrational frequencies, and for the two molecules some modes are observed in both the infra-red and the Raman. The bands in N_2O show only simple PR structure (no Q branches) while those in NO_2 show complex rotational structure. What can be deduced about the structure of each molecule?

CHAPTER FIVE

ELECTRONIC SPECTROSCOPY OF ATOMS

5.1 THE STRUCTURE OF ATOMS

5.1.1 Electronic Wave Functions

It is well known that an atom consists of a central, positively charged nucleus, which contributes nearly all the mass to the system, surrounded by negatively charged electrons in sufficient number to balance the nuclear charge. Hydrogen, the smallest and simplest atom, has a nuclear charge of $+1$ units (where the unit is the electronic charge, 1.60×10^{-19} coulomb) and one electron; each succeeding atom increases the nuclear charge and electron total by unity, up to atoms with 100 or more electrons.

Modern theories have long ceased to regard the electron as a particle which obeys the laws of classical mechanics applicable to massive, everyday objects; instead, in common with all entities of subatomic size, we consider that it obeys the laws of quantum mechanics (or wave mechanics) as embodied in the Schrödinger wave equation. In principle, this equation may be used to determine many things: e.g., the way in which electrons group themselves about a nucleus when forming an atom, the energy which each electron may have, the way in which it can undergo transitions between energy states, etc. In practice the application of the Schrödinger equation to these problems presents difficulties which can only be overcome in the case of the simplest atoms or by the use of gross approximations. Here, however, we shall be concerned only with the results obtained—and then only in qualitative terms—rather than the mathematical theory of the process.

The Schrödinger theory can be used to predict the *probability* of an electron with a particular energy being at a particular point in space, and it expresses this probability in terms of a very important algebraic expression called the *wave function* of the electron. The wave function is given the Greek symbol ψ. Quite simply, the probability of finding an electron, whose wave function is ψ, within unit volume at a given point in space, is proportional to the value of ψ^2 at that point:

$$\text{Relative probability density} = \psi^2 \qquad (5.1)$$

Let us see what this means. Electronic wave functions consist of three elements: (1) some fundamental *physical constants* (π, h, c, m, e, etc.—where m and e are the mass and charge, respectively, of the electron); (2) *parameters* peculiar to the system under discussion—e.g., for atoms, distance from the nucleus, either radially (r) or along some coordinate axes (x, y, z); and (3) one or more *quantum numbers*. These latter are by no means arbitrarily introduced into the problem in order to make the predictions match experiment; they *belong* to the solution of the Schrödinger equation in the sense that ψ represents a sensible physical situation *only if* the quantum numbers have certain values.

As an example we may quote here the expression for a set of wave functions, ψ_n, which are solutions to the Schrödinger equation for the hydrogen atom:

$$\psi_n = f\left(\frac{r}{a_0}\right) \exp\left(-\frac{r}{na_0}\right) \qquad (5.2)$$

where $a_0 = h^2/4\pi^2 me^2$, r is radial distance from the nucleus, $f(r/a_0)$ is a power series of degree $(n-1)$ in r/a_0, and n is the *principal quantum number*, which can have only integral values, 1, 2, 3, ..., ∞. The constant a_0 has dimensions of length (and is, in fact, about 53 nm) and so the quantity r/a_0 is a pure number. Thus for particular values of r and n, ψ_n and ψ_n^2 are also simply numbers, and ψ_n^2 represents the probability of finding the electron at our chosen distance r from the nucleus when it is in the state represented by the given n value.

It is found that the electronic wave functions of all atoms require the introduction of only four quantum numbers. We shall describe these briefly here, leaving a more thorough discussion to later sections.

5.1.2 The Shape of Atomic Orbitals; Atomic Quantum Numbers

Table 5.1 lists the four quantum numbers, gives the allowed values of each, and states what is the function of each. The *principal quantum number*, as stated earlier, can take integral values from one to infinity. It governs the

Table 5.1 The atomic quantum numbers

Quantum No.	Allowed values	Function
Principal, n	1, 2, 3, ...	Governs the energy and size of the orbital
Orbital, l	$(n-1), (n-2), ..., 0$	Governs the shape of the orbital and the electronic angular momentum
Magnetic, m	$\pm l, \pm(l-1), ..., 0$	Governs the direction of an orbital and the electrons' behaviour in a magnetic field
Spin, s	$+\frac{1}{2}$	Governs the axial angular momentum of the electron

energy of the electron mainly (although we shall see later that the other quantum numbers also affect this energy to some extent). The table shows that n also governs the size of the electronic *orbital*; this latter is a term used to represent the space within which an electron can move according to the Schrödinger theory—it corresponds approximately to the earlier idea of Bohr that electrons move in circular or elliptical *orbits* like planets round a sun. Energy and size of the orbital are connected in that the smaller the orbital the closer to the nucleus the electron will be and hence the more firmly bound.

The *orbital* (or *azimuthal*) *quantum number l*, also has integral values only, but these must be less than n. Thus for $n = 3$, l can be 2, 1, or 0. It governs the shape of the orbital (cf. Fig. 5.1) and the angular momentum of the electron as it circulates about the nucleus in its orbital.

The *magnetic quantum number m* takes integral values which depend on l. Thus for $l = 2$, m can be $+2$, $+1$, 0, -1, -1, or -2; in general there are $2l + 1$ values of m. Besides denoting the behaviour of electrons in orbitals when the atom is placed in a magnetic field, the m quantum number can also be used to specify the direction of a particular orbital.

The *spin quantum number s* is of magnitude $+\frac{1}{2}$ only (but cf. Sec. 5.2.2). It measures the spin angular momentum which the electron possesses whether it is present in an atom or in free space.

Since wave functions represent only a probability distribution of an electron it is difficult to define precisely the shape and size of an orbital. From Eq. (5.2) we see that even at very large values of r, ψ_n (and hence ψ_n^2) still has a value, even though small. Thus an orbital tails off to infinity (although, because of the smallness of a_0, 'infinity' on the atomic scale might be taken as 10^{-4} or 10^{-5} cm) in all directions. However, the difficulty can be overcome if we agree to draw a three-dimensional shape *within* which the electron spends, say, 95 per cent or some other fraction of its time. This can be taken as the effective boundary of the electron's domain and it can be called *the orbital*.

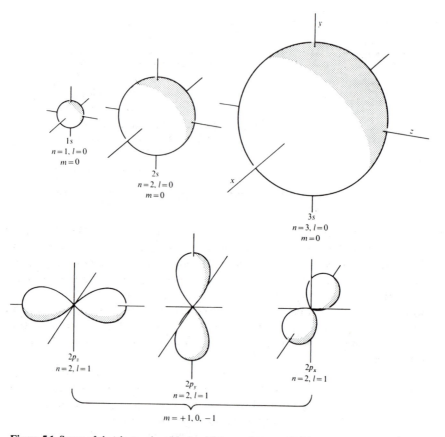

Figure 5.1 Some of the electronic orbitals which may be occupied by the electron in a hydrogen atom.

Considering still the wave function of Eq. (5.2) we see that the corresponding orbital must be spherical, for at any given distance r from the nucleus ψ_n has the same value irrespective of direction. Thus the 95 per cent boundary will be spherical. For larger n the function tails off less rapidly with distance and so the electron can spend proportionately more of its time further from the nucleus; thus the 95 per cent sphere will increase in size with n. We have drawn the cases $n = 1$, 2, and 3 at the top of Fig. 5.1. These spherical orbitals, it so happens, are associated with an l value of zero (and hence $m = 0$) and they are referred to as s orbitals. (Although it is perhaps helpful to connect s with 'spherical'—in fact the label arose historically because of the alleged particular *sharpness* of spectral lines arising from transition of electrons occupying s orbitals—the connection which should be remembered is between s orbitals and $l = 0$.) The s orbitals are labelled according to their n quantum numbers: $1s, 2s, \ldots, ns$.

Orbitals with $l = 1$ (and hence $n \geq 2$) also arise as solutions to the Schrödinger equation for the hydrogen atom. These are twin-lobed and have the approximate shape shown for $n = 2$ in the lower half of Fig. 5.1. Orbitals with $n = 3$, $l = 1$ are larger but have the same shape. Such orbitals are labelled p (historically their transitions were thought to be 'principal') and we see that, for a given n, there are *three* of them, one along each coordinate axis. They can be distinguished as np_x, np_y, and np_z, if necessary. The fact that there are three of them is connected with the three values of m, $m = +1$, 0, and -1, allowed for $l = 1$ states. It is conventional to associate the value $m = 0$ with the np_z orbitals but, for good physical reasons which lie outside the scope of this book, it is not then possible to associate the other m values with either np_x or np_y. Other representations of these orbitals can be drawn, however, in such a way that there is a one-to-one correspondence between each m value and an orbital; these representations are less convenient for the descriptive purposes of this book and we shall not discuss them here.

We can go further: for $l = 2$ (hence $n \geq 3$) we have a set of d orbitals (historically 'diffuse'), and $l = 3(n \geq 4)$ f orbitals (historically 'fundamental'): there are five of the former ($m = \pm 2$, ± 1 or 0) and seven of the latter ($m = \pm 3$, ± 2, ± 1 or 0). Sketches of d orbitals show that they have four lobes, while the f have six, but we shall not attempt to reproduce these here. Orbitals with higher l values, $l = 4, 5, 6, \ldots$, are of less importance and we shall not consider them further; if necessary they are labelled alphabetically after f, that is, $l = 4$, g; $l = 5$, h, etc.

5.1.3 The Energies of Atomic Orbitals; Hydrogen Atom Spectrum

However large an atom its electrons take up orbitals of the s, p, d, $\ldots$ type (according to very specific laws which we shall discuss later) and so the overall shape of each electron's domain is unaltered. The *energy* of each orbital, on the other hand, varies considerably from atom to atom. There are two main contributions to this energy: (1) attraction between electrons and nucleus, (2) repulsion between electrons in the same atom.

We consider first the case of hydrogen in some detail: this is the simplest because, having only one electron, factor (2) is completely absent. We shall later see how the picture should be modified for larger atoms.

Because of the absence of interelectronic effects all orbitals with the same n value have the same energy in hydrogen. Thus the $2s$ and $2p$ orbitals, for instance, are degenerate, as are the $3s$, $3p$, and $3d$. However the energies of the $2s$, $3s$, $4s$, $\ldots$ orbitals differ considerably. For the s orbitals given by Eq. (5.2):

$$\psi_{ns} = f\left(\frac{r}{a_0}\right) \exp\left(-\frac{r}{na_0}\right)$$

the Schrödinger equation shows that the energy is:

$$E_n = -\frac{me^4}{8h^2\varepsilon_o^2 n^2} \quad \text{J}$$

$$\varepsilon_n = -\frac{me^4}{8h^3 c\varepsilon_o^2 n^2} = -\frac{R}{n^2} \quad \text{cm}^{-1} \qquad (n = 1, 2, 3, \ldots) \qquad (5.3)$$

where ε_o is the vacuum permittivity, and where the fundamental constants have been collected together and given the symbol R, called the Rydberg constant. Since p, d, ... orbitals have the same energies as the corresponding s (for hydrogen *only*), Eq. (5.3) represents *all* the electronic energy levels of this atom.

The lowest value of ε_n is plainly $\varepsilon_n = -R$ cm^{-1} (when $n = 1$), and so this represents the most stable (or ground) state. ε_n increases with increasing n, reaching a limit, $\varepsilon_n = 0$ for $n = \infty$. This represents complete removal of the electron from the nucleus, i.e., the state of ionization. We sketch these energy levels for $n = 1$ to 5 and $l = 0$, 1, and 2 only in Fig. 5.2. (Some

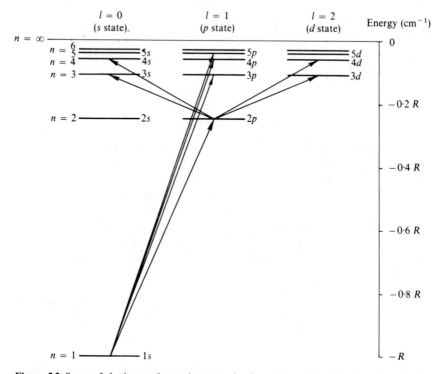

Figure 5.2 Some of the lower electronic energy levels and transitions between them for the single electron of the hydrogen atom.

possible transitions, also shown, will be discussed shortly.) The three p states and five d states for each n are degenerate and not shown separately.

Equation (5.3) and Fig. 5.2, then, represent the energy levels of the atom; in order to discuss the spectra which may arise we need the selection rules governing transitions. The Schrödinger equation shows these to be:

$$\Delta n = \text{anything} \quad \text{and} \quad \Delta l = \pm 1 \text{ only} \quad (5.4)$$

From these selection rules we see immediately that an electron in the ground state (the $1s$) can undergo a transition into any p state:

$$1s \rightarrow np \quad (n \geq 2)$$

while a $2p$ electron can have transitions either into an s state or a d state:

$$2p \rightarrow ns \quad \text{or} \quad nd$$

Since s and d orbitals are here degenerate the energy of both these transitions will be identical. These transitions are sketched in Fig. 5.2.

In general an electron in a lower state n'', can undergo a transition into a higher state n', with absorption of energy:

$$\Delta \varepsilon = \varepsilon_{n'} - \varepsilon_{n''} \quad \text{cm}^{-1}$$

$$\therefore \; \bar{\nu}_{\text{spect.}} = -\frac{R}{n'^2} - \left(-\frac{R}{n''^2}\right) = R\left\{\frac{1}{n''^2} - \frac{1}{n'^2}\right\} \quad \text{cm}^{-1} \quad (5.5)$$

An identical spectral line will be produced in emission if the electron falls from state n' to state n''. In both cases l must change by unity. Let us consider a few of these transitions, restricting ourselves to absorption for simplicity.

Transitions $1s \rightarrow n'p$, $n' = 2, 3, 4, \ldots$. For these

$$\bar{\nu}_{\text{Lyman}} = R\left\{\frac{1}{1} - \frac{1}{n'^2}\right\} = R - \frac{R}{n'^2} \quad \text{cm}^{-1}$$

$$= \frac{3R}{4}, \frac{8R}{9}, \frac{15R}{16}, \frac{24R}{25}, \ldots \quad \text{cm}^{-1} \quad (\text{for } n' = 2, 3, 4, 5, \ldots)$$

Hence we expect a series of lines at the wavenumbers given above. Just such a series is indeed observed in the atomic hydrogen spectrum, and it is called the Lyman series after its discoverer. The appearance of this spectrum is sketched in Fig. 5.3 together with a scale in units of R and in wavenumbers. We can see that the spectrum converges to the point R cm^{-1}, and from the observed spectrum the very precise value $R = 109\,677 \cdot 581$ cm^{-1} is obtained. This convergence limit, which arises when $n' = \infty$, is shown dashed on the figure. It plainly represents complete removal of the electron—i.e., ionization—and the energy required to ionize the atom is given, in cm^{-1}, by the value of R. Using the conversion factor 1 cm$^{-1} = 1 \cdot 987 \times 10^{-23}$ J,

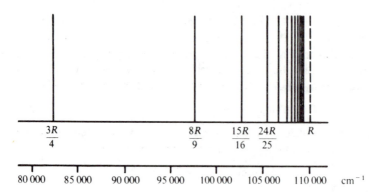

Figure 5.3 Representation of part of the Lyman series of the hydrogen atom showing the convergence (ionization) point.

we have a very precise measure of the *ionization potential* from the ground (1s) state: $2\cdot1781 \times 10^{-18}$ J (which may be more familiar in non-SI units as 13·595 eV).

Another set of transitions arises from an electron initially in the 2s or 2p states: $2s \rightarrow n'p$ or $2p \rightarrow n's, n'd$. For these we write:

$$\bar{\nu}_{\text{Balmer}} = R\left\{\frac{1}{4} - \frac{1}{n'^2}\right\} \quad \text{cm}^{-1}$$

$$= \frac{5R}{36}, \frac{3R}{16}, \frac{21R}{100}, \ldots \quad \text{cm}^{-1} \qquad (\text{for } n' = 3, 4, 5, \ldots)$$

Thus we expect another series of lines converging to $\frac{1}{4}R$ cm^{-1} ($n' = \infty$); this series, called the Balmer series after its discoverer, is observed and the value of $\frac{1}{4}R$ obtained from its convergence limit—which represents the ionization potential from the first excited state—is in excellent agreement with the value of R from the Lyman series.

Other similar line series (called the Paschen, Brackett, Pfund, etc., series) are observed for $n'' = 3, 4, 5, \ldots$; indeed these spectra were observed long before the modern theory of atomic structure had been developed. The spectral lines were correlated empirically by Rydberg, and he showed that an equation of the form given in Eq. (5.5) described the wavenumbers of each. It is after him that the Rydberg constant is named.

It should be mentioned that each line series discussed above shows a *continuous absorption* or *emission* to high wavenumbers of the convergence limits. The convergence limit represents the situation where the atomic electron has absorbed just sufficient energy from radiation to escape from the nucleus with zero velocity. It can, however, absorb more energy than this and hence escape with higher velocities and since the kinetic energy of an electron moving in free space is *not* quantized, *any* energy above the ionization energy can be absorbed. Hence the spectrum in this region is continuous.

This completes our discussion of what might be termed the coarse structure of the hydrogen atom spectrum. In order to consider the fine structure we need to know how the other quantum numbers, besides n, affect the electronic energy levels.

5.2 ELECTRONIC ANGULAR MOMENTUM

5.2.1 Orbital Angular Momentum

An electron moving in its orbital about a nucleus possesses *orbital* angular momentum, a measure of which is given by the l value corresponding to the orbital. This momentum is, of course, quantized, and it is usually expressed in terms of the unit $h/2\pi$, where h is Planck's constant. We may write:

$$\text{Orbital angular momentum} = \sqrt{l(l+1)} \cdot \frac{h}{2\pi} = \sqrt{l(l+1)} \text{ units} \quad (5.6)$$

Now angular momentum is a *vector* quantity, by which we mean that its *direction* is important as well as its magnitude—the axis of a spinning top, for instance, points in a particular direction. Conventionally, vectors may be represented by arrows, and the angular momentum vector is represented by an arrow based at the centre of the top, along the top axis, and of length proportional to the magnitude of the angular momentum. Such an arrow can lie in two different directions, at 180° to each other; these directions are associated, depending on the sign convention used, with clockwise and anticlockwise rotations of the top. Mathematically we can ignore the spinning body and deal merely with the properties of the arrow.

It is usual to distinguish vector quantities by the use of *bold-face* type and we shall accordingly represent orbital angular momentum by the symbol **l** where:

$$\mathbf{l} = \sqrt{l(l+1)} \text{ units} \quad (5.7)$$

In this equation l is always zero or positive and hence so is **l**. Since **l** and l are so closely connected they are often loosely used interchangeably: thus we speak of an electron having 'an angular momentum of 2' when we strictly mean that $l = 2$ and $\mathbf{l} = \sqrt{2 \times 3} = 2.44$ units.

We might at first think that the angular momentum vector of an electron could point in an infinite number of different directions. This, however, would be to reckon without the quantum theory. In fact, once a reference direction has been specified (and this may be done in many ways, either externally, such as by applying an electric or magnetic field, or internally, perhaps in terms of the angular momentum vector of one particular

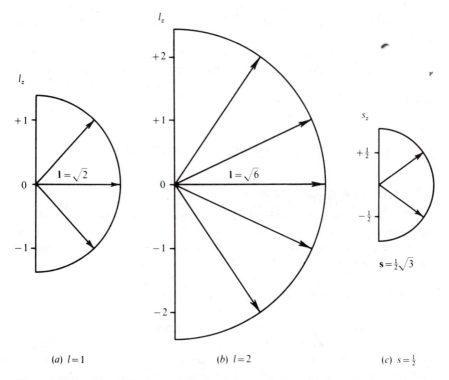

Figure 5.4 The allowed directions of the electronic angular momentum vector for an electron in (a) a *p* state (1 = 1), (b) a *d* state (1 = 2), and (c) the allowed directions of the electronic spin angular momentum vector. The reference direction is taken arbitrarily as upwards in the plane of the paper.

electron), the angular momentum vector can point *only so that its components along the reference direction are integral multiples of $h/2\pi$*. Figure 5.4(a) and (b) show the situation for an electron with $l = 1$ and $l = 2$ respectively (i.e., a *p* and a *d* electron). The reference direction, here taken to be vertical in the figure, is conventionally used to define the *z* axis, and so we can write the *components* of l in this direction as $\mathbf{l}_z$ (note the use of bold-face type for $\mathbf{l}_z$, since the components of angular momentum are clearly vectors, having both magnitude and direction). Alternatively, since we know that the $\mathbf{l}_z$ are integral multiples of $h/2\pi$, we can represent the components in terms of an integral number l_z (not bold-face), where

$$\mathbf{l}_z = l_z \cdot \frac{h}{2\pi} \tag{5.8}$$

This latter notation is used in Fig. 5.4, and we see there that for $l = 1$, l_z takes values of $+1$, 0, and -1, while for $l = 2$ the values are $+2$, $+1$, 0, -1, and -2. In general we see that l_z has values:

$$l_z = l, l - 1, \ldots, 0, \ldots, -(l - 1), l \tag{5.9}$$

and that there are $2l + 1$ values of l_z for a given l. Plainly l_z is to be identified with the magnetic quantum number m introduced in Sec. 5.1.2:

$$l_z \equiv m$$

and this justifies our previous assertion that m governs essentially the *direction* of an orbital.

Before proceeding further let us reiterate the distinction between $l, \mathbf{l}, \mathbf{l}_z$, and l_z. The quantum number l is an integer, positive or zero, representing the state of an electron in an atom and determining its orbital angular momentum. The vector $\mathbf{l}$ designates the magnitude and direction of this momentum as shown by the vector arrows of Fig. 5.4. When expressed in units of $h/2\pi$, $\mathbf{l}$ is numerically equal to $\sqrt{l(l + 1)}$. Once a reference direction is specified (and this is often arbitrary) $\mathbf{l}$ can point only so as to have components $\mathbf{l}_z = l_z h/2\pi$ (with l_z an integer or zero) along that direction.

Usually the orbital energy of the electron depends only on the magnitude and not the direction of its angular momentum; thus the $2l + 1$ values of l_z are all *degenerate*. But we should note that it is possible to *lift the degeneracy* (cf. Sec. 5.6) so that levels with different l_z have different energy.

5.2.2 Electron Spin Angular Momentum

Every electron in an atom can be considered to be spinning about an axis as well as orbiting about the nucleus. Its spin motion is designated by the *spin quantum number s*, which can be shown to have a value of $\frac{1}{2}$ only. Thus the spin angular momentum is given by:

$$\mathbf{s} = \sqrt{s(s + 1)} \frac{h}{2\pi} = \sqrt{\frac{1}{2} \times \frac{3}{2}} \text{ units}$$

$$= \tfrac{1}{2}\sqrt{3} \text{ units} \tag{5.10}$$

The quantization law for spin momentum is that the vector can point so as to have components in the reference direction which are *half-integral* multiples of $h/2\pi$, that is, so that $\mathbf{s}_z = s_z h/2\pi$ with s_z taking the values $+\frac{1}{2}$ or $-\frac{1}{2}$ only. The two (that is, $2s + 1$) allowed directions are shown in Fig. 5.4(c); they are normally degenerate.

5.2.3 Total Electronic Angular Momentum

We now need to discover some means whereby the orbital and spin contributions to the electronic angular momentum may be combined. Formally we can write:

$$\mathbf{j} = \mathbf{l} + \mathbf{s} \tag{5.11}$$

where $\mathbf{j}$ is the *total angular momentum*. Since $\mathbf{l}$ and $\mathbf{s}$ are vectors, Eq. (5.11) must be taken to imply *vector addition*. Also formally, we can express $\mathbf{j}$ in terms of a total angular momentum quantum number j:

$$\mathbf{j} = \sqrt{j(j+1)}\,\frac{h}{2\pi} = \sqrt{j(j+1)} \text{ units} \tag{5.12}$$

where j is *half-integral* (since s is half-integral for a one-electron atom), and a quantal law applies equally to $\mathbf{j}$ as to $\mathbf{l}$ and $\mathbf{s}$: $\mathbf{j}$ *can have z-components which are half-integral only*, i.e.,

$$j_z = \pm j, \ \pm(j-1), \ \pm(j-2), \ldots, \tfrac{1}{2} \tag{5.13}$$

There are two methods by which we can deduce the various allowed values of $\mathbf{j}$ for particular $\mathbf{l}$ and $\mathbf{s}$ values. We shall consider them both briefly.

1. *Vector summation.* In ordinary mechanics two forces in different directions may be added by a graphical method in which vector arrows are drawn to represent the magnitude and direction of the forces, the 'parallelogram is completed', and the magnitude and direction of the resultant given by the diagonal of the parallelogram. Exactly the same method can be used to find the resultant ($\mathbf{j}$) of the vectors $\mathbf{l}$ and $\mathbf{s}$. The important difference is that quantum mechanical laws restrict the angle between $\mathbf{l}$ and $\mathbf{s}$ to values such that $\mathbf{j}$ is given by Eq. (5.12) with *half-integral j*. Thus $\mathbf{j}$ can take values

$$\tfrac{1}{2}\sqrt{3}, \ \tfrac{1}{2}\sqrt{15}, \ \tfrac{1}{2}\sqrt{35}, \ldots \qquad \text{corresponding to} \qquad j = \tfrac{1}{2}, \tfrac{3}{2}, \tfrac{5}{2}, \ldots$$

The method is illustrated in Fig. 5.5(a) and (b) for the case $l = 1$ (that is, $\mathbf{l} = \sqrt{2}$) and $s = \tfrac{1}{2}$ ($\mathbf{s} = \tfrac{1}{2}\sqrt{3}$). In ($a$) the summation yields $\mathbf{j} = \tfrac{1}{2}\sqrt{15}$, which corresponds to a j value of $\tfrac{3}{2}$, while in (b) $\mathbf{j} = \tfrac{1}{2}\sqrt{3}$ or $j = \tfrac{1}{2}$. Construction or calculation shows that $\mathbf{l}$ and $\mathbf{s}$ may not be combined in any other way to give an allowed value of $\mathbf{j}$.

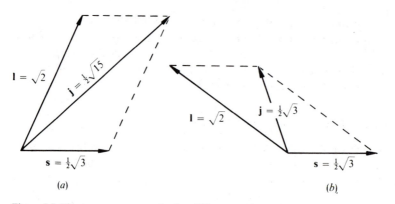

(a)   (b)

Figure 5.5 The two energy states having different total angular momentum which can arise as a result of the vector addition of $\mathbf{l} = \sqrt{2}$ and $\mathbf{s} = \tfrac{1}{2}\sqrt{3}$.

Note that we can get exactly the same answer by summing the *quantum numbers* l and s to get the *quantum number* j. In this example $l = 1$, $s = \frac{1}{2}$, and hence:

$$j = l + s = \tfrac{3}{2} \quad \text{or} \quad j = l - s = \tfrac{1}{2}$$

This simple approach, although adequate for systems with one electron only, is not readily extended to multi-electron systems. For these we must use the rather more fundamental method outlined below.

2. *Summation of z components.* If the components along a common direction of two vectors are added, the summation yields the component in that direction of their resultant. We have seen (cf. Eq. (5.9)) that the z components of $l = 1$ are ± 1, and 0, while those of $s = \frac{1}{2}$ are $\pm \frac{1}{2}$ only. Taking all possible sums of these quantities we have:

$$j_z = l_z + s_z$$

$$\therefore\ j_z = 1 + \tfrac{1}{2},\ 1 - \tfrac{1}{2},\ 0 + \tfrac{1}{2},\ 0 - \tfrac{1}{2},\ -1 + \tfrac{1}{2},\ -1 - \tfrac{1}{2}$$

$$= \tfrac{3}{2},\ \tfrac{1}{2},\ \tfrac{1}{2},\ -\tfrac{1}{2},\ -\tfrac{1}{2},\ -\tfrac{3}{2}$$

In this list of six j_z components, the maximum value is $\frac{3}{2}$, which we know (cf. Eq. (5.13)) must belong to $j = \frac{3}{2}$. Other components of $j = \frac{3}{2}$ are $\frac{1}{2}$, $-\frac{1}{2}$, and $-\frac{3}{2}$ and, striking these from the above six, we are left with $j_z = +\frac{1}{2}$ and $-\frac{1}{2}$. These values are plainly consistent with $j = \frac{1}{2}$.

Thus all the six components are accounted for if we say that the states $j = \frac{3}{2}$ and $j = \frac{1}{2}$ may be formed from $l = 1$ and $s = \frac{1}{2}$. This is, of course, in agreement with the vector summation method.

Both these methods show that for a p electron (that is, $l = 1$), the orbital and spin momenta may be combined to produce a total momentum of $\mathbf{j} = \frac{1}{2}\sqrt{15}$ when $\mathbf{l}$ and $\mathbf{s}$ reinforce (physically we would say that the angular momenta have the same *direction*) or to give $\mathbf{j} = \frac{1}{2}\sqrt{3}$ when $\mathbf{l}$ and $\mathbf{s}$ oppose each other. Thus the total momentum is different in *magnitude* in the two cases and hence we have arrived at two *different energy states* depending on whether $\mathbf{l}$ and $\mathbf{s}$ reinforce or oppose. Both energy states are p states, however (since l is 1 for both), and they may be distinguished by writing the j quantum number value as a subscript to the *state symbol P*, thus $P_{3/2}$ or $P_{1/2}$. (We here use a capital letter for the state of a whole atom and a small letter for the state of an individual electron; in the hydrogen atom, which contains only one electron, the distinction is trivial.) States such as these, split into two energies, are termed *doublet states*; their doublet nature is usually indicated by writing a superscript 2 to the state symbol, thus: $^2P_{3/2}$, $^2P_{1/2}$. The state (or term) symbols produced are to be read 'doublet P three halves' or 'doublet P one half' respectively.

All other higher l values for the electron will obviously produce doublet states when combined with $s = \frac{1}{2}$; for instance, $l = 2$, 3, 4, ... will yield

$^2D_{5/2, 3/2}$, $^2F_{7/2, 5/2}$, $^2G_{9/2, 7/2}$, etc. The student should satisfy himself of this, preferably by using the z-component summation method outlined above. There is, however, a slight difficulty with s states ($l = 0$). Here, since $l = 0$, it can make no contribution to the vector sum, and the only possible resultant is $s = \frac{1}{2}\sqrt{3}$ or $s = \frac{1}{2}$. Remember $s = -\frac{1}{2}$ is not allowed, since the *quantum number* cannot be negative; it is only the z component of the vector which can have negative values. Thus for an s electron we would have the state symbol $S_{1/2}$ only. This is nonetheless formally written as a doublet state ($^2S_{1/2}$) for reasons which should become clear during the discussion of multiplicity in Sec. 5.4.3.

We can now consider the relevance of this discussion to atomic spectroscopy.

5.2.4 The Fine Structure of the Hydrogen Atom Spectrum

The hydrogen atom contains but one electron and so the coupling of orbital and spin momenta and consequent splitting of energy levels will be exactly as described above. We summarize the essential details of the energy levels

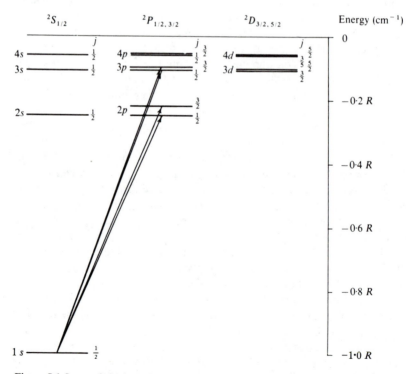

Figure 5.6 Some of the lower energy levels of the hydrogen atom showing the inclusion of j-splitting. The splitting is greatly exaggerated for clarity.

in Fig. 5.6. Each level is labelled with its n quantum number on the extreme left, and its j value on the right; the l value is indicated by the state symbols S, P, D, ... at the top of each column. There is no attempt to show the energy-level splitting of the P and D states to scale in this diagram—the separation between levels differing only in j is many thousands of times smaller than the separation between levels of different n. However, we do indicate that the j-splitting *decreases* with increasing n and with increasing l. The F, G, ... states, not shown on the diagram, follow the same pattern.

The selection rules for n and l are the same as before:

$$\Delta n = \text{anything} \qquad \Delta l = \pm 1 \text{ only} \tag{5.14}$$

but now there is a selection rule for j:

$$\Delta j = 0, \pm 1 \tag{5.15}$$

These selection rules indicate that transitions are allowed between any S level and any P level:

$$^2S_{1/2} \to {}^2P_{1/2} \qquad (\Delta j = 0)$$

$$^2S_{1/2} \to {}^2P_{3/2} \qquad (\Delta j = +1)$$

Thus the spectrum to be expected from the ground ($1s$) state will be identical with the Lyman series (cf. Sec. 5.1.3) except that *every line will be a doublet*. In fact the separation between the lines is too small to be readily resolved but we shall shortly consider the spectrum of sodium in which this splitting is easily observed.

Transitions between the ^2P and ^2D states are rather more complex; Fig. 5.7 shows four of the energy levels involved. Plainly the transition at lowest frequency will be that between the closest pair of levels, the $^2P_{3/2}$ and $^2D_{3/2}$. This, corresponding to $\Delta j = 0$ is allowed. The next transition, $^2P_{3/2} \to {}^2D_{5/2}$ ($\Delta j = +1$), is also allowed and will occur close to the first because the separation between the doublet D states is very small. Thirdly, and more widely spaced, will be $^2P_{1/2} \to {}^2D_{3/2}$ ($\Delta j = +1$), but the fourth transition (shown dotted) $^2P_{1/2} \to {}^2D_{5/2}$, is not allowed since for this $\Delta j = +2$.

Thus the spectrum will consist of the three lines shown at the foot of the figure. This, arising from transitions between doublet levels, is usually referred to as a 'compound doublet' spectrum.

We see, then, that the inclusion of coupling between orbital and spin momenta has led to a slight increase in the complexity of the hydrogen spectrum. In practice, the complexity will be observed only in the spectra of heavier atoms, since for them the j-splitting is larger than for hydrogen. In principle, however, all the lines in the hydrogen spectrum should be close doublets if the transitions involve s levels, or 'compound doublets' if s electrons are not involved.

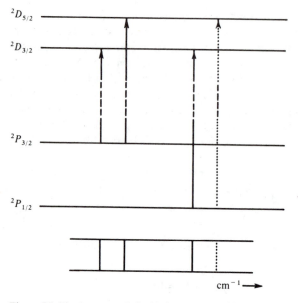

Figure 5.7 The 'compound doublet' spectrum arising as the result of transitions between ^2P and ^2D levels in the hydrogen atom.

5.3 MANY-ELECTRON ATOMS

5.3.1 The Building-up Principle

The Schrödinger equation shows that electrons in atoms occupy orbitals of the same type and shape as the s, p, d, ... orbitals discussed for the hydrogen atom, but that the energies of these electrons differ markedly from atom to atom. There is no general expression for the energy levels of a many-electron atom comparable to Eq. (5.3) for hydrogen; each atom must be treated as a special case and its energy levels either tabulated or shown on a diagram similar to Fig. 5.2 or Fig. 5.6.

There are three basic rules, known as the building-up rules, which determine how electrons in large atoms occupy orbitals. These may be summarized as:

1. Pauli's principle: no two electrons in an atom may have the same set of values for n, l, l_z ($\equiv m$), and s_z.
2. Electrons tend to occupy the orbital with lowest energy available.
3. Hund's principle: electrons tend to occupy degenerate orbitals singly with their spins parallel.

Rule 1 effectively limits to two the number of electrons in each orbital.

An example may make this clear: we may characterize both an orbital and an electron occupying it by specifying the n, l, and m quantum numbers. Thus a $1s$ orbital or $1s$ electron has $n = 1$, $l = 0$, and $m = l_z = 0$; the electron (but not the orbital) is further characterized by a statement of its spin direction, i.e., by specifying $s_z = +\frac{1}{2}$ or $s_z = -\frac{1}{2}$. Two electrons can together occupy the $1s$ orbital provided, according to rule 1, that one has the set of values $n = 1$, $l = 0$, $l_z = 0$, $s_z = +\frac{1}{2}$, and the other $n = 1$, $l = 0$, $l_z = 0$, $s_z = -\frac{1}{2}$. We talk, rather loosely, of two electrons occupying the same orbital only if their spins are *paired* (or *opposed*). A third electron cannot exist in the same orbital without repeating a set of values for n, l, l_z, and s_z already taken up. It would have to be placed into some other orbital and the choice is determined by rule 2: it would go into the next higher vacant or half-vacant orbital. In general, orbital energies in many-electron atoms increase with increasing n, as they do for hydrogen, but they also increase with increasing l, whereas we noted for hydrogen that all s, p, d, ... orbitals with the same n were degenerate. In fact the order of the energy levels for most atoms is as follows:

$$1s < 2s < 2p < 3s < 3p < 4s < 3d < 4p < 5s < 4d \cdots \qquad (5.16)$$

Thus when the $1s$ orbital is full (i.e., contains two electrons) the next available orbital is the $2s$, and after this the $2p$. Now we remember that there are *three* $2p$ orbitals, one along each coordinate axis, and *each* of these can contain two electrons. We may write the n, l, l_z, and s_z values as:

$$\left. \begin{array}{llll} n = 2 & l = 1 & l_z = 1 & s_z = \pm\frac{1}{2} \\ n = 2 & l = 1 & l_z = 0 & s_z = \pm\frac{1}{2} \\ n = 2 & l = 1 & l_z = -1 & s_z = \pm\frac{1}{2} \end{array} \right\} \text{ total six electrons}$$

All three p orbitals remain degenerate (as do the five d orbitals, seven f, etc.) for a given n. It is rule 3 which tells us how electrons occupy these degenerate orbitals. Hund's rule states that when, for example, the $2p_x$ orbital contains an electron, the next electron will go into a *different* $2p$, say $2p_y$, orbital, and a third into the $2p_z$. This may be looked upon as a consequence of repulsion between electrons. A fourth electron has no choice but to pair its spin with an electron already in one $2p$ orbital, while a fifth and sixth will complete the filling of the three $2p$'s.

On this basis we can build up the *electronic configurations* of the 10 smallest atoms, from hydrogen to neon. This is shown in Table 5.2 where each box represents an orbital occupied by one or two electrons with spin directions shown by the arrows. A convenient notation for the electronic configuration is also shown in the table.

When a set of orbitals of given n and l is filled it is referred to as a *closed shell*. Thus the $1s^2$ set of helium, the $2s^2$ set of beryllium, and the $2p^6$ set of neon are all closed shells. The convenience of this is that closed shells

Table 5.2 Electronic structure of some atoms

	1s	2s	2p	
Hydrogen	↑			$1s^1$
Helium	↑↓			$1s^2$
Lithium	↑↓	↑		$1s^2 2s^1$
Beryllium	↑↓	↑↓		$1s^2 2s^2$
Boron	↑↓	↑↓	↑			$1s^2 2s^2 2p_x^1$
Carbon	↑↓	↑↓	↑	↑		$1s^2 2s^2 2p_x^1 2p_y^1$
Nitrogen	↑↓	↑↓	↑	↑	↑	$1s^2 2s^2 2p_x^1 2p_y^1 2p_z^1$
Oxygen	↑↓	↑↓	↑↓	↑	↑	$1s^2 2s^2 2p_x^2 2p_y^1 2p_z^1$
Fluorine	↑↓	↑↓	↑↓	↑↓	↑	$1s^2 2s^2 2p_x^2 2p_y^2 2p_z^1$
Neon	↑↓	↑↓	↑↓	↑↓	↑↓	$1s^2 2s^2 2p_x^2 2p_y^2 2p_z^2$

make no contribution to the orbital or spin angular momentum of the whole atom and hence they may be ignored when discussing atomic spectra. This represents a considerable simplification.

5.3.2 The Spectrum of Lithium and Other Hydrogen-like Species

The alkali metals, lithium, sodium, potassium, rubidium, and cesium, all have a single electron outside a closed-shell core (cf. lithium in Table 5.2). Superficially, then, they resemble hydrogen and this resemblance is augmented by the fact that we can ignore the angular momentum of the core and deal merely with the spin and orbital momentum of the outer electron. Thus we immediately expect the p, d, ... levels to be split into doublets because of coupling between **l** and **s**.

The energy levels of lithium are sketched in Fig. 5.8, which figure should be compared with the corresponding Fig. 5.6 for hydrogen. The two diagrams are similar except for the energy difference between the s, p, and d orbitals of given n in the case of lithium, and the fact that, for this metal, the $1s$ state is filled with electrons which do not generally take part in spectroscopic transitions, it requiring much less energy to induce the $2s$ electron to undergo a transition. Under high energy conditions, however, one or both of the $1s$ electrons may be promoted.

The selection rules for alkali metals are the same as for hydrogen, that is, $\Delta n = $ anything, $\Delta l = \pm 1$, $\Delta j = 0, \pm 1$, and so the spectra will be similar also. Thus transitions from the ground state ($1s^2 2s$) can occur to p levels: $2S_{1/2} \to nP_{1/2, 3/2}$, and a series of doublets similar to the Lyman series will be formed converging to some point from which the ionization potential can be found. From the $2p$ state, however, two separate series of lines will be seen:

$$2\,^2P_{1/2, 3/2} \to n\,^2S_{1/2}$$

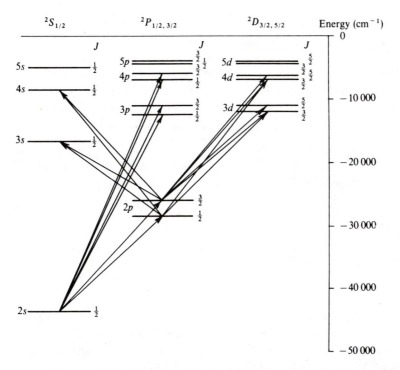

Figure 5.8 Some of the lower energy levels of the lithium atom showing the difference in energy of s, p, and d states with the same value of n. Some allowed transitions are also shown. The j-splitting is greatly exaggerated.

and

$$2\ ^2P_{1/2,\ 3/2} \rightarrow n\ ^2D_{3/2,\ 5/2}$$

The former will be doublets, the latter compound doublets, but their frequencies will differ because the s and d orbital energies are no longer the same.

The same remarks apply to the other alkali metals, the differences between their spectra and that of lithium being a matter of scale only. For instance the j-splitting due to coupling between **l** and **s** increases markedly with the atomic number. Thus the doublet separation of lines in the spectral series, which is scarcely observable for hydrogen, is less than 1 cm^{-1} for the 2p level of lithium, about 17 cm^{-1} for sodium, and over 5000 cm^{-1} for cesium.

Any atom which has a single electron moving outside a closed shell will exhibit a spectrum of the type discussed above. Thus ions of the type He$^+$, Be$^+$, B^{2+}, etc., should, and indeed do, show what are termed 'hydrogen-like spectra'.

5.4 THE ANGULAR MOMENTUM OF MANY-ELECTRON ATOMS

We turn now to consider the contribution of two or more electrons in the outer shell to the total angular momentum of the atom. There are two different ways in which we might sum the orbital and spin momentum of several electrons:

1. First sum the orbital contributions, and then the spin contributions separately, finally add the total orbital and total spin contributions to reach the grand total. Symbolically:

$$\sum \mathbf{l}_i = \mathbf{L} \qquad \sum \mathbf{s}_i = \mathbf{S} \qquad \mathbf{L} + \mathbf{S} = \mathbf{J}$$

   where we use bold-face capital letters to designate total momentum.

2. Sum the orbital and spin momenta of each electron separately, finally summing the individual totals to form the grand total:

$$\mathbf{l}_i + \mathbf{s}_i = \mathbf{j}_i \qquad \sum \mathbf{j}_i = \mathbf{J}$$

The first method, known as Russell–Saunders coupling, gives results in accordance with the spectra of small and medium-sized atoms, while the second (called j, j coupling, since individual j's are summed) applies better to large atoms. We shall consider only the former in detail.

5.4.1 Summation of Orbital Contributions

The orbital momenta, $\mathbf{l}_1$, $\mathbf{l}_2$, ... of several electrons may be added by the same methods as were discussed in Sec. 5.2.3 for the summation of the orbital and spin momenta of a single electron. Thus we could:

1. Add the vectors $\mathbf{l}_1$, $\mathbf{l}_2$, ... graphically remembering that their resultant $\mathbf{L}$ must be expressible by

$$\mathbf{L} = \sqrt{L(L + 1)} \qquad (L = 0, 1, 2, \ldots) \tag{5.17}$$

   where L is the total orbital momentum quantum number. Thus $\mathbf{L}$ can have values $0, \sqrt{2}, \sqrt{6}, \sqrt{12}, \ldots$ only. Figure 5.9 illustrates the method for a p and a d electron, $l_1 = 1$, $l_2 = 2$, hence $\mathbf{l}_1 = \sqrt{2}$, $\mathbf{l}_2 = \sqrt{6}$. There are three, and only three, ways in which the two vectors may be combined to give $\mathbf{L}$ consistent with Eq. (5.17). The three values of $\mathbf{L}$ are seen to be $\sqrt{12}$, $\sqrt{6}$, and $\sqrt{2}$, corresponding to the quantum number $L = 3$, 2, and 1 respectively.

2. Alternatively we could add the individual *quantum numbers* l_1 and l_2 to obtain the total quantum number L according to:

$$L = l_1 + l_2 \qquad l_1 + l_2 - 1, \ldots \qquad |l_1 - l_2| \tag{5.18}$$

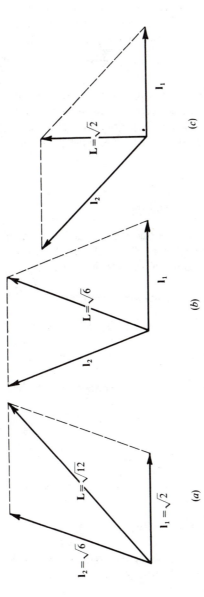

Figure 5.9 Summation of orbital angular momenta for a p and a d electron.

where the modulus sign $|...|$ indicates that we are to take $l_1 - l_2$ or $l_2 - l_1$, whichever is positive. For two electrons, there will plainly be $2l_i + 1$ different values of L, where l_i is the smaller of the two l values.

3. Finally we could add the z components of the individual vectors, picking out from the result sets of components corresponding to the various allowed **L** values. Symbolically this process is:

$$L_z = \sum l_{i_z}$$

Of these methods, (2) is the simplest but it is only applicable when the individual electrons concerned have different n or different l values (these are termed *non-equivalent* electrons). If n and l are the same for two or more electrons they are termed *equivalent* and method 3 must be used. Examples will be given later.

5.4.2 Summation of Spin Contributions

The same methods may be used here as in Sec. 5.4.1. Briefly, if we write the total spin angular momentum as **S**, and the total spin quantum number as S (which is often simply called the total spin), we can have:

1. Graphical summation, provided the resultant is

$$\mathbf{S} = \sqrt{S(S + 1)} \qquad (5.19)$$

where S is either *integral or zero* only, if the number of contributing spins is *even*, or *half-integral* only, if the number is *odd*.

2. Summation of individual quantum numbers; for N spins we have:

$$S = \sum s_i, \sum s_i - 1, \sum s_i - 2, \ldots$$

$$= \frac{N}{2}, \frac{N}{2} - 1, \ldots, \tfrac{1}{2} \qquad \text{(for } N \text{ odd)} \qquad (5.20)$$

$$= \frac{N}{2}, \frac{N}{2} - 1, \ldots, 0 \qquad \text{(for } N \text{ even)}$$

3. Summation of individual s_z to give S_z.

Method 2, which is *always* applicable, is the simplest. Thus for two electrons we have the two possible spin states:

$$S = \tfrac{1}{2} + \tfrac{1}{2} = 1 \qquad \text{or} \qquad S = \tfrac{1}{2} + \tfrac{1}{2} - 1 = 0$$

In the former the spins are called *parallel* and the state may be written (↑↑), while in the latter they are *paired* or *opposed* and written (↑↓).

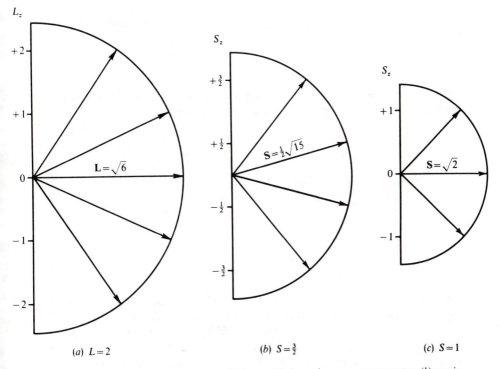

(a) L = 2   *(b) S = 3/2*   *(c) S = 1*

Figure 5.10 Showing the z components of (*a*) an orbital angular momentum vector, (*b*) a spin vector for which S is half-integral, and (*c*) a spin vector for which S is integral.

Again, for three electrons we may have:

$$S = \tfrac{1}{2} + \tfrac{1}{2} + \tfrac{1}{2} = \tfrac{3}{2} \quad (\uparrow\uparrow\uparrow)$$

$$S = \tfrac{1}{2} + \tfrac{1}{2} + \tfrac{1}{2} - 1 = \tfrac{1}{2} \quad (\uparrow\uparrow\downarrow, \uparrow\downarrow\uparrow \text{ or } \downarrow\uparrow\uparrow)$$

where we see that there are three ways in which the $S = \tfrac{1}{2}$ state may be realized, one in which $S = \tfrac{3}{2}$.

As we have implied above, both **L** and **S** have z components along a reference direction. For **L** these components are limited to integral values by quantum laws and, as we can see from Fig. 5.10(*a*), there are, in general, $2L + 1$ of them, while for **S** the S_z will be integral only or half-integral only, depending on whether S is integral or half-integral. We show examples in Fig. 5.10(*b*) and (*c*). In both cases there are $2S + 1$ components.

5.4.3 Total Angular Momentum

The addition of the total orbital momentum **L** and the total spin momentum **S** to give the grand total momentum **J** can be carried out in the same ways as the addition of **I** and **s** to give **j**, for a single electron. The only additional point is that the quantum number J in the expression

$$\mathbf{J} = \sqrt{J(J + 1)} \cdot \frac{h}{2\pi} \qquad (5.21)$$

must be *integral* if S is integral, and *half-integral* if S is half-integral. In terms of the quantum numbers we can write immediately:

$$J = L + S, L + S - 1, \ldots, |L - S| \qquad (5.22)$$

where, as before, the *positive* value of $L - S$ is the lowest limit of the series of values.

For example, if $L = 2$, $S = \frac{3}{2}$, we would have

$$J = \tfrac{7}{2}, \tfrac{5}{2}, \tfrac{3}{2}, \text{ and } \tfrac{1}{2}$$

while if $L = 2$, $S = 1$, the J values are:

$$J = 3, 2, \text{ or } 1 \text{ only}$$

In general we see that there are $2S + 1$ different values of J and hence $2S + 1$ states with different total momentum. The energy of a state depends on its total momentum, so we arrive at $2S + 1$ different energy levels, the energy of each depending on the way in which **L** and **S** are combined. The quantity $2S + 1$, which occupies a special place in atomic spectroscopy, is called the *multiplicity* of the system.

We recall that, when discussing the total angular momentum of a single electron (Sec. 5.2.3), we found that each state, except those with $l = 0$, consisted of two very slightly different energy levels owing to j-splitting; we called these 'doublet states'. We also called the $l = 0$ states doublets, although each has but one energy level since $j = \frac{1}{2}$ only. We now see that the concept of multiplicity justifies us in labelling *all* one-electron states as doublets, since for all of them $S = \frac{1}{2}$, hence $2S + 1 = 2$, and they all have a multiplicity of two.

It is a general rule that in states with $L \geq S$, whether consisting of one electron or of many, the multiplicity is equal to the actual number of levels with different J, whereas if $L < S$, then there are only $2L + 1$ different J values, which is less than the multiplicity. As an example of the latter, if $L = 1$ and $S = 2$, there are only three different J values, $J = 3, 2,$ or 1, whereas the multiplicity is $2S + 1 = 5$.

It is in fact highly convenient to label states with their multiplicities rather than to show the number of different J values because, as we shall see shortly, there is a selection rule which forbids transitions between states with different multiplicities; thus designating the multiplicity of a particular state immediately indicates to which other levels in the system a transition may take place.

5.4.4 Term Symbols

In the whole of this section we have been describing the way in which the total angular momentum of an atom is built up from its various components. Using one sort of coupling only (the Russell–Saunders coupling) we arrive at vector quantities **L**, **S**, and **J** for a system which may be expressed in terms of quantum numbers L, S, and J:

$$\mathbf{L} = \sqrt{L(L + 1)} \qquad \mathbf{S} = \sqrt{S(S + 1)} \qquad \mathbf{J} = \sqrt{J(J + 1)} \qquad (5.23)$$

where the integral L and integral or half-integral S and J are themselves combinations of individual electronic quantum numbers.

In any particular atom, then, we see that the individual electronic angular momenta may be combined in various ways to give different states each having a different total angular momentum (**J**) and hence a different energy (unless some states happen to be degenerate). Before discussing the effect of these states on the spectrum of an atom we require some symbolism which we may use to describe states conveniently. We have already introduced such *state symbols* or *term symbols* in Sec. 5.2.6, but we now consider them rather more fully.

The term symbol for a particular atomic state is written as follows:

$$\text{Term symbol} = {}^{2S+1}L_J \qquad (5.24)$$

where the numerical superscript gives the *multiplicity* of the state, the numerical subscript gives the total angular momentum quantum number J, and the value of the oribital quantum number L is expressed by a *letter*:

For $\quad L = 0, 1, 2, 3, 4, \ldots$

Symbol $= S, P, D, F, G, \ldots$

which symbolism is comparable with the $s, p, d, \ldots$ already used for single-electron states with $l = 0, 1, 2, \ldots$.

Let us now see some examples.

1. $S = \frac{1}{2}$, $L = 2$; hence $J = \frac{5}{2}$ or $\frac{3}{2}$ and $2S + 1 = 2$. Term symbols: ${}^2D_{5/2}$ and ${}^2D_{3/2}$, which are to be read 'doublet D five halves' and 'doublet D three halves' respectively.
2. $S = 1$, $L = 1$; hence $J = 2$, 1, or 0, and $2S + 1 = 3$. Term symbols: ${}^3P_2$, ${}^3P_1$, or ${}^3P_0$ (read 'triplet P two', etc.).
   In both these examples we see that (since $L \geq S$), the multiplicity is the same as the number of different energy states.
3. $S = \frac{3}{2}$, $L = 1$; hence $J = \frac{5}{2}, \frac{3}{2}$, or $\frac{1}{2}$ and $2S + 1 = 4$. Term symbols: ${}^4P_{5/2}$, ${}^4P_{3/2}$, ${}^4P_{1/2}$ (read 'quartet P five halves', etc.) where, since $L < S$, there are only three different energy states but each is nonetheless described as *quartet* since $2S + 1 = 4$.

The reverse process is equally easy; given a term symbol for a particular atomic state we can immediately deduce the various total angular momenta of that state. Some examples:

4. $^3S_1$: we read immediately that $2S + 1 = 3$, hence $S = 1$, and that $L = 0$ and $J = 1$.
5. $^2P_{3/2}$: $L = 1$, $J = \frac{3}{2}$, $2S + 1 = 2$, hence $S = \frac{1}{2}$.

Note, however, that the term symbol tells us only the total spin, total orbital, and grand total momenta of the whole atom—it tells us nothing of the states of the individual electrons in the atoms, nor even how many electrons contribute to the total. Thus in example 5 above, the fact that $S = \frac{1}{2}$ implies that the atom has an odd number of contributing electrons, all except one of which have their spins paired. Thus a single electron (↑), three electrons (↑↑↓), five electrons (↑↓↑↓↑), etc., all form a doublet state. Similarly, the value $L = 1$ implies, perhaps, one p electron, or perhaps one p and two s electrons, or one of many other possible combinations.

Normally this is not important; the spectroscopist is interested only in the energy state of the atom as a whole. Should we wish to specify the energy states of individual electrons, however, we can do so by including them in the term symbol as a prefix. Thus in example 5 we might have $2p\ ^2P_{3/2}$, or $1s2p3s\ ^2P_{3/2}$, etc.

We can now apply our knowledge of atomic states to the discussion of the spectra of some atoms with two or more electrons. We start with the simplest, that of helium.

5.4.5 The Spectrum of Helium and the Alkaline Earths

Helium, atomic number two, consists of a central nucleus and two outer electrons. Clearly there are only two possibilities for the relative spins of the two electrons:

(1) their spins are paired; in which case if s_{1_z} is $+\frac{1}{2}$, s_{2_z} must be $-\frac{1}{2}$, hence $S_z = s_{1_z} + s_{2_z} = 0$, and so $S = 0$ and we have *singlet* states; or
(2) their spins are parallel; now $s_{1_z} = s_{2_z} = +\frac{1}{2}$, say, so that $S_z = 1$ and the states are *triplet*.

The lowest possible energy state of this atom is when both electrons occupy the $1s$ orbital; this, by Pauli's principle, is possible only if their spins are paired, so the ground state of helium must be a singlet state. Further $L = l_1 + l_2 = 0$, and hence J can only be zero. The ground state of helium, therefore, is $^1S_0$.

The relevant selection rules for many-electron systems are:

$$\Delta S = 0 \qquad \Delta L = \pm 1 \qquad \Delta J = 0, \pm 1 \qquad (5.25)$$

(There is a further rule that a state with $J = 0$ cannot make a transition to another $J = 0$ state, but this will not concern us here.) We see immediately that, since S cannot change during a transition, the singlet ground state can undergo transitions only to other singlet states. The selection rules for L and J are the same as those for l and j considered earlier.

For the moment we shall imagine that only one electron undergoes transitions, leaving the other in the $1s$ orbital, and the left-hand side of Fig. 5.11 shows the energy levels for the various singlet states which arise.

Initially the $1s^2\ {}^1S_0$ state can undergo a transition only to $1s^1\ np^1$ states (abbreviated to $1snp$); in the latter $L = 1$, $S = 0$, and hence $J = 1$ only, so the transition may be symbolized:

$$1s^2\ {}^1S_0 \rightarrow 1snp\ {}^1P_1$$

or, briefly:

$${}^1S_0 \rightarrow {}^1P_1$$

From the ${}^1P_1$ state the system could either revert to ${}^1S_0$ states, as shown in the figure, or undergo transitions to the higher ${}^1D_2$ states (for these $S = 0$, $L = 2$, hence $J = 2$ only). In general, then, all these transitions will give rise to spectral series very similar to those of lithium except that here transitions are between *singlet* states only and all the spectral lines will be single.

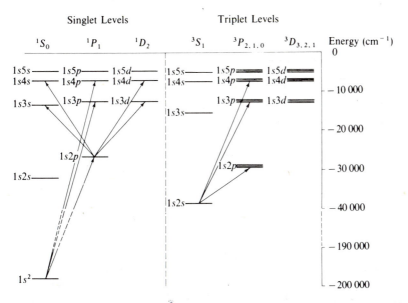

Figure 5.11 Some of the energy levels of the electrons in the helium atom together with a few allowed transitions.

Returning now to the situation in which the electron spins are parallel (case (2) the triplet states) we see that, since the electrons are now forbidden by Pauli's principle from occupying the same orbital, the lowest energy state is $1s2s$. This and other triplet energy levels are shown on the right of Fig. 5.11. The $1s2s$ state has $S = 1$, $L = 0$ and hence $J = 1$ only, and so it is $^3S_1$; by the selection rules of Eq. (5.25) it can only undergo transitions into the $1snp$ triplet states: these, with $S = 1$, $L = 1$, have $J = 2$, 1 or 0, and so the transitions may be written:

$$^3S_1 \to {}^3P_2, {}^3P_1, {}^3P_0$$

All three transitions are allowed, since $\Delta J = 0$ or ± 1, so the resulting spectral lines will be *triplets*.

Transitions from the ^3P states may take place either to ^3S states (spectral series of triples) or to ^3D states. In the latter case the spectral series may be very complex if completely resolved. For ^3D we have $S = 1$, $L = 2$, hence $J = 3$, 2, or 1 and we show in Fig. 5.12 a transition between ^3P and ^3D states, bearing in mind the selection rule $\Delta J = 0, \pm 1$. We note that $^3P_2$ can go to each of $^3D_{3, 2, 1}$, $^3P_1$ can go only to $^3D_{2, 1}$, and $^3P_0$ can go only to $^3D_1$. Thus the complete spectrum (shown at the foot of the figure) should consist of six lines. Normally, however, the very close spacing is not resolved, and only three lines are seen; for this reason the spectrum is referred to as a *compound triplet*.

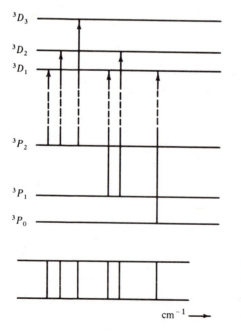

Figure 5.12 The 'compound triplet' spectrum arising from transitions between ^3P and ^3D levels in the helium atom. The separation between levels of different J is much exaggerated.

cm^{-1} ⟶

We might note in passing that Fig. 5.12 shows that levels with higher J have a higher energy in helium, and that the separation decreases from top to bottom. This is not the case with all atoms, however. If higher J is equivalent to lower energy then the separation *increases* from top to bottom and the multiplet is described as *inverted*. In helium, and other atoms with similar behaviour, the multiplet is *normal* or *regular*.

We see, then, that the spectrum of helium consists of spectral series grouped into two types which overlap each other in frequency. In one type, involving transitions between singlet levels, all the spectral lines are themselves singlets, while in the other the transitions are between triplet states and each 'line' is at least a close triplet and possibly even more complex. Because of the selection rule $\Delta S = 0$ there is a strong prohibition on transitions between singlet and triplet states, and transitions *cannot occur* between the right- and left-hand sides of Fig. 5.11. Early experimenters, noting the difference between the two types of spectral series, suggested that helium exists in two modifications, ortho- and para-helium. This is not far from the truth, although we know now that the difference between the two forms is very subtle: it is merely that one has its electron spins always opposed, and the other always parallel.

Other atoms containing two outer electrons exhibit spectra similar to that of helium. Thus the alkaline earths, beryllium, magnesium, calcium, etc., fall in this category, as do ionized species with just two remaining electrons, for example, B^+, C^{2+}, etc.

We should remind the reader at this point that the above discussion on helium has been carried through on the assumption that one electron remains in the $1s$ orbital all the time. This is a reasonable assumption since a great deal of energy would be required to excite *two* electrons simultaneously, and this would not happen under normal spectroscopic conditions. However, not all atoms have only s electrons in their ground state configuration, and we consider next some of the consequences.

5.4.6 Equivalent and Non-Equivalent Electrons; Energy Levels of Carbon

The ground state electronic configuration of carbon is $1s^2 2s^2 2p^2$, which indicates that both the $1s$ and $2s$ orbitals are filled (and hence contribute nothing to the angular momentum of the atom) while the $2p$ orbitals are only partially filled. The $2p$ electrons, also, are most easily removed, so it is these which normally undergo spectroscopic transitions.

Two or more electrons are referred to as *equivalent* if they have the same value of n and of l. Thus the two $2p$ electrons in the ground state of carbon are equivalent ($n_1 = n_2 = 2$, $l_1 = l_2 = 1$) while the set of $1s2s$ are non-equivalent ($n_1 \neq n_2$ although $l_1 = l_2$) as are, for example, $2s2p$ ($n_1 = n_2$ but $l_1 \neq l_2$). Special care is necessary when considering the total angular momentum of equivalent electrons since restrictions are placed on the values of the quantum numbers which each may have. Let us consider the case of $2p^2$ in some detail.

The first restriction arises from Pauli's principle (Sec. 5.3.1). Since we have $n_1 = n_2$ and $l_1 = l_2$ we cannot simultaneously choose $l_{1_z} = l_{2_z}$ and $s_{1_z} = s_{2_z}$.

Further restrictions follow from physical considerations. The basic principle is that electrons cannot be distinguished from each other and so if the energies of two electrons are exchanged we have no way of discovering experimentally that such exchange has taken place. The implication is that if the values of all four numbers n, l, l_z, and s_z for each of two electrons are exchanged, the initial situation is identical in every way with the final. When considering total momentum and the term symbols of atoms we are interested only in different situations and we must not count twice those systems which are interconvertible merely by an exchange of all four numbers n, l, l_z, and s_z. Consider some examples chosen from the $2p^2$ case:

1. We have $n_1 = n_2$, $l_1 = l_2$ and if we also choose $l_{1_z} = l_{2_z}$ then we know (Pauli's principle) that if $s_{1_z} = +\frac{1}{2}$, then $s_{2_z} = -\frac{1}{2}$; alternatively if $s_{1_z} = -\frac{1}{2}$ then $s_{2_z} = +\frac{1}{2}$. Now these two cases are completely identical because one can be reached from the other by exchanging n, l, l_z and s_z. Thus, while we can consider either set alone as typical of the state, we must not consider both together.

2. Similarly, if we assume $s_{1_z} = s_{2_z}$ then we know $l_{1_z} \neq l_{2_z}$. For p electrons $l = 1$ and hence $l_z = 1, 0$ or -1. So we might have:

$$s_{1_z} = s_{2_z}: \quad l_{1_z} = 1 \qquad l_{2_z} = 0 \text{ or } -1$$
$$l_{1_z} = 0 \qquad l_{2_z} = 1 \text{ or } -1$$
$$l_{1_z} = -1 \qquad l_{2_z} = 1 \text{ or } 0$$

Note, however, that the system represented by the pair of values $(1, 0)$ for (l_{1_z}, l_{2_z}) is identical with that for $(0, 1)$; $(0, -1)$ is identical with $(-1, 0)$ and $(1, -1)$ with $(-1, 1)$. Thus we reduce the above six pairs to only three *different* sets:

$$s_{1_z} = s_{2_z}: \quad l_{1_z} = 1 \qquad l_{2_z} = 0$$
$$l_{1_z} = 1 \qquad l_{2_z} = -1$$
$$l_{1_z} = 0 \qquad l_{2_z} = -1$$

3. Finally we note that if $l_{1_z} \neq l_{2_z}$ and $s_{1_z} \neq s_{2_z}$, then interchange of only *one* pair of values (for example, $s_{1_z} = +\frac{1}{2}$, $s_{2_z} = -\frac{1}{2}$, $\rightarrow s_{1_z} = -\frac{1}{2}$, $s_{2_z} = +\frac{1}{2}$) does produce a *different* situation; physically an electron

Table 5.3 Sub-states of two equivalent p electrons ($n_1 = n_2 = 2; l_1 = l_2 = 1$)

l_{1_z}	l_{2_z}	s_{1_z}	s_{2_z}	L_z $(l_{1_z}+l_{2_z})$	S_z $(s_{1_z}+s_{2_z})$	Substate
$+1$	$+1$	$+\frac{1}{2}$	$-\frac{1}{2}$	$+2$	0	(a)
$+1$	0	$+\frac{1}{2}$	$+\frac{1}{2}$	$+1$	$+1$	(b)
$+1$	0	$+\frac{1}{2}$	$-\frac{1}{2}$	$+1$	0	(c)
$+1$	0	$-\frac{1}{2}$	$+\frac{1}{2}$	$+1$	0	(d)
$+1$	0	$-\frac{1}{2}$	$-\frac{1}{2}$	$+1$	-1	(e)
$+1$	-1	$+\frac{1}{2}$	$+\frac{1}{2}$	0	$+1$	(f)
$+1$	-1	$+\frac{1}{2}$	$-\frac{1}{2}$	0	0	(g)
$+1$	-1	$-\frac{1}{2}$	$+\frac{1}{2}$	0	0	(h)
$+1$	-1	$-\frac{1}{2}$	$-\frac{1}{2}$	0	-1	(i)
0	0	$+\frac{1}{2}$	$-\frac{1}{2}$	0	0	(j)
0	-1	$+\frac{1}{2}$	$+\frac{1}{2}$	-1	$+1$	(k)
0	-1	$+\frac{1}{2}$	$-\frac{1}{2}$	-1	0	(l)
0	-1	$-\frac{1}{2}$	$+\frac{1}{2}$	-1	0	(m)
0	-1	$-\frac{1}{2}$	$-\frac{1}{2}$	-1	-1	(n)
-1	-1	$+\frac{1}{2}$	$-\frac{1}{2}$	-2	0	(o)

already distinguishable by its l_z value is being reversed in spin. All four numbers n, l, l_z, and s_z must be exchanged to produce and indistinguishable state.

Keeping these rules in mind we can construct Table 5.3, in which the four columns list combinations of l_{1_z}, l_{2_z}, s_{1_z}, and s_{2_z} leading to different energy states (let us call them substrates). We are interested in the *total* energy and so we show in the next two columns the values of $l_{1_z} + l_{2_z} = L_z$ and $s_{1_z} + s_{2_z} = S_z$ respectively. The final column merely supplies a convenient label to each substate for the following discussion. The 15 substates in the table constitute the z components of the various **L** and **S** vectors which may be formed from two equivalent p electrons. We can find the term symbols in the following way:

1. Note first that the largest L_z value in the table is $L_z = +2$ (substate (a)), and this is associated with $S_z = 0$. $L_z = +2$ must be a z component of the state $L = 2$, that is, one component of a D state, the other components of which are $L_z = +1, 0, -1,$ and -2. In the table we can find several substates of the requisite L_z values, all associated with $S_z = 0$; it is immaterial which of the alternatives we choose, so let us take substates (c), (g), (l), and (o). Since $S_z = 0$ the state must be singlet, hence we have:

$$^1D = \text{substates } (a), (c), (g), (l), \text{ and } (o)$$

With $L = 2, S = 0$ this can only be a $^1D_2$ state.

2. Of the remaining substates the largest L_z is $+1$ associated with $S_z = +1$ (substate (b)). This is plainly one component of a ^3P state, the other components of which may be selected as:

^3P state: $S_z = +1$: $L_z = 1, 0,$ and -1, that is, (b), (f), (k)

                   $S_z = 0$: $L_z = 1, 0,$ and -1, that is, (d), (h), (m)

                   $S_z = -1$: $L_z = 1, 0,$ and -1, that is, (e), (j), (n)

Here we have considered the three components of $S = 1, S_z = +1, 0,$ and -1, to be associated in turn with the components of $L = 1$. The three states listed correspond to term symbols $^3P_2$, $^3P_1$, and $^3P_0$.

3. Finally the one remaining substate, (j), has $L_z = S_z = 0$ and this plainly comprises a $^1S_0$ state.

Overall, then, two equivalent p electrons give rise to the three different energy states ^1D, ^1S, and ^3P, of which the latter, being a triplet state, has three close energy levels $^3P_2$, $^3P_1$, and $^3P_0$. Hund's rule, which we quoted in Sec. 5.3.1, may be expressed for equivalent electrons as: 'The state of lowest energy for a given electronic configuration is that having the *greatest multiplicity*. If more than one state has the same multiplicity then the lowest of these is that with the greatest L value'.

Thus for carbon the ground state is the ^3P, the next in energy is the ^1D, and finally the ^1S. We note that this new expression of Hund's rule implies that electrons in degenerate orbitals tend to have their spins parallel (since this gives the greatest multiplicity, and hence lowest energy); this in turn means that electrons tend to go into separate orbitals since in the same orbital they must have paired spins. Thus we are justified in writing the electronic structures of carbon, nitrogen, and oxygen as in Table 5.2.

If now one of the $2p$ electrons of carbon is promoted to the $3p$ state we have the configuration $1s^2 2s^2 2p3p$. This is an excited state in which the p electrons are non-equivalent. The interested student should show, by the method of Table 5.3, that six different term symbols can be found for this configuration, that is, ^1S, ^1P, ^1D, ^3S, ^3P, and ^3D. In this case, since $n_1 \neq n_2$, neither the Pauli principle nor the principle of indistinguishability offers restrictions to l_z and s_z values, and hence more terms result.

For non-equivalent electrons, however, it is simpler to deal directly with L and S values. Thus we have $s_1 + s_2 = 1$ or 0 depending upon whether the electron spins are parallel or opposed, while for $l_1 = l_2 = 1$ we can have $L = 2, 1$ or 0. We can then tabulate L, S, and J directly, together with their term symbols:

L	S	J	Term symbol
2	1	3, 2, or 1	$^3D_{3,2,1}$
2	0	2	$^1D_2$
1	1	2, 1 or 0	$^3P_{2,1,0}$
1	0	1	$^4P_1$
0	1	1	$^3S_1$
0	0	0	$^1S_0$

and we arrive at the six states listed previously. Note that this direct method is not applicable to equivalent electrons because summation of l to give L implies that all l_z are allowed: this, we have seen, is not true when the electrons are equivalent.

Many other electronic configurations occur, both for carbon and other atoms, in which two or more equivalent electrons contribute to the total energy. We shall not discuss these further, however, except to state that their total energies and term symbols may be discovered by the same process as exemplified above for $2p^2$ electrons.

We can, however, now conveniently discuss rather more fully the operation of the helium–neon laser which was mentioned briefly in Sec. 1.10. This is an example of a continuous laser; by means of an electric discharge the helium atoms in a mixture of helium and neon are excited and ionized. Those which are excited into singlet states decay by emitting radiation until they arrive in the ground state once more, ready for re-excitation; those excited to triplet states, however (see Fig. 5.11), can decay only as far as the $1s2s$ $^3S_1$ state, which is metastable, since the selection rule $\Delta S = 0$ prevents its reversion to the ground $1s^2$ $^1S_0$ state. The $1s2s$ $^3S_1$ state is about 160 000 cm^{-1} above the ground state.

Turning now to the other component of the mixture, neon, this has a ground state configuration $1s^2 2s^2 2p^6$ $^1S_0$; it happens that one of its excited states, the $1s^2 2s^2 2p^5 4s^1$, where one of the $2p$ electrons has been promoted to the $4s$ orbital, is very nearly 160 000 cm^{-1} above the ground state, so collisions between excited helium atoms and ground state neon atoms can result in a resonance exchange of energy:

$$He^* + Ne \rightarrow He + Ne^*$$

Thus the electric discharge essentially pumps *neon* atoms into an excited state. This state can undergo spontaneous decay to lower singlet states, but here the induced decay is quite important and, if radiation of about 8700 cm^{-1} is present, the decay to $1s^2 2s^2 2p^5 3p^1$ is induced, while radiation of 15 800 cm^{-1} results in decay to $1s^2 2s^2 2p^5 3s^1$—it is this latter which gives the usual 632.8 nm radiation from this laser.

The presence of radiation of the appropriate frequency is ensured by keeping the helium–neon mixture at low pressure in a tube placed between a pair of highly efficient mirrors. Thus the majority of the radiation is repeatedly reflected up and down the tube, and it is only the one per cent or so which 'escapes' through the mirrors that constitutes the useful output from the laser. Nonetheless, because all the available power is concentrated into a very narrow, highly monochromatic and coherent beam, these lasers are increasingly used as sources of light and power.

5.5 PHOTOELECTRON SPECTROSCOPY AND X-RAY FLUORESCENCE SPECTROSCOPY

5.5.1 Photoelectron Spectroscopy

The recently developed technique of Photoelectron Spectroscopy (PES) offers an excellent way of studying the electronic energy levels of atoms and—as we shall see in Sec. 6.2.5—of molecules too. Basically the method is equivalent to measuring the ionization energy of both outer (valence) electrons and the more firmly bound inner electrons; electrons are ejected from atoms or molecules essentially by giving them an energy 'kick' of known size and measuring the amount of excess energy with which each emerges from the atom or molecule.

For valence electrons the energy is usually provided by radiation from a helium lamp, in which low-pressure helium gas is excited by an electric discharge. The radiation arises when atoms excited into the $1s2p\ ^1P_1$ state (cf. Fig. 5.11) by the discharge return to the ground $1s^2\ ^1S_0$ state, and has a frequency of $5\cdot14 \times 10^{15}$ Hz and a wavelength of $58\cdot4$ nm. For our purposes, however, it is more convenient to think of this radiation as consisting of a stream of photons, each of energy $h\nu$, that is, $3\cdot41 \times 10^{-18}$ J (for those who still prefer non-SI units, this figure is equivalent to $21\cdot2$ eV). When such radiation falls on to a sample atom, one can think of a photon 'colliding' with an electron in the atom and giving up its energy. If the electron is held very firmly by its nucleus, $3\cdot14 \times 10^{-18}$ J may well not be enough to dislodge it, but if it is one of the outer electrons, a kick of this magnitude will certainly eject it from the atom. If its binding energy is small, say $0\cdot8 \times 10^{-18}$ J (or 5 eV), then it will leave with a kinetic energy equivalent to $(3\cdot4 - 0\cdot8) \times 10^{-18}$ J $= 2\cdot6 \times 10^{-18}$ J, whereas if it is held with an energy of, say $3\cdot2 \times 10^{-18}$ J (20 eV), it will leave with a much smaller kinetic energy, only $0\cdot2 \times 10^{-18}$ J. We may write:

$$\text{Kinetic energy of ejection} = 3\cdot4 \times 10^{-18} \text{ J} - \text{binding energy}$$

or, more generally:

$$\text{Kinetic energy} = h\nu - I \tag{5.26}$$

where $h\nu$ is the energy of the exciting radiation and I the binding energy (ionization potential) of the electron. In order to measure binding energies we clearly need a technique to estimate the kinetic energy, or velocity, of the ejected electrons.

There are two main approaches to this, both shown in Fig. 5.13. In (*a*) the radiation passes along the axis of a metal can, which contains a concentric metal grid, and which also holds the sample. Electrons ejected from sample atoms may pass through the grid and are then collected by the can, to be registered as a current by a sensitive electrometer. While the grid is uncharged, *all* the electrons emitted by the sample reach the can and the current is at its maximum. If now a negative potential is applied to the grid, slow-moving electrons will be repelled and will not reach the can, so the current falls. As the grid potential is increased from zero a curve of current (i.e., relative number of electrons reaching the can) versus grid potential (i.e., kinetic energy) can be plotted.

Figure 5.13(*b*) shows the second method. Here the electrons ejected from the sample are collimated into a beam by a slit in the can and then,

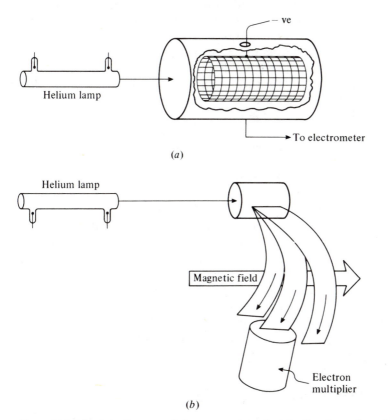

(*a*)

(*b*)

Figure 5.13 Schematic diagrams of the apparatus required for the observation of photoelectron spectra using (*a*) the electron retardation method, and (*b*) the velocity spectrometer method.

since all electrons have the same mass, they can be sorted into their velocities by a magnetic field—a simple form of velocity spectrometer akin to a mass spectrometer. The path of moving electrons is curved into a circle whose radius is directly proportional to their velocity and inversely proportional to the magnitude of the field, so that increasing the field from zero causes electrons of successively higher velocities to be recorded by the cascade electron detector.

The grid method of Fig. 5.13(a) has the advantage of higher sensitivity in that electrons emitted in all directions from sample atoms are collected by the can, whereas in the velocity spectrometer of Fig. 5.13(b) only those electrons passing through the slit are utilized. The latter method is advantageous, however, when experiments concerning the directional nature (or polarization) of the ejected electrons are undertaken.

While the 58·4 nm helium radiation is sufficiently energetic to eject valence electrons from all other elements, it cannot disturb *inner* electrons. For these the lamp may be replaced by an X-ray source; for example with X-rays of 10 nm wavelength, the photon energy is some 2×10^{-17} J (100 eV), while 1 nm X-rays carry energy of 2×10^{-16} J (1000 eV) and are sufficient to eject electrons from the inner shells of most atoms.

By means of a combination of radiation sources, then, the energy levels of atomic electrons can be measured with some precision. At present electrons ejected with energies closer than about 10^{-21} J (0·01 eV) cannot be distinguished, so it is not yet possible to investigate the finer details of atomic energy levels such as j-splitting in smaller atoms. However, the broad pattern of observed energies is, of course, quite consistent with current atomic theories.

In addition, however, the technique has useful potential in chemical analysis. Although during compound formation the valence electrons of atoms take up quite new energy levels (see Chapter 6), all the inner orbitals, and even some outer ones, are unaffected by bonding. Thus the energies of the so-called 'lone pair' electrons of elements such as chlorine, oxygen, or nitrogen are scarcely changed by compound formation since they play no part in normal bonds, and the characteristic binding energy of these electrons in each case makes their recognition easy. Equally for inner electrons, X-ray sources give spectra from which the characteristic energy level pattern of each atom present can be recognized. On the whole, however, it is simpler to use the closely related technique of X-ray fluorescence spectroscopy for this type of analysis, a method described very briefly in Sec. 5.5.2.

5.5.2 X-ray Fluorescence Spectroscopy

When inner electrons are ejected from an atom by an X-ray beam, the 'holes' left in the inner orbitals are rapidly filled by electrons dropping down from higher levels; much of the energy which they emit during this process appears in the X-ray region, but to lower frequency than the original exciting beam—the substance is said to *fluoresce*. The emitted secondary X-rays may be readily analysed into their component frequencies by diffraction from a crystal surface, since this forms a grating with appropriately sized spacings for the wavelengths involved. Now since the inner orbital energies are quite unaffected by external influences, and since each atom has its own quite characteristic pattern of energy levels, X-ray fluorescence spectroscopy is an excellent, rapid, and quite unambiguous method of detecting and estimating particular atoms in a sample. The technique is, for instance, widely used to estimate the composition of alloys, particularly since the analysis is sufficiently rapid for a sample to be withdrawn from the melt and analysed in time for adjustments to be made to the bulk alloy composition before casting.

5.6 THE ZEEMAN EFFECT

We have been concerned in this chapter with two sorts of electronic energy. Firstly, there is energy of *position*—energy arising by virtue of interaction between electrons and the nucleus and between electrons and other electrons in the same atom. This energy can be described in terms of the n and l quantum numbers, although we have discussed it, rather less precisely, by drawing energy level diagrams. Secondly, there is energy of *motion*—energy arising from the summed orbital and spin momenta of the electrons in the atom which depends on the l_z and s_z values of each electron and the way in which these are coupled. This gives rise to the fine structure of spectroscopic lines discussed earlier.

Angular momentum can be considered as arising from a physical movement of electrons about the nucleus and, since electrons are charged, such motion constitutes a circulating electric current and hence a magnetic field. This field can, indeed, be detected, and it is its interaction with exterior fields which is the subject of this section.

We can represent the angular momentum field by a vector $\boldsymbol{\mu}$—the magnetic dipole of the atom—and it is readily shown that $\boldsymbol{\mu}$ is directly proportional to the angular momentum $\mathbf{J}$ and has the same direction. If the electron is considered as a point of mass m and charge e, then we have:

$$\boldsymbol{\mu} = -\frac{e}{2m} \mathbf{J} \quad \text{JT}^{-1}$$

(Here we use the SI unit of magnetic field, the tesla (T), which is equivalent to 10 000 gauss in electromagnetic units.) But quantum mechanics indicates that the electron is not a point charge and a more exact expression for $\boldsymbol{\mu}$ is:

$$\boldsymbol{\mu} = -\frac{ge}{2m} \mathbf{J} = -\frac{ge}{2m} \sqrt{J(J+1)} \frac{h}{2\pi} \quad \text{J T}^{-1} \tag{5.27}$$

where g is a purely numerical factor, called the Landé splitting factor. This factor depends on the state of the electrons in the atom and is given by:

$$g = \frac{3}{2} + \frac{S(S + 1) - L(L + 1)}{2J(J + 1)} \tag{5.28}$$

In general g lies between 0 and 2.

We now recall (cf. Eq. (5.13) for one electron) that $\mathbf{J}$ can have either integral or half-integral components J_z along a reference direction, depending upon whether the quantum number J is integral or half-integral. Figure 5.14(a) shows this for a state with $J = \frac{3}{2}$, the $2J + 1$ components being given in general by:

$$J_z = J, J - 1, \ldots, \tfrac{1}{2} \text{ or } 0, \ldots, -J \tag{5.29}$$

Further, since $\boldsymbol{\mu}$ is proportional to $\mathbf{J}$, $\boldsymbol{\mu}$ will also have components in the z direction which are given by:

$$\mu_z = -\frac{ge}{2m}\frac{h}{2\pi} J_z \tag{5.30}$$

These are shown diagrammatically at Fig. 5.14(b). If now an external field is applied to the atom, thus specifying the previously arbitrary z direction, the atomic dipole $\boldsymbol{\mu}$ will interact with the applied field to an extent depending on its component in the field direction. If the strength of the applied field is B_z then the extent of the interaction is simply $\mu_z B_z$:

$$\text{Interaction} = \Delta E = \mu_z B_z = -\frac{heg}{4\pi m} B_z \quad \text{J} \tag{5.31}$$

In this equation we have expressed the interaction as ΔE since the application of the field splits the originally degenerate energy levels corresponding to the $2J + 1$ values of J_z into $2J + 1$ *different* energy levels. This is shown for $J = \frac{3}{2}$ in Fig. 5.14(c). It is this splitting, or lifting of the degeneracy on the application of an external magnetic field, which is called the *Zeeman effect* after its discoverer.

The energy splitting is very small; the factor $he/4\pi m$ in Eq. (5.31), known as the *Bohr magneton*, has a value of 9.27×10^{-24} J T^{-1}; thus for $g = 1$, and for an applied field B_z of one tesla (that is, 10 000 gauss), the interaction energy is only some 10^{-23} joules, which in turn is of the order 0.5 cm^{-1}. This small splitting is, of course, reflected in a splitting of the spectral transitions observed when a magnetic field is applied to an atom. In order to discuss the effect on the spectrum we need one further selection rule:

$$\Delta J_z = 0, \pm 1$$

Let us consider the doublet lines in the sodium spectrum produced, as we have discussed in Sec. 5.3.2, by transitions between the $^2S_{1/2}$ states and

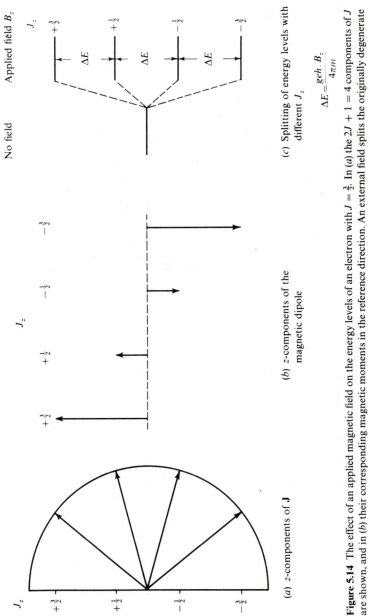

(a) z-components of **J**

(b) z-components of the magnetic dipole

(c) Splitting of energy levels with different J_z

$$\Delta E = \frac{geh . B_z}{4\pi m}$$

Figure 5.14 The effect of an applied magnetic field on the energy levels of an electron with $J = \frac{3}{2}$. In (a) the $2J + 1 = 4$ components of J are shown, and in (b) their corresponding magnetic moments in the reference direction. An external field splits the originally degenerate levels into four separate levels as in (c).

the $^2P_{1/2}$ and $^2P_{3/2}$ states. When a field B_z is applied to the atom, the $^2S_{1/2}$ and $^2P_{1/2}$ states are both split into two (since $J = \frac{1}{2}$, $2J + 1 = 2$), while the $^2P_{3/2}$ is split into four. The extent of the splitting (Eq. (5.31)) is proportional to the g factor in each state and, from Eq. (5.28) we can easily calculate:

$$^2S_{1/2}: \quad S = \tfrac{1}{2}, L = 0, J = \tfrac{1}{2}, \text{ hence } g = 2$$

$$^2P_{1/2}: \quad S = \tfrac{1}{2}, L = 1, J = \tfrac{1}{2}, \text{ hence } g = \tfrac{2}{3}$$

$$^2P_{3/2}: \quad S = \tfrac{1}{2}, L = 1, J = \tfrac{3}{2}, \text{ hence } g = 1\tfrac{1}{3}$$

and we see that the $^2S_{1/2}$, $^2P_{1/2}$, and $^2P_{3/2}$ levels are split in the ratio of $3:1:2$. We show the situation in Fig. 5.15. On the left of the figure we see the energy levels and transitions *before* the field B_z is applied; the levels are unsplit and the spectrum is a simple doublet. On the right we see the effect of the applied field. The spectrum shows that the original line due to the $^2S_{1/2} \rightarrow {}^2P_{1/2}$ transition disappears and is replaced by *four* new lines, while the $^2S_{1/2} \rightarrow {}^2P_{3/2}$ transition is replaced by *six* new lines.

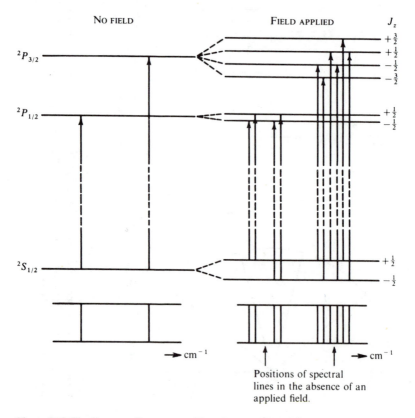

Positions of spectral
lines in the absence of an
applied field.

Figure 5.15 The Zeeman effect on transitions between ^2S and ^2P states. The situation before the field is applied is shown on the left, that after on the right.

The effect described above is usually referred to as the *anomalous Zeeman effect*—although, in fact, most atoms show the effect in this form. The *normal Zeeman effect* applies to transitions between singlet states only (e.g., the transitions of electrons in the helium atom shown on the left of Fig. 5.11). For singlet states we have:

$$2S + 1 = 1 \qquad \text{hence} \qquad S = 0$$

$$\therefore \; J = L \qquad \text{and} \qquad g = 1 \qquad (\text{cf. Eq. (5.28)})$$

Thus the splitting between all singlet levels is identical for a given applied field and the corresponding Zeeman spectrum is considerably simplified.

In general, the Zeeman effect can give very useful information about the electronic states of atoms. In the first place, the number of lines into which each transition becomes split when a field is applied depends on the J value of the states between which transitions arise. Next the g value, deduced from the splitting for a known applied field, gives information about the L and S values of the electron undergoing transitions. Overall, then, the term symbols for various atomic states can be deduced by Zeeman experiments. In this way all the details of atomic states, term symbols, etc., discussed above, have been amply confirmed experimentally.

5.7 THE INFLUENCE OF NUCLEAR SPIN

The nuclei of many atoms are known to be spinning about an axis. We shall discuss at some length in Chapter 7 the spectrum which this spin may give rise to in the radiofrequency region, but it is pertinent here to consider very briefly what effect such spin may have on the electronic spectra of atoms.

The nuclear spin quantum number I may be zero, integral, or half-integral depending on the particular nucleus considered. Thus the nuclear angular momentum, given by

$$\mathbf{I} = \sqrt{I(I + 1)} \, \frac{h}{2\pi} = \sqrt{I(I + 1)} \text{ units} \qquad (5.32)$$

can have values $0, \sqrt{3}/2, \sqrt{2}, \sqrt{15}/2$, etc.

The effect of $\mathbf{I}$ on the spectrum can be understood if we define the total momentum (electronic + nuclear) of an atom by $\mathbf{F}$:

$$\mathbf{F} = \sqrt{F(F + 1)} \, \frac{h}{2\pi} = \sqrt{F(F + 1)} \text{ units} \qquad (5.33)$$

where F is the *total momentum quantum number*. If, as before, J is the total electronic quantum number, then we may write

$$F = J + I, J + I - 1, \ldots, |J - I| \qquad (5.34)$$

thus giving $2J + 1$ or $2I + 1$ different energy states, whichever is the less.

The energy-level splitting due to nuclear spin is of the order 10^{-3} that due to electron spin; thus extremely fine resolving power is necessary for its observation and it is normally referred to as *hyperfine structure*.

5.8 CONCLUSION

This completes all we have to say about atomic spectroscopy. In the next chapter we extend the ideas introduced here to cover the electronic spectra of simple molecules, and we shall briefly discuss the techniques of electronic spectroscopy.

BIBLIOGRAPHY

See the Bibliography to Chapter 6 with the following important additions:

Herzberg, G.: *Atomic Spectra and Atomic Structure*, Dover, 1944.
Kuhn, M. G.: *Atomic Spectra*, Academic Press, 1962.
Shore, B. W., and D. H. Menzel: *Principles of Atomic Spectra*, Wiley, 1968.

PROBLEMS

(Useful constants: $R = 109\ 677 \cdot 581\ \text{cm}^{-1}$; $1\ \text{cm}^{-1} \equiv 11 \cdot 958\ \text{J mol}^{-1}$.)

5.1 Calculate the first three lines in the absorption spectrum arising from transitions from the $3s$ level of the hydrogen atom; what is the ionization energy of this level?

5.2 The term symbol for a particular atomic state is quoted as $^4D_{5/2}$. What are the values of L, S, and J for this state? What is the minimum number of electrons which could give rise to this? Suggest a possible electronic configuration.

5.3 What are the term symbols for the following pairs of non-equivalent electrons: (*a*) ss, (*b*) pp, (*c*) sd, and (*d*) pd?

5.4 What are the term symbols for the following pairs of equivalent electrons: (*a*) s^2, (*b*) p^2, and (*c*) d^2?

5.5 The term symbols for particular states of three different atoms are quoted as $^4S_1$, $^2D_{7/2}$, and $^0P_1$; explain why these are erroneous.

5.6 Figure 5.7 (p. 170) shows the three transitions arising between ^2P and ^2D states; into how many lines would each of these transitions split if a magnetic field were applied? (Assume that the g-value is different for each energy level.)

5.7 Show that the p^3 configuration with equivalent electrons contains the states ^2D, ^2P, and ^4S. Which state is lowest in energy? What additional states arise if the electrons are non-equivalent?

CHAPTER SIX

ELECTRONIC SPECTROSCOPY
OF MOLECULES

In the first section of this chapter we shall discuss, in some detail, the electronic spectra of diatomic molecules. We shall find that the overall appearance of such spectra can be considered without assuming any knowledge of molecular structure, without reference to any particular electronic transition, and indeed, with little more than a formal understanding of the nature of electronic transitions within molecules. In Sec. 6.2 we shall summarize modern ideas of molecular structure and show how these lead to a classification of electronic states analogous to the classification of atomic states discussed in the previous chapter. Section 6.3 will extend the ideas of Secs 6.1 and 6.2 to polyatomic molecules and Sec. 6.4 will deal briefly with experimental techniques.

6.1 ELECTRONIC SPECTRA OF DIATOMIC MOLECULES

6.1.1 The Born–Oppenheimer Approximation

As a first approach to the electronic spectra of diatomic molecules we may use the Born–Oppenheimer approximation previously mentioned in Sec. 3.2; in the present context this may be written:

$$E_{\text{total}} = E_{\text{electronic}} + E_{\text{vibration}} + E_{\text{rotation}} \qquad (6.1)$$

which implies that the electronic, vibrational, and rotational energies of a molecule are completely independent of each other. We shall see later to what extent this approximation is invalid. A change in the total energy of a molecule may then be written:

$$\Delta E_{total} = \Delta E_{elec.} + \Delta E_{vib.} + \Delta E_{rot.} \quad J$$

or

$$\Delta \varepsilon_{total} = \Delta \varepsilon_{elec.} + \Delta \varepsilon_{vib.} + \Delta \varepsilon_{rot.} \quad cm^{-1} \tag{6.2}$$

The approximate orders of magnitude of these changes are:

$$\Delta \varepsilon_{elec.} \approx \Delta \varepsilon_{vib.} \times 10^3 \approx \Delta \varepsilon_{rot.} \times 10^6 \tag{6.3}$$

and so we see that vibrational changes will produce a 'coarse structure' and rotational changes a 'fine structure' on the spectra of electronic transitions. We should also note that whereas pure rotation spectra (Chapter 2) are shown only by molecules possessing a permanent electric dipole moment, and vibrational spectra (Chapter 3) require a change of dipole during the motion, electronic spectra are given by *all* molecules since changes in the electron distribution in a molecule are always accompanied by a dipole change. This means that homonuclear molecules (for example, H_2 or N_2), which show no rotation or vibration–rotation spectra, *do* give an electronic spectrum and show vibrational and rotational structure in their spectra from which rotational constants and bond vibration frequencies may be derived.

Initially we shall ignore rotational fine structure and discuss the appearance of the vibrational coarse structure of spectra.

6.1.2 Vibrational Coarse Structure: Progressions

Ignoring rotational changes means that we rewrite Eq. (6.1) as

$$E_{total} = E_{elec.} + E_{vib.} \quad J$$

or

$$\varepsilon_{total} = \varepsilon_{elec.} + \varepsilon_{vib.} \quad cm^{-1} \tag{6.4}$$

From Eq. (3.12) we can write immediately:

$$\varepsilon_{total} = \varepsilon_{elec.} + (v + \tfrac{1}{2})\bar{\omega}_e - x_e(v + \tfrac{1}{2})^2\bar{\omega}_e \quad cm^{-1} \qquad (v = 0, 1, 2, \ldots) \tag{6.5}$$

The energy levels of this equation are shown in Fig. 6.1 for two arbitrary values of $\varepsilon_{elec.}$. As in previous chapters the lower states are distinguished by a *double* prime (v'', $\varepsilon''_{elec.}$), while the upper states carry only a *single* prime (v', $\varepsilon'_{elec.}$). Note that such a diagram cannot show correctly the relative separations between levels of different $\varepsilon_{elec.}$, on the one hand, and those with different v' or v'' on the other (cf. Eq. (6.3)), but that the spacing between the upper vibrational levels is deliberately shown to be rather smaller than that between the lower; this is the normal situation since an excited electronic state usually corresponds to a weaker bond in the molecule and hence a smaller vibrational wavenumber $\bar{\omega}_e$.

There is essentially no selection rule for v when a molecule undergoes an electronic transition, i.e., every transition $v'' \rightarrow v'$ has some probability, and a great many spectral lines would, therefore, be expected. However, the situation is considerably simplified if the *absorption* spectrum is considered from the electronic ground state. In this case, as we have seen in Sec. 3.1.3, virtually all the molecules exist in the lowest vibrational state, that is, $v'' = 0$, and so the only transitions to be observed with appreciable intensity are those indicated in Fig. 6.1. These are conventionally labelled according to their (v', v'') numbers (note: upper state *first*), that is, (0, 0), (1, 0), (2, 0), etc. Such a set of transitions is called a *band* since, under low resolution, each line of the set appears somewhat broad and diffuse, and is more particularly called a v' *progression*, since the value of v' increases by unity for each line in the set. The diagram shows that the lines in a band crowd together more closely at high frequencies; this is a direct consequence of the anharmonicity of the upper state vibration which causes the excited vibrational levels to converge.

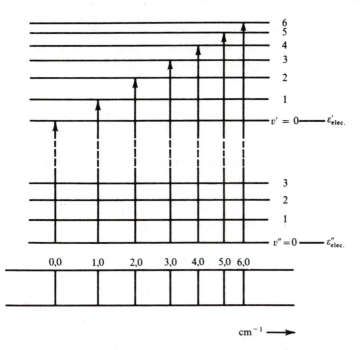

Figure 6.1 The vibrational 'coarse' structure of the band formed during electronic absorption from the ground ($v'' = 0$) state to a higher state.

An analytical expression can easily be written for this spectrum. From Eq. (6.5) we have immediately:

$$\Delta\varepsilon_{total} = \Delta\varepsilon_{elec.} + \Delta\varepsilon_{vib.}$$

$$\therefore \bar{\nu}_{spec.} = (\varepsilon' - \varepsilon'') + \{(v' + \tfrac{1}{2})\bar{\omega}'_e - x'_e(v' + \tfrac{1}{2})^2\bar{\omega}'_e\}$$

$$- \{(v'' + \tfrac{1}{2})\bar{\omega}''_e - x''_e(v'' + \tfrac{1}{2})^2\bar{\omega}''_e\} \quad \text{cm}^{-1} \quad (6.6)$$

and, provided some half-dozen lines can be observed in the band, values for $\bar{\omega}'_e$, x'_e, $\bar{\omega}''_e$, and x''_ε, as well as the separation between electronic states, $(\varepsilon' - \varepsilon'')$, can be calculated. Thus the observation of a band spectrum leads not only to values of the vibrational frequency and anharmonicity constant in the ground state ($\bar{\omega}''_e$ and x''_e), but also to these parameters in the excited electronic state ($\bar{\omega}'_e$ and x'_e). This latter information is particularly valuable since such excited states may be extremely unstable and the molecule may exist in them for very short times; nonetheless the band spectrum can tell us a great deal about the bond strength of such species.

We shall see later that molecules normally have many excited electronic energy levels, so that the whole absorption spectrum of a diatomic molecule will be more complicated than Fig. 6.1 suggests: the ground state can usually undergo a transition to several excited states, and each such transition will be accompanied by a band spectrum similar to Fig. 6.1.

Further, in *emission spectra* the previously excited molecule may be in one of a large number of available (ε', v') states, and has a similar multitude of (ε'', v'') states to which it may revert. Thus emission spectra are usually extremely complicated, and a great deal of care and patience is needed for a complete analysis.

6.1.3 Intensity of Vibrational–Electronic Spectra: the Franck–Condon Principle

Although quantum mechanics imposes no restrictions on the change in the vibrational quantum number during an electronic transition, the vibrational lines in a progression are not all observed to be of the same intensity. In some spectra the (0, 0) transition is the strongest, in others the intensity increases to a maximum at some value of v', while in yet others only a few vibrational lines with high v' are seen, followed by a continuum. All these types of spectrum are readily explicable in terms of the *Franck–Condon principle* which states that *an electronic transition takes place so rapidly that a vibrating molecule does not change its internuclear distance appreciably during the transition.*

We have already seen in Chapter 3 how the energy of a diatomic molecule varies with internuclear distance (cf. Fig. 3.3). We recall that this figure, the Morse curve, represents the energy when one atom is considered fixed on the $r = 0$ axis and the other is allowed to oscillate between the

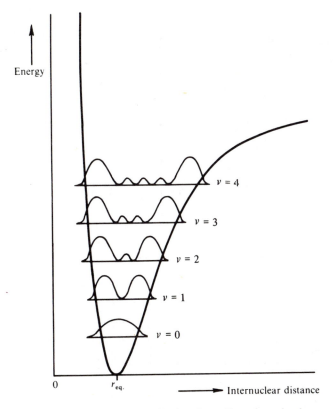

Figure 6.2 The probability distribution for a diatomic molecule according to the quantum theory. The nuclei are most likely to be found at distances apart given by the maxima of the curve for each vibrational state.

limits of the curve. Classical theory would suggest that the oscillating atom would spend most of its time *on* the curve at the turning point of its motion, since it is moving most slowly there; quantum theory, while agreeing with this view for high values of the vibrational quantum number, shows that for $v = 0$ the atom is most likely to be found at the *centre* of its motion, i.e., at the equilibrium internuclear distance $r_{eq.}$. For $v = 1, 2, 3, \ldots$ the most probable positions steadily approach the extremities until, for high v, the quantal and classical pictures merge. This behaviour is shown in Fig. 6.2 where we plot the probability distribution in each vibrational state against internuclear distance. Those who have studied quantum mechanics will realize that Fig. 6.2 shows the variation of ψ^2 with internuclear distance, where ψ is the vibrational wave function.

If a diatomic molecule undergoes a transition into an upper electronic state in which the excited molecule is stable with respect to dissociation into its atoms, then we can represent the upper state by a Morse curve similar in

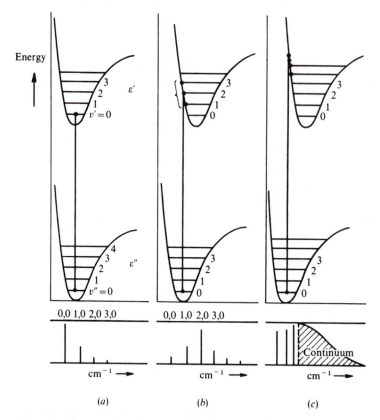

Figure 6.3 The operation of the Franck–Condon principle for (a) internuclear distances equal in upper and lower states, (b) upper-state internuclear distance a little greater than that in the lower state, and (c) upper-state distance considerably greater.

outline to that of the ground electronic state. There will probably (but not necessarily) be differences in such parameters as vibrational frequency, equilibrium internuclear distance, or dissocation energy between the two states, but this simply means that we should consider each excited molecule as a new, but rather similar, molecule with a different, but also rather similar, Morse curve.

Figure 6.3 shows three possibilities. In (a) we show the upper electronic state having the same equilibrium internuclear distance as the lower. Now the Franck–Condon principle suggests that a transition occurs *vertically* on this diagram, since the internuclear distance does not change, and so if we consider the molecule to be initially in the ground state both electronically (ε'') and vibrationally ($v'' = 0$), then the most probable transition is that indicated by the vertical line in Fig. 6.3(a). Thus the strongest spectral line of the $v'' = 0$ progression will be the (0, 0). However, the quantum theory only says that the *probability* of finding the oscillating atom is greatest at the equilibrium distance in the $v = 0$ state—it allows some, although small, chance of the atom being near the extremities of its vibrational motion. Hence there is some chance of the transition starting from the ends of the $v'' = 0$ state and finishing in the $v' = 1, 2$, etc., states. The (1, 0), (2, 0), etc., lines diminish rapidly in intensity, however, as shown at the foot of Fig. 6.3(a).

In Fig. 6.3(b) we show the case where the excited electronic state has a *slightly* greater internuclear separation than the ground state. Now a vertical transition from the $v'' = 0$ level will most likely occur into the upper vibrational state $v' = 2$, transitions to lower and higher v' states being less likely; in general the upper state most probably reached will depend on the difference between the equilibrium separations in the lower and upper states. In Fig. 6.3(c) the upper state separation is drawn as *considerably* greater than that in the lower state and we see that, firstly, the vibrational level to which a transition takes place has a high v' value. Further, transitions can now occur to a state where the excited molecule has energy in excess of its own dissociation energy. From such states the molecule will dissociate without any vibrations and, since the atoms which are formed may take up any value of kinetic energy, the transitions are not quantized and a continuum results. This is shown at the foot of the figure. We consider the phenomenon of dissociation more fully in the next section.

The situation is rather more complex for emission spectra and for absorption from an excited vibrational state, for now transitions take place from *both ends* of the vibrational limits with equal probability; hence each progression will show two maxima which will coincide only if the equilibrium separations are the same in both states.

6.1.4 Dissociation Energy and Dissociation Products

Figure 6.4(a) and (b) shows two of the ways in which electronic excitation can lead to dissociation (a third way called *predissociation*, will be considered in Sec. 6.1.7). Part (a) of the figure represents the case, previously discussed, where the equilibrium nuclear separation in the upper state is considerably greater than that in the lower. The dashed line limits of the Morse curves represent the dissociation of the normal and excited molecule into atoms, the dissociation energies being D_0'' and D_0' from the $v = 0$ state in each case. We see that the total energy of the dissociation products (i.e., atoms) from the upper state is greater by an amount called $E_{ex.}$ than that of the products of dissociation in the lower state. This energy is the *excitation energy* of one (or rarely both) of the atoms produced on dissociation.

We saw in the previous section that the spectrum of this system consists of some vibrational transitions (quantized) followed by a continuum (nonquantized transitions) representing dissociation. The lower wavenumber

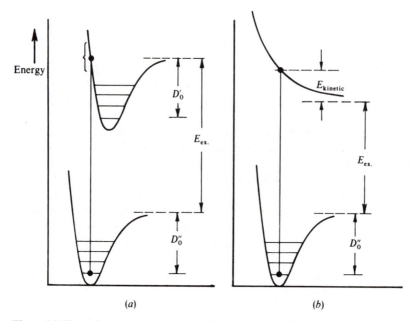

Figure 6.4 Illustrating dissociation by excitation into (a) a stable upper state, and (b) a continuous upper state.

limit of this continuum must represent just sufficient energy to cause dissociation and no more (i.e., the dissociation products separate with virtually zero kinetic energy) and thus we have

$$\bar{\nu}_{(\text{continuum limit})} = D_0'' + E_{\text{ex.}} \quad \text{cm}^{-1} \tag{6.7}$$

and we see that we can measure D_0'', the dissociation energy, if we know $E_{\text{ex.}}$, the excitation energy of the products, whatever they may be. Now, although the excitation energy of atoms to various electronic states is readily measurable by atomic spectroscopy (cf. Chapter 5), the precise *state* of dissociation products is not always obvious. There are several ways in which the total energy $D_0'' + E_{\text{ex.}}$ may be separated into its components, however; here we shall mention just two.

Firstly, thermochemical studies often lead to an approximate value of D_0'' and hence, since $D_0'' + E_{\text{ex.}}$ is accurately measurable spectroscopically, a rough value for $E_{\text{ex.}}$ is obtained. When the spectrum of the atomic products is studied, it usually happens that only one value of excitation energy corresponds at all well with $E_{\text{ex.}}$. Thus the state of the products is known, $E_{\text{ex.}}$ measured accurately, and a precise value of D_0'' deduced.

Secondly, if more than one spectroscopic dissociation limit is found, corresponding to dissociation into two or more different states of products with different excitation energies, the separations between the excitation energies are often found to correspond closely with the separations between only one set of excited states of the atoms observed spectroscopically. Thus the nature of the excited products and their energies are immediately known.

In Fig. 6.4(*b*) we illustrate the case in which the upper electronic state is *unstable*: there is no minimum in the energy curve and, as soon as a molecule is raised to this state by excitation, the molecule dissociates into products with total excitation energy $E_{\text{ex.}}$. The products fly apart with kinetic energy E_{kinetic} which represents (as shown on the figure) the excess energy in the final state above that needed just to dissociate the molecule. Since E_{kinetic} is not quantized the whole spectrum for this system will exhibit a continuum the lower limit of which (if observable) will be precisely the energy $D_0'' + E_{\text{ex.}}$. As before, if $E_{\text{ex.}}$ can be found from a knowledge of the dissociation products, D_0'' can be measured with great accuracy.

We shall see in Sec. 6.2.1 what sort of circumstances lead to the minimum in the upper state (Fig. 6.4(*a*)) on the one hand, or the *continuous* upper state (Fig. 6.4(*b*)) on the other.

In many electronic spectra no continua appear at all—the internuclear distances in the upper and lower states are such that transitions near to the dissociation limit are of negligible probability—but it is still possible to derive a value for the dissociation energy by noting how the vibrational lines converge. We have already seen in Chapter 3 (cf. Eq. (3.12)), that the vibrational energy levels may be written:

$$\varepsilon_v = (v + \tfrac{1}{2})\bar{\omega}_e - x_e(v + \tfrac{1}{2})^2\bar{\omega}_e \quad \text{cm}^{-1} \tag{6.8}$$

and so the separation between neighbouring levels, $\Delta\varepsilon$, is plainly:

$$\Delta\varepsilon = \varepsilon_{v+1} - \varepsilon_v$$
$$= \bar{\omega}_e\{1 - 2x_e(v + 1)\} \quad \text{cm}^{-1} \tag{6.9}$$

This separation obviously decreases linearly with increasing v and the dissociation limit is reached when $\Delta\varepsilon \to 0$. Thus the maximum value of v is given by $v_{\text{max.}}$, where:

$$\bar{\omega}_e\{1 - 2x_e(v_{\text{max.}} + 1)\} = 0$$

i.e.,

$$v_{\text{max.}} = \frac{1}{2x_e} - 1 \tag{6.10}$$

We recall that the anharmonicity constant, x_e, is of the order 10^{-2}, hence $v_{\text{max.}}$ is about 50.

We saw in Sec. 3.1.3, that two vibrational transitions (in the infra-red) were sufficient to determine x_e and $\bar{\omega}_e$. Thus, an example given there for HCl yielded $\bar{\omega}_e = 2990$ cm^{-1}, $x_e = 0.0174$. From Eq. (6.10) we calculate $v_{\text{max.}} = 27.74$ and the next lowest integer is $v = 27$. Replacing $v = 27$, $\bar{\omega}_e = 2990$ cm^{-1} and $x_e = 0.0174$ into Eq. (6.8) gives the maximum value of the vibrational energy as $42\,890$ cm^{-1} or $513\cdot0$ kJ mol^{-1}. This is to be compared with a more accurate value of $427\cdot2$ kJ mol^{-1} evaluated thermochemically.

The discrepancy between these two figures arises from two causes. Firstly, the infra-red data only allows us to consider two or three vibrational transitions (the fundamental plus the first and second overtones). The electronic spectrum, as we have seen, shows many more vibrational lines (in fact the number is limited not by quantum restrictions, but by the Franck–Condon principle) and we shall get a better value of D_0'' if we make use of this extra data. Secondly, we have assumed that Eq. (6.8) applies exactly even at high values of v; this is not true because cubic and even quartic terms become important at this stage. Because of these, $\Delta\varepsilon$ decreases more rapidly than Eq. (6.9) suggests.

Both these points may be met if we plot the *separation* between vibrational transitions, $\Delta\varepsilon$, as observed in the electronic spectrum, against the

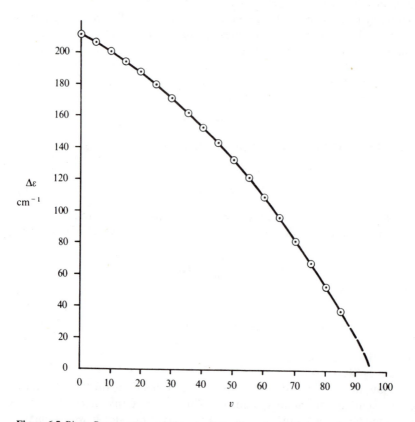

Figure 6.5 Birge–Sponer extrapolation to determine the dissociation energy of the iodine molecule, I$_2$. *(Taken from the data of R. D. Verma, J. Chem. Phys., vol. 32, p. 738 (1960), by kind permission of the author.)*

vibrational quantum number. Initially, Eq. (6.9) will apply quite accurately and the graph will be a straight line which may be extrapolated either to find v_{max}. or, since the dissociation energy itself is simply the sum of all the increments $\Delta\varepsilon$ from $v = 0$ to $v = v_{max}$., the *area* under the $\Delta\varepsilon$ versus v graph gives this energy directly. Such a linear extrapolation was first suggested by Birge and Sponer and is usually given their name.

On the other hand, if extensive data are available about a set of electronic–vibration transitions, the graph of $\Delta\varepsilon$ versus v will, at high v, begin to fall off more sharply as cubic and quartic terms become significant. In this case the most accurate determination of dissociation energy is obtained by extrapolating the smooth curve and finding the area beneath it. Figure 6.5 shows this process for data on iodine vapour given by R. D. Verma, *J. Chem. Phys.*, **32**, 738 (1960).

In absorption spectra it is normally the series of lines originating at $v'' = 0$ which is observed (cf. Fig. 6.1). Thus the convergence of the levels in the upper state and hence the dissociation energy of that state is normally found. While this in itself is of great interest, particularly since molecules in excited states usually revert to the ground state within fractions of a microsecond, the dissociation energy in the ground state can be found quite easily provided, as before, the dissociation products and their excitation energy are known. Thus, in Fig. 6.4(a), if we know $E_{ex.}$ (from atomic spectroscopy), and D'_0 (from Birge–Sponer extrapolation), and if we can measure the energy of the (0, 0) transition either directly or by calculation from the observed energy levels, we have:

$$D''_0 = \text{energy of } (0, 0) + D'_0 - E_{ex.} \quad \text{cm}^{-1} \qquad (6.11)$$

6.1.5 Rotational Fine Structure of Electronic–Vibration Transitions

So far we have seen that the electronic spectrum of a diatomic molecule consists of one or more series of convergent lines constituting the vibrational coarse structure on each electronic transition. Normally each of these 'lines' is observed to be broad and diffuse or, if the resolution is sufficiently good, each appears as a cluster of many very close lines. This is, of course, the rotational fine structure.

To a very good approximation we can ignore centrifugal distortion and we have the energy levels of a rotating diatomic molecule (cf. Eqs (2.11) and (2.12)) as:

$$\varepsilon_{rot.} = \frac{h}{8\pi^2 Ic} J(J + 1) = BJ(J + 1) \text{ cm}^{-1} \qquad (J = 0, 1, 2, \ldots) \quad (6.12)$$

where I is the moment of inertia, B the rotational constant, and J the rotational quantum number. Thus, by the Born–Oppenheimer approximation, the total energy (excluding kinetic of translation) of a diatomic molecule is:

$$\varepsilon_{total} = \varepsilon_{elec.} + \varepsilon_{vib.} + BJ(J + 1) \text{ cm}^{-1} \tag{6.13}$$

Changes in the total energy may be written:

$$\Delta\varepsilon_{total} = \Delta\{\varepsilon_{elect.} + \varepsilon_{vib.}\} + \Delta\{BJ(J + 1)\} \text{ cm}^{-1} \tag{6.14}$$

and the wavenumber of a spectroscopic line corresponding to such a change becomes simply:

$$\bar{\nu}_{spect.} = \bar{\nu}_{(v', v'')} + \Delta\{BJ(J + 1)\} \text{ cm}^{-1} \tag{6.15}$$

where we write $\bar{\nu}_{(v'', v'')}$ to represent the wavenumber of an electronic–vibrational transition. This plainly corresponds to any *one* of the transitions, for example, (0, 0) or (1, 0), etc., considered in previous sections. Here we are mainly concerned with $\Delta\{BJ(J + 1)\}$.

The selection rule for J depends upon the type of electronic transition undergone by the molecule. We shall discuss these in more detail in Sec. 6.2.2; for the moment we must simply state that if both the upper and lower electronic states are $^1\Sigma$ states (i.e., states in which there is no electronic angular momentum about the internuclear axis), this selection rule is:

$$\Delta J = \pm 1 \text{ only} \qquad \text{for } ^1\Sigma \rightarrow {}^1\Sigma \text{ transitions} \tag{6.16}$$

whereas for all other transitions (i.e., provided either the upper or the lower states (or both) have angular momentum about the bond axis) the selection rule becomes:

$$\Delta J = 0, \text{ or } \pm 1 \tag{6.17}$$

For this latter case there is the added restriction that a state with $J = 0$ cannot undergo a transition to another $J = 0$ state:

$$J = 0 \nleftrightarrow J = 0 \tag{6.18}$$

Thus we see that for transitions between $^1\Sigma$ states, P and R branches only will occur, while for other transitions Q branches will appear in addition.

We can expand Eq. (6.15) as follows:

$$\bar{\nu}_{spect.} = \bar{\nu}_{(v', v'')} + B'J'(J' + 1) - B''J''(J'' + 1) \text{ cm}^{-1} \tag{6.19}$$

where B' and J' refer to the upper electronic state, B'' and J'' to the lower. When we considered vibration–rotational spectra in Chapter 3, we saw (cf. Sec. 3.4) that the difference between B values in different vibrational levels was very small and could be ignored except in explaining finer details of the spectra. But this is by no means the case in electronic spectroscopy: here we have seen, when discussing the Franck–Condon principle in Sec. 6.1.3, that equilibrium internuclear distances in the lower and upper electronic states may differ considerably, in which case the moments of inertia, and hence B values, in the two states will also differ. We cannot say a priori which of the two B values will be greater. Quite often the electron excited is one of those forming the bond between the nuclei; if this is so, the bond in the upper state will be weaker and probably longer (cf. Fig. 6.3(b) or (c)) so that the equilibrium moment of inertia increases during the transition and B decreases. Thus $B' < B''$. The reverse is sometimes true, however, e.g., when the electron is excited from an antibonding orbital (see Sec. 6.2.2).

We can discuss the rotational fine structure quite generally by applying the selection rules of Eqs (6.16), (6.17), and (6.18) to the expression for spectral lines, Eq. (6.19). We may note, in passing, that the treatment given here for the P and R branch lines is identical with that given in Sec. 3.4 for the vibration–rotation spectrum, except that there we were concerned with B_0 and B_1—B values in lower and upper *vibrational* states. Here our concern is with B values in lower and upper *electronic* states, B'' and B', and we also consider the formation of a Q branch.

Taking the P, R and Q branches in turn:

1. P branch: $\Delta J = -1$, $J'' = J' + 1$

$$\Delta \varepsilon = \bar{\nu}_P = \bar{\nu}_{(v', v'')} - (B' + B'')(J' + 1) + (B' - B'')(J' + 1)^2 \text{ cm}^{-1}$$

$$\text{where } J' = 0, 1, 2, \ldots \quad (6.20a)$$

2. R branch: $\Delta J = +1$, $J' = J'' + 1$

$$\Delta \varepsilon = \bar{\nu}_R = \bar{\nu}_{(v', v'')} + (B' + B'')(J'' + 1) + (B' - B'')(J'' + 1)^2 \text{ cm}^{-1}$$

$$\text{where } J'' = 0, 1, 2, \ldots \quad (6.20b)$$

These two equations can be combined into:

$$\bar{\nu}_{P, R} = \bar{\nu}_{(v', v'')} + (B' + B'')m + (B' - B'')m^2 \quad \text{cm}^{-1}$$

$$\text{where } m = \pm 1, \pm 2, \ldots \quad (6.20c)$$

positive m values comprising the R branch (i.e., corresponding to $\Delta J = +1$) and negative values the P branch ($\Delta J = -1$). Note that m cannot be zero (this would correspond in, e.g., the P branch, to $J' = -1$ which is impossible) so that no line from the P and R branch appears at the *band origin* $\bar{\nu}_{(v', v'')}$. We draw the appearance of the R and P branches separately in Fig. 6.6(a) and (b) respectively, taking a 10 per cent difference between the upper and lower B values and choosing $B' < B''$. Note that, *with this choice*, P branch lines occur on the *low* wavenumber side of the band origin and the spacing between the lines increases with m. On the other hand the R branch appears on the *high* wavenumber of the origin and the line spacing decreases rapidly with m—so rapidly that the lines eventually reach a maximum wavenumber and then begin to return to low wavenumbers with increasing spacing.† It will be remembered that

---

† The returning lines of the R branch coincide with earlier lines if Eq. (6.20b) is obeyed exactly. For real molecules cubic and quartic terms become important at high values of m.

in Sec. 3.4, a similar decrease in spacing was observed in the R branch but this was much too slow for a convergence limit to be reached; the rapid convergence here is due simply to the magnitude of $B' - B''$. The point at which the R branch separation decreases to zero is termed the *band head*.

3. Q branch: $\Delta J = 0$, $J' = J''$

$$\Delta \varepsilon = \bar{v}_Q = \bar{v}_{(v', v'')} + (B' - B'')J'' + (B' - B'')J''^2 \quad \mathrm{cm}^{-1}$$

$$\text{where } J'' = 1, 2, 3, \ldots \quad (6.21)$$

Note that here $J'' = J' \neq 0$ since we have the restriction shown in Eq. (6.18). Thus again *no line will appear at the band origin*. We sketch the Q branch in Fig. 6.6(c), again for $B' < B''$ and a 10 per cent difference between the two. We see that the lines lie to *low* wavenumber of the origin and their spacing increases. The first few lines of this branch are not usually resolved.

The complete rotational spectrum is shown in Fig. 6.6(d). We have seen in Sec. 2.3.2 that many rotational levels are populated even at room temperature; consequently, a large number of the P and R (and Q, where appropriate) lines will appear in the spectrum with comparable intensity. The spectrum is usually dominated by the band head, since here several of

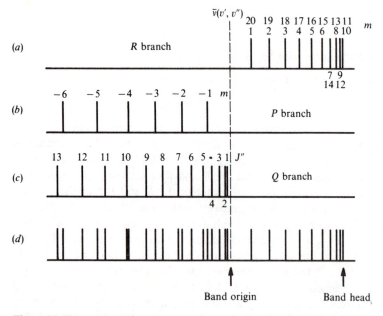

Figure 6.6 The rotational fine structure of a particular vibrational–electronic transition for a diatomic molecule. The R, P, and Q branches are shown separately at (a), (b), and (c) respectively, with the complete spectrum at (d).

the R branch lines crowd together; for this reason, the Q branch is not very apparent if it occurs.

In the situation we have been discussing ($B' < B''$), the band head appears in the R branch on the high wavenumber side of the origin; such a band is said to be *degraded* (or *shaded*) *towards the red*—i.e., the tail of the band where the intensity falls off points towards the red (low-frequency) end of the spectrum. If, on the other hand, $B' > B''$, then all our previous arguments are reversed. Briefly: (1) the Q branch spreads to *high* wavenumber, (2) the R branch (still, of course, on the *high* wavenumber side) consists of a series of lines with *increasing* separation, and (3) the band head appears in the P branch to *low* frequency of the origin. Such a band is *shaded to the violet*.

Normally, all the vibrational bands in any one electronic transition (e.g., the set of bands shown as a line spectrum in Fig. 6.1) are shaded in the same direction, while different electronic transitions in the same molecule may well show different shadings. Thus, observation of the shading may assist in the analysis of a complex spectrum. However, it may happen that different shadings are observed in bands belonging to the same electronic transition. This is because the B' and B'' values are not altogether independent of the vibrational state (as we have already seen in Sec. 3.4) so that, if $B' - B''$ is small, it may reverse sign for some higher vibrational levels. This behaviour is observed, for example, in the molecular fragment AlF, but is rare.

6.1.6 The Fortrat Diagram

We may rewrite the expressions for the P, R, and Q lines, Eq. (6.20c) and (6.21), with *continuously variable* parameters p and q:

$$\bar{\nu}_{P,\,R} = \bar{\nu}_{(v',\,v'')} + (B' + B'')p + (B' - B'')p^2 \qquad (6.22a)$$

$$\bar{\nu}_Q = \bar{\nu}_{(v',\,v'')} + (B' - B'')q + (B' - B'')q^2 \qquad (6.22b)$$

when we see that they each represent a parabola, p taking both positive and negative values, while q is positive only. We sketch these parabolae in Fig. 6.7 choosing, as before, $B' < B''$ and a difference of 10 per cent between them, and labelling regions of positive p with $\bar{\nu}_R$ and negative p with $\bar{\nu}_P$. These parabolae are usually referred to as the Fortrat parabolae. If we now illustrate the fact that p and q may in fact take only integral values (but not zero) by drawing circles round the allowed points on the parabolae, we can then read off the $\bar{\nu}$ values of the spectral lines directly from the graph. We show at the foot of the figure the first few lines of each branch with dotted leader lines connecting spectrum and Fortrat diagram at intervals.

A useful property of the Fortrat diagram is that the band head is plainly at the vertex of the P, R parabola. We may calculate the position of

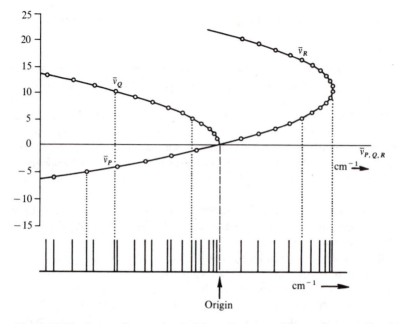

Figure 6.7 The Fortrat diagram sketched for a 10 per cent difference between B' and B'' (with $B' < B''$). The spectrum illustrated at the foot is identical with that of Fig. 6.6(d).

the vertex by differentiation of Eq. (6.22a):

$$\frac{d\bar{\nu}_{P,R}}{dp} = B' + B'' + 2(B' - B'')p = 0$$

or

$$p = -\frac{B' + B''}{2(B' - B'')} \quad \text{for band head} \tag{6.23}$$

Thus if $B' < B''$ (upper state has longer equilibrium bond length) the band head occurs at *positive* p values (i.e., in the R branch), the line at maximum wavenumber being given by the nearest positive integer to p. Conversely, for $B' > B''$ the band head occurs in the region of p negative, i.e., in the P branch. A simple calculation shows that for a 10 per cent difference between B' and B'' the band head occurs at $p \approx 10$.

6.1.7 Predissociation

If a large number of vibrational transitions are observed for a particular molecule, it sometimes happens that the vibrational and rotational structure are quite distinct within a progression for large and small changes in the vibrational quantum number, but either the rotational structure is blurred or a complete continuum is observed for intermediate changes. A

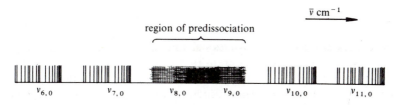

Figure 6.8 Diagrammatic illustration of the appearance of predissociation. The rotational fine structure is clearly defined for vibrational transitions both above and below the predissociation region, but in this region the fine structure becomes blurred and lost.

diagram showing the appearance of such a band is sketched in Fig. 6.8. A continuum at high wavenumber would correspond to ordinary dissociation (cf. Sec. 6.1.4) but the central continuum, occurring at energies well below the true dissociation limit, is referred to as *predissociation*.

Predissociation can arise when the Morse curves of a particular molecule in two different excited states intersect; one such possibility is shown in Fig. 6.9. One of the excited states is stable, since it has a minimum in the curve, and the other is continuous. Some of the vibrational levels are also shown, and let us suppose a transition takes place from some lower state

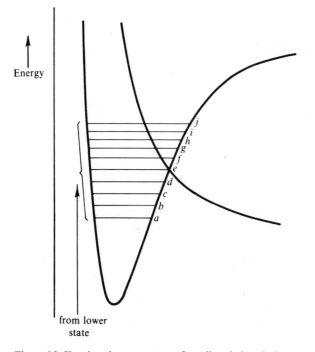

Figure 6.9 Showing the occurrence of predissociation during transitions into a stable upper state intersected by a continuous state.

into the vibrational levels shown bracketed on the left. Now if a transition takes place into the levels labelled a, b, or c, a normal vibrational–electronic spectrum occurs complete with rotational fine structure; two such bands appear at the left of Fig. 6.8. If the transition is to levels d, e, or f there is a possibility that the molecule will 'cross over' on to the continuous curve and thus dissociate. In general, transition from one curve to another in this way (a so-called *radiationless transfer* since no energy is absorbed or emitted in the process) is faster than the time taken by the molecule to rotate ($\sim 10^{-10}$ s) but usually slower than the vibrational time ($\sim 10^{-13}$ s). Thus predissociation will occur before the molecule rotates (and thus all rotational fine structure will be destroyed in the spectrum), while the vibrational structure is usually not destroyed. If the cross-over is faster than the vibrational time, then a complete continuum will occur in the spectrum as shown in Fig. 6.8.

On the other hand, transitions into levels g, h, . . . will give rise to a normal vibrational–electronic spectrum including rotational fine structure once more. As we have seen previously (Sec. 6.1.3) the molecule spends most time at the extreme ends of its vibrational motion when v is large, and very little time in between. When moving in the vibrational states g, h, . . ., the molecule spends insufficient time near the cross-over point for appreciable dissociation to occur and a normal spectrum results.

6.1.8 Diatomic Molecules: a Summary

When the rotational fine structure of electronic spectra can be resolved—as it normally can for diatomic molecules—we see that a great deal of useful information becomes available. We can immediately determine the rotational constant, and hence calculate the moment of inertia and bond length, for both the lower and the upper electronic states. Isotopic species in the molecule will cause a slight difference in the rotational constant, so such isotopes may be detected, and their concentrations measured from the band intensity. Equally, the vibrational levels of the electronic states can be determined from the position of band origins; these lead to the evaluation of fundamental vibration frequencies, of bond force constants and, perhaps, of dissociation energies too. The latter, however, are more accurately determined if a continuum is observed at the end of a band spectrum.

Where data obtained from such spectra can be checked independently, e.g., by microwave or infra-red spectroscopy, by X-ray or neutron diffraction, or by thermochemical methods, perfectly satisfactory agreement is found. Thus we can use electronic spectroscopic methods with great confidence to determine bond lengths and strengths in those molecules to which such independent methods are not applicable.

Probably the most important application of this type of spectroscopy is to the study of excited states and unstable radicals. Thus we have seen that

B values and dissociation energies are obtained for both the upper and lower electronic states—and data for the upper state are not obtainable by other means. Further, considerable amounts of energy are involved in the production of electronic spectra, and complex molecules are frequently disrupted into fragments during the process, the fragments, or free radicals, normally being very short-lived. Examples are legion, a few of the more important diatomic ones being CH, NH, C_2, OH, CN, etc. Spectra arising from these radicals can be recognized and studied, leading to the determination of bond lengths, force constants, dissociation energies, etc. Further, if the variation of the intensity of such spectra over short periods of time is studied—as in the techniques of flash photolysis—information can be obtained about the rate at which the radicals are produced and destroyed. Since the length of time during which some radicals have an independent existence is measured in microseconds or less, it is remarkable that many such 'diatomic molecules' are as well characterized as, for example, the rather more stable H_2 or CO.

6.2 ELECTRONIC STRUCTURE OF DIATOMIC MOLECULES

6.2.1 Molecular Orbital Theory

Several theories have been suggested to account for the formation of molecules from atoms. All, if taken to a sufficiently high degree of approximation, seem to agree with observed data, but the calculation involved is so extensive that complete agreement is seldom reached and then only in the simplest examples. Here we shall discuss just one of these theories—the *molecular orbital theory*; we choose this, not because it is better or simpler than others (such considerations depend upon the particular problem in hand and are, in any case, largely subjective), but because it gives a convenient pictorial representation of molecule formation which is particularly suited to the discussion of electronic transitions, and because the ideas it uses are entirely analogous to those of atomic structure which we have discussed in the previous chapter.

Thus we have seen that electrons in atoms do not occupy space haphazardly or have arbitrary energies, but that their distribution and energy are governed by well-defined natural laws. These characteristics may be calculated from the Schrödinger equation and expressed in terms of a three-dimensional wave function, or orbital, ψ, which depends on the values of three quantum numbers, n, l, and m (or l_z); the spin of the electron also contributes to the energy. Definite rules determine which orbitals are occupied in the ground state and what transitions may take place between orbitals.

Molecular orbital theory supposes orbitals to extend about, and embrace, *two or more nuclei*, the shape and energy of these orbitals being calculable from the Schrödinger equation in terms of three quantum numbers. Essentially the same rules (i.e., lowest energy first, maximum of two (paired) electrons per orbital, parallel spins in degenerate orbitals) apply to their filling as to the filling of atomic orbitals.

The situation is relatively simple for diatomic molecules where the molecular orbital embraces two nuclei only and we shall discuss these molecules in some detail first. The extension to polyatomic molecules will be outlined in Sec. 6.3.

6.2.2 The Shapes of Some Molecular Orbitals

As in atomic orbital theory (cf. Sec. 5.1.2) the shape of a molecular orbital is the space within which an electron belonging to that orbital will spend 95 per cent (or some other arbitrary fraction) of its time. While detailed computation of these shapes from the Schrödinger theory may be extremely difficult, a very good qualitative idea of their approximate shape may be obtained by considering molecular orbitals to be made up of sums and differences of the atomic orbitals of the constituent atoms—the so-called *linear combination of atomic orbitals* (LCAO) approximation. Thus for a diatomic molecule we could imagine the formation of two different molecular orbitals whose wave functions would be:

$$\psi_{\text{mo.}} = \psi_1 + \psi_2 \quad \text{or} \quad \psi_{\text{mo.}} = \psi_1 - \psi_2 \tag{6.24}$$

where ψ_1 and ψ_2 are the relevant atomic orbitals of the two atoms. Note that the function $\psi_2 - \psi_1$ is identical with $\psi_1 - \psi_2$, since it is $\psi_{\text{mo.}}^2$ which represents the probability of finding an electron in a particular place.

Let us consider the hydrogen molecule, H_2, as an example; the obvious atomic orbitals to use are the $1s$ orbitals of each atom. Figure 6.10(a) shows the situation:

$$\psi_{H_2} = \psi_{1s} + \psi_{1s} \tag{6.25}$$

We recall from the previous chapter (Eq. (5.2)) that ψ_{1s} is everywhere positive in value and so, where the atomic orbitals overlap, the value of ψ_{H_2} will be increased. This suggests (and detailed calculation confirms) that the molecular orbital of Eq. (6.25) is a simple ellipsoid, symmetrical in shape. The concentration of electronic charge between the nuclei acts as a sort of cement keeping the nuclei together and thus this orbital represents the formation of a bond between the atoms. It is called a *bonding orbital* and given the label $1s\sigma$ since it is produced from two $1s$ orbitals (we shall see later that σ has a similar significance regarding the orbital angular momenta of molecular electrons as s has for atoms).

End-view

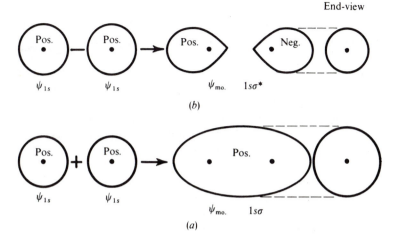

(b)

(a)

Figure 6.10 The formation of (a) a bonding $1s\sigma$ orbital and (b) an antibonding $1s\sigma^*$ orbital from two atomic $1s$ orbitals.

On the other hand, Fig. 6.10(b) shows the situation for

$$\psi_{H_2} = \psi_{1s} - \psi_{1s} \tag{6.26}$$

Since, again, ψ_{1s} is everywhere positive, where the two separate ψ_{1s} orbitals overlap, they will cancel each other. Thus between the nuclei $\psi_{mo.}$ will be zero, while it will be positive near one nucleus and negative near the other (remember that it is ψ^2 which determines probability and this is, in either case, positive). Now, however, the shape of the molecular orbital shows that electronic charge, far from being concentrated *between* the nuclei, is greatest *outside* them; thus the nuclear repulsion is enhanced and the orbital is described as *antibonding*. It leads to a state of higher energy than two separate atoms and is labelled $1s\sigma^*$, the asterisk representing high energy. Figure 6.10 also shows, at the extreme right, an end-view of these orbitals. They are both seen to have cylindrical symmetry about the bond axis; it is this property which leads to their both being described as σ orbitals although in other respects they have quite different appearances.

Another facet of orbital symmetry should be mentioned here. If the molecule considered is homonuclear (i.e., made of two *identical* atoms) then the midpoint of the bond between them is a *centre of symmetry*—starting from any point in the molecule, on or off the internuclear axis, exactly similar surroundings are encountered by proceeding to the point diagonally opposite the centre. Such a process is known as *inversion*, and such molecular properties as electron density, force fields, etc., are quite unchanged by inversion. However, we note that ψ (as opposed to ψ^2) may or may not be changed *in sign* by inversion. Thus inversion of the $1s\sigma$ molecular orbital of

Fig. 6.10(*a*) plainly causes no change in ψ since it is everywhere positive; this orbital, in which ψ is completely symmetrical, is described as *even* and usually given the symbol *g* (German: *gerade* = even) as a suffix: $1s\sigma_g$. On the other hand the $1s\sigma^*$ orbital in (*b*) of the figure is antisymmetrical, since inversion reverses the sign of ψ. This orbital is thus *odd* and given the subscript *u*, $1s\sigma_u$ or $1s\sigma_u^*$ (German: *ungerade* = odd). In the case of molecular hydrogen, then, the bonding orbital is even, the antibonding is odd; this situation may be reversed for other molecular orbitals as we shall see.

If the molecule is heteronuclear (for example, CO, HCl, etc.) then there is no centre of symmetry and the odd–even classification of orbitals does not arise.

Before turning to the shapes of other molecular orbitals, it is instructive to consider how the energy of the $1s\sigma_g$ and $1s\sigma_u^*$ orbitals varies with the distance between the nuclei. This variation may be calculated from the Schrödinger equation and the result is shown in Fig. 6.11. The $1s\sigma_g$, the

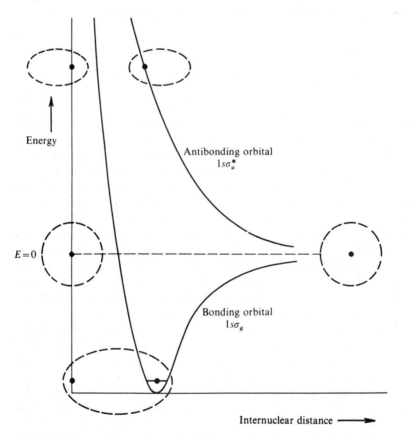

Figure 6.11 The variation of energy with internuclear distance in the bonding and antibonding orbitals, $1s\sigma_g$ and $1s\sigma_u^*$.

bonding orbital, shows a typical Morse curve for a diatomic molecule, the minimum in the curve showing that a bond is formed between the atoms. The $1s\sigma_u^*$, on the other hand, shows no minimum, but is one of the 'continuous' curves already discussed in Sec. 6.1.4. In this case the dissociation limits of the two curves coincide since the dissociation products are identical—two hydrogen atoms. The relationship of the orbitals sketched in Fig. 6.10 to the energy curve is shown by superimposing the molecular orbitals at their appropriate equilibrium internuclear distance and the separate atomic orbitals at a large distance.

Two $2s$ atomic orbitals can form $2s\sigma_g$ and $2s\sigma_u^*$ bonding and antibonding orbitals with identical shape to (but larger and with higher energy than) the $1s\sigma_g$ and $1s\sigma_u^*$ orbitals. Two $2p$-type orbitals can overlap in two different ways depending on their relative orientation. If we label the internuclear axis the z direction, then we may consider first the $2p$ orbital which lies along this axis for each atom, i.e., the $2p_z$ orbitals. Now the expression for the wave function of a $2p_z$ orbital has the form $\psi_{2p_z} = zf(r)$, where $f(r)$ is a positive function of distance from the nucleus. We see then, that for $+z$ directions ψ is also positive, while it is negative for $-z$. The two lobes of a $2p$ orbital thus have opposite signs of ψ (although, of course, ψ^2 is everywhere positive).

We draw these orbitals and indicate their signs on the left of Fig. 6.12 and it is evident that in the molecular orbital

$$\psi_{\text{mo.}} = \psi_{2p_z} + \psi_{2p_z} \tag{6.27}$$

the electron density between the nuclei is cancelled to zero and the orbital will have the shape shown at the top right of the figure. Similarly in

$$\psi_{\text{mo.}} = \psi_{2p_z} - \psi_{2p_z} \tag{6.28}$$

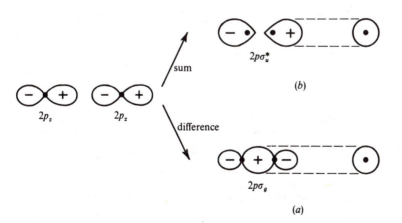

Figure 6.12 The formation of (a) a bonding $2p\sigma_g$ and (b) an antibonding $2p\sigma_u^*$ orbital from two atomic $2p_z$ orbitals, where the z axis is taken as the internuclear axis.

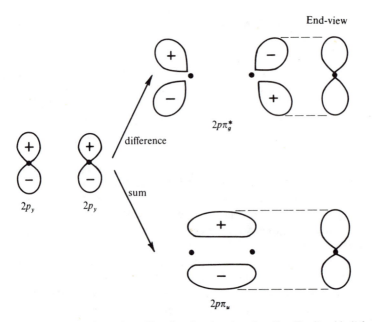

Figure 6.13 The formation of bonding $(2p\pi_u)$ and antibonding $(2p\pi_g^*)$ orbitals from two atomic $2p_y$ orbitals, the z axis being the internuclear axis. Atomic $2p_x$ orbitals would form identical molecular orbitals except that all lobes would be rotated through a right angle about the z axis.

the electron density increases between the nuclei and the shape is shown at the bottom right of the figure. Plainly the latter is *bonding*, and consideration of its symmetry shows that it is *even* (*g*), while the former is plainly *antibonding* and *odd* (*u*) in character. The end-view of both is the same, however, and shows symmetry about the bond axis; for this reason both are referred to as σ orbitals and they may be labelled $2p\sigma_g$ and $2p\sigma_u^*$ respectively to indicate their origin.

The overlap of two $2p_y$ orbitals is shown in Fig. 6.13 ($2p_x$ are exactly similar but rotated out of the plane of the paper through 90°). The summed orbitals, which are *bonding*, are sketched at bottom right and we see that the molecular orbital formed consists of two streamers, one above and one below the nuclei. The end-view of this orbital is shown at the extreme right; it has the appearance of an atomic p orbital and hence it is labelled π. In this case the bonding orbital is evidently *odd*, so we have a $2p\pi_u$ state.

On the other hand, if the atomic orbitals are subtracted the orbital picture shown at top right of Fig. 6.13 is produced. This has a similar end-view to $2p\pi_u$, is *antibonding* and *even*, hence it is labelled π_g^*.

More complex orbitals exist—δ, ϕ, etc., formed by interaction between d, f, etc., atomic orbitals—but they need not concern us; the simple molecules with which we shall deal use σ and π orbitals only. We do need 'to

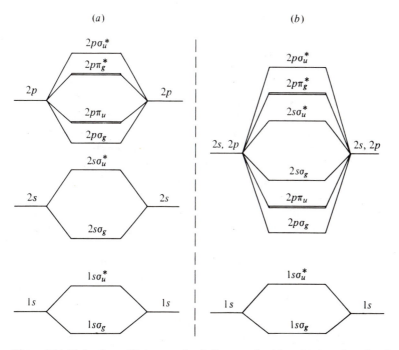

Figure 6.14 Molecular orbital energy level diagrams for (a) molecules other than hydrogen, and (b) the hydrogen molecule.

know, however, the order of increasing energy of the molecular orbitals so far discussed so that we can consider the ground states of some atoms. This we show diagrammatically (and by no means to scale) in Fig. 6.14(a), where we also indicate, to the right and left, the atomic orbitals which combine to form each molecular state. As drawn here the diagram represents the situation for the larger molecules such as N_2, O_2, and F_2; for smaller molecules such as Li_2, the $2p\pi_u$, $2p\sigma_g$, and $2s\sigma_u^*$ become overlapped, while for the lightest molecule of all, H_2, the situation is as shown in Fig. 6.14(b). Here, because in the separate atoms the $2s$ and $2p$ orbitals have identical energies, the molecular orbitals take up the order shown. Using these diagrams and the Pauli principle (not more than two electrons to each orbital) we can build up the electronic configurations of simple molecules. Some examples follow:

Firstly hydrogen, H_2: in the ground state of this molecule the two $1s$ atomic electrons, one from each atom, can both occupy the molecular $1s\sigma_g$ provided their spins are opposed. The energy of the electrons in this state, as we can see from Figs. 6.11 and 6.14(b), is lower than that of two separate atoms, hence the molecule is stable. We can write its configuration $1s\sigma_g^2$.

Next helium, He_2: if this molecule were to form it would have a total of four electrons to place into orbitals, two from each atom. Only two of these could go into the $1s\sigma_g$; the other two would have to go into $1s\sigma_u^*$. But we can see from Figs 6.11 and 6.14(a) that more energy would be absorbed by the latter than evolved by the former, hence the molecule is unstable with respect to the atoms.

Thirdly, nitrogen, N_2: the two atoms each have an electronic configuration $1s^2 2s^2 2p^3$ (cf. Table 5.2), so we have a total of 14 electrons to dispose of, in pairs, into the molecular orbitals of Fig. 6.14(a); clearly the configuration will be $1s\sigma_g^2 1s\sigma_u^2 2s\sigma_g^2 2s\sigma_u^2 2p\sigma_g^2 2p\pi_u^4$ (where we omit the asterisks to avoid confusion). As a good approximation we can allow the bonding and antibonding contributions of the $1s\sigma_g^2$ and $1s\sigma_u^2$ to cancel, and similarly with the $2s\sigma_g^2$ and $2s\sigma_u^2$. We are left, then, with three pairs of electrons in bonding orbitals, the $2p\sigma_g$ and the $2p\pi_u$, so we conclude that the molecule has a triple bond.

Finally, oxygen, O_2: each atom has one electron more than the nitrogen atom (Table 5.2), and clearly the lowest available orbital in Fig. 6.14(a) is the *antibonding* $2p\pi_g$. Two electrons in this orbital effectively cancel the bonding contribution of a pair in the $2p\pi_u$, and we are left with a net bonding of four electrons, i.e., a double bond. One unusual characteristic of the structure of O_2 concerns the two electrons in the antibonding $2p\pi_g$. Since there are here two degenerate orbitals, the electrons will occupy one each to satisfy electron repulsion; however, according to Hund's rule (p. 186) which applies to molecules just as to atoms, they will preferentially have their spins *parallel* rather than paired. Thus O_2 is a molecule containing some unpaired electrons—a structural characteristic which gives it its most unusual magnetic properties.

6.2.3 Electronic Angular Momentum in Diatomic Molecules; Classification of States

We found in Sec. 5.2, that the total energy of an electron, while depending mainly on its average distance from the nucleus (represented by the quantum number n) also depends on its orbital and spin angular momenta (quantum numbers l and s) and on the way in which these are coupled together (quantum number j). For several electrons in an atom we found that their separate energies can be combined in different ways to produce a variety of states; simple rules allow the ground state to be predicted in any particular case.

Much the same comments apply to electrons in molecules. Thus a single electron in a molecule has a quantum number n specifying the size of its orbital and mainly determining its energy, and a number l specifying its orbital angular momentum. Small letters s, p, d, ... are used, as before, to designate l values of 0, 1, 2, ... However, it will be remembered that in order to discuss the components of $\mathbf{l}$ we required to invoke some reference direction called the z direction; in a diatomic molecule a reference direction is already quite obviously specified—the inernuclear axis, or bond—and it would be perverse (not to say *wrong*) to discuss the components of $\mathbf{l}$ along any other direction. Furthermore, a force-field exists along this direction due to the presence of two nuclear charges; therefore different $\mathbf{l}$ components are not degenerate but represent *different energies*.

The axial component of orbital angular momentum is of more importance in molecules than the momentum itself and for this reason it is given the special symbol λ. Formally $\lambda \equiv |l_z|$, so that λ takes *positive* integral values or is zero, and we designate the λ state of an electron in a molecule by using the small Greek letters corresponding to the $s, p, d, \ldots$ of atomic nomenclature. Thus we have, for

$$l_z = 0, \pm 1, \pm 2, \pm 3, \ldots$$

$$\lambda = 0, 1, 2, 3, \ldots$$

and the symbols are $\qquad \sigma, \pi, \delta, \phi, \ldots$

Since λ has positive values only, each λ state with $\lambda > 0$ is *doubly degenerate*, because it corresponds to l_z being both positive and negative. The significance of λ is that the *axial component of orbital angular momentum* $= \lambda h/2\pi$ or λ *units*.

The total orbital angular momentum of several electrons in a molecule can be discussed, as for atoms, in terms of the quantum number $L = \Sigma l$, $\Sigma l - 1$, etc., with $\mathbf{L} = \sqrt{L(L + 1)}h/2\pi$, but again the axial component, denoted by Λ, is of greatest significance. Since, by definition, all individual λ_i lie along the internuclear axis, their summation is particularly simple. We have

$$\Lambda = |\Sigma \lambda_i| \qquad (6.29)$$

and states are designated by capital Greek letters Σ, Π, Δ, etc., for $\Lambda = 0, 1,$ 2, ... We must take into account, when using Eq. (6.29), that the individual λ_i may have the same or opposite directions and all possible combinations which give a positive Λ should be considered. Thus for a π and a δ electron ($\lambda_1 = 1$, $\lambda_2 = 2$) we could have $\Lambda = 1$ or 3 (but not -1), that is, a Π or a Φ state.

Electron *spin* momentum, on the other hand, is not greatly affected by the electric field of the two nuclei—we say the spin–axial coupling is weak, whereas the orbital–axial coupling is usually strong. Normally, therefore, we use the same notation for electronic spin in molecules as in atoms; the total spin momentum $\mathbf{S}$ is given by $\sqrt{S(S + 1)}$ where the total spin quantum number S is:

$$S = \Sigma S_i, \Sigma S_i - 1, \Sigma S_i - 2, \ldots, \tfrac{1}{2} \text{ or } 0 \qquad (6.30)$$

The multiplicity of a molecular state is, as for atoms, $2S + 1$ and this is usually indicated as an upper prefix to the state symbol. Thus for the Π and Φ states discussed in the previous paragraph, the states would be written $^3\Pi$ or $^3\Phi$ if the individual π and δ electron spins are parallel ($S = \frac{1}{2} + \frac{1}{2} = 1$, $2S + 1 = 3$), or as $^1\Pi$ or $^1\Phi$ if the spins are paired.

When the *axial* component of a spin is required, however, it is often designated by σ for a single electron or Σ for several (corresponding to s and S for the atomic case). In this case the multiplicity is $2\Sigma + 1$.

Finally, we consider the axial component of the *total* electronic angular momentum, i.e., the sum of the axial components of spin and orbital motion. In general the total momentum is strongly coupled to the axis and its axial component is more significant than the momentum itself. If we write the axial component as Ω we have simply:

$$\Omega = |\Lambda + \Sigma| \qquad (6.31)$$

but we must remember that Λ and Σ may have the same or opposite directions along the internuclear axis. Thus in the $^3\Pi$ state described above we have $\Lambda = 1$, $\Sigma = 1$, hence $\Omega = 2$ or 0. The $^1\Pi$ state has $\Lambda = 1$, $\Sigma = 0$, hence we have $\Omega = 1$ only. These states would be indicated by writing their Ω values as subscripts: $^3\Pi_2, ^3\Pi_0, ^3\Pi_1$.

Perhaps it will assist the student if we draw up a table (Table 6.1) showing the symbols used to designate the various sorts of angular momentum in atoms and molecules, together with their axial components, for one or more electrons.

Table 6.1 Comparison of symbols used for electronic angular momenta

	Orbital momentum	Spin momentum	Total momentum
For atoms			
Single electron	l (symbol s, p, d for $l = 0, 1, 2, \ldots$)	s	j
Single electron (z component)	l_z	s_z	j_z
Several electrons	L (symbol S, P, D for $L = 0, 1, 2, \ldots$)	S	J
Several electrons (z component)	L_z	S_z	J_z
For molecules			
Single electron	l	s	j_a (seldom used)
Single electron (axial component)	λ (symbol σ, π, δ, for $\lambda = 0, 1, 2, \ldots$)	σ	ω
Several electrons	L	S	J_a (seldom used)
Several electrons (axial component)	Λ (symbol Σ, Π, Δ, for $\Lambda = 0, 1, 2, \ldots$)	Σ	Ω

6.2.4 An Example: the Spectrum of Molecular Hydrogen

Before turning to polyatomic molecules, let us see how the above ideas may be applied to the simplest molecule, H_2. We shall consider first the nature of the ground state and some excited states of the molecule and how these relate to occupancy of the molecular orbitals of Fig. 6.14(b); then the energy of these states and, finally, what transitions may arise between them. The student may find it helpful to discover the many points of similarity between the discussion which follows and that given in Sec. 5.4.5 on helium.

The hydrogen molecule contains two electrons, one contributed to by each of the atoms. We would thus expect to find singlet and triplet states, depending on whether the electron spins are paired or parallel. In the *ground state* both electrons will occupy the same lowest orbital, i.e., the $1s\sigma_g$ of Fig. 6.14(b) and, by Pauli's principle, they must then form a singlet state. Both electrons are σ electrons (since both are in a σ orbital) hence $\lambda_1 = \lambda_2 = 0$ and $\Lambda = 0$ also; the state is thus $^1\Sigma$. We could indicate the value of Ω as a subscript ($\Omega = \Lambda + \Sigma = 0 + 0 = 0$, since $\Sigma = 0$ for singlet states) but it is more informative to specify the symmetry (g or u) of the orbital. In this case both electrons are in the same g orbital, hence the total state is $^1\Sigma_g$.

A further subdivision of Σ states is normally made, representing another facet of molecular symmetry. In any diatomic molecule (whether homo- or heteronuclear) any plane drawn through both nuclei is a *plane of symmetry*, i.e., electron density, shape, force fields, etc., are quite unchanged by reflection in the plane. However, the wave function of the electron, ψ, may either be completely unchanged (symmetrical) or changed in sign only (antisymmetrical) with respect to such a reflection (in either case, of course, ψ^2 is unchanged). The former states are distinguished by a superscript $+$ and the latter by $-$. For several reasons this division is made for Σ states only and nearly all such states are in fact $+$. Certainly all the states of hydrogen are symmetric.

Thus the ground state of molecular hydrogen can be written:

$$\text{Ground state:} \quad (1s\sigma_g)^2 \; {}^1\Sigma_g^+$$

A large number of excited singlet states also exist; let us consider some of the lower ones for which one electron only has been raised from the ground state into some higher molecular orbital, i.e., singly excited states. We can ignore promotion into any of the starred states of Fig. 6.14(b), since this would lead to the formation of an unstable molecule and immediate dissociation (cf. Fig. 6.11, where the placing of an electron in each of the σ_g and σ_u^* orbitals produces dissociation into two H atoms). Thus we may consider the three possible excited states $(1s\sigma_g 2s\sigma_g)$, $(1s\sigma_g 2p\sigma_g)$, and $(1s\sigma_g 2p\pi_u)$.

Taking $(1s\sigma_g 2s\sigma_g)$ first: here both electrons are σ electrons, hence $\Lambda = \lambda_1 + \lambda_2 = 0$ and, since we are considering only singlet states, $S = 0$ also.

Further, since both constituent orbitals are *even* and *symmetrical*, the overall state will be the same, and we have $(1s\sigma_g 2s\sigma_g)^1\Sigma_g^+$.

Now $(1s\sigma_g 2p\sigma_g)$: here we again have a $^1\Sigma$ state since both electrons are σ, but the overall state is now *odd* (u); this may be rationalized if we think of one electron as rising from a hydrogen atom in the *even* $1s$ state, and the other from an *odd* $2p$ state—the combination of an odd and an even state leading to an overall odd state. Thus $(1s\sigma_g 2p\sigma_g)\ ^1\Sigma_u^+$.

Finally the $(1s\sigma_g 2p\pi_u)$: now $\Lambda = \lambda_1 + \lambda_2 = 1$, since one electron is in a π state and, again since one electron originates from a $2p$ orbital, the combined state is u: $^1\Pi_u$.

The energies of these three states increase in the order of the constituent molecular orbitals, as shown in Fig. 6.14(b), i.e.,

$$^1\Sigma_u^+ < {}^1\Pi_u < {}^1\Sigma_g^+$$

Similar states are obtained by excitation to the $3s$ and $3p$ states, to the $4s$ and $4p$ states, etc. Also for $n = 3, 4, \ldots$ there exists the possibility of excitation to the nd orbital. It may be shown by methods similar to those above that interaction between $1s$ and nd electrons can lead to the three configurations and state symbols in increasing energy:

$$(1s\sigma\ nd\sigma)\ ^1\Sigma_g^+ < (1s\sigma\ nd\pi)\ ^1\Pi_g < (1s\sigma\ nd\delta)\ ^1\Delta_g$$

Some of these energy levels are shown at the left of Fig. 6.15. Transitions between them can occur according to the *selection rules*:

1. $$\Delta\Lambda = 0, \pm 1 \tag{6.32}$$

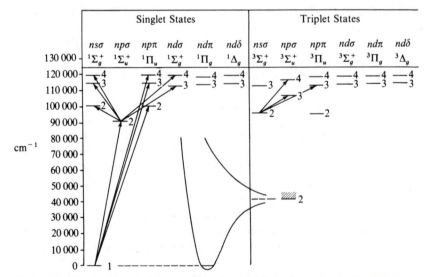

Figure 6.15 The singlet and triplet energy levels of the hydrogen molecule. One electron only is assumed to undergo transitions, the other remaining in the $1s\sigma$ state.

Thus transitions $\Sigma \leftrightarrow \Sigma$, $\Sigma \leftrightarrow \Pi$, $\Pi \leftrightarrow \Pi$, etc., are allowed, but $\Sigma \leftrightarrow \Delta$, for example, is not.

2. $$\Delta S = 0 \qquad (6.33)$$

For the present we are concerned only with singlet states so this rule does not arise.

3. $$\Delta \Omega = 0, \ \pm 1 \qquad (6.34)$$

This follows directly from 1 and 2 above.

4. There are also restrictions on symmetry changes. Σ^+ states can undergo transitions only into other Σ^+ states (or, of course, into Π states) while Σ^- go only into Σ^- (or Π). Symbolically:

$$\Sigma^+ \leftrightarrow \Sigma^+ \qquad \Sigma^- \leftrightarrow \Sigma^- \qquad \Sigma^+ \nleftrightarrow \Sigma^- \qquad (6.35)$$

And finally:

$$g \leftrightarrow u \qquad g \nleftrightarrow g \qquad u \nleftrightarrow u \qquad (6.36)$$

We show a few allowed transitions from the ground state and the lowest excited state in Fig. 6.15.

Let us now consider some of the triplet states of molecular hydrogen, i.e., those states in which the electron spins are parallel and hence $S = 1$. Plainly both electrons cannot now occupy the same orbital so the state of lowest energy will be either $(1s\sigma_g 2s\sigma_g)$, $(1s\sigma_g 2p\sigma_g)$, or $(1s\sigma_g 2p\pi_u)$. The first two are evidently $^3\Sigma$ states, the third is $^3\Pi$, and, following the rules outlined above, we can write down their state symbols and order of energies as:

$$(1s\sigma_g 2p\sigma_g) \ ^3\Sigma_u^+ < (1s\sigma_g 2p\pi_u) \ ^3\Pi_u < (1s\sigma_g 2s\sigma_g) \ ^3\Sigma_g^+$$

These energy levels are shown on the right of Fig. 6.15, together with those formed by the introduction of $3d$ and $4d$ orbitals. (The very small splitting of the levels into states with different $\Omega = \Lambda + S$ is ignored in the figure.) A few of the allowed transitions are shown from the $(1s\sigma 2s\sigma)$ state, but it should be particularly noted that, because of the selection rule $\Delta S = 0$ given in Eq. (6.33), transitions are not allowed between singlet and triplet states, i.e., between the two halves of Fig. 6.15.

Transitions are not shown from the lowest $^3\Sigma_u^+$ state on Fig. 6.15, that is, the lowest triplet state. This is not because transitions are forbidden, but because the state is the continuous one shown in the upper half of Fig. 6.11, the $1s\sigma_u$. Thus in this state the molecule immediately dissociates into atoms before further transitions can occur. The energy level shown for this state in Fig. 6.15 is the lower limit, i.e., the dissociation limit, and in fact the state extends continuously from this limit up to the top of the diagram. Part of Fig. 6.11 is reproduced on Fig. 6.15 to underline the relationship between them.

Thus although the hydrogen spectrum will be complicated by the presence of vibrational and rotational structure on each of the transitions sketched in the figure, basically the overall pattern consists of sets of Rydberg-like line series from which the positions of the energy levels can be found.

6.2.5 Molecular Photoelectron Spectroscopy

Photoelectron spectroscopy and its application to atoms has been described in Sec. 5.5; here we want to discuss the information which this technique can give us about diatomic molecules. To do this we shall use the oxygen molecule as an example.

Photoelectron spectroscopy relies on the ejection of an electron, and consequent formation of an ion, under the influence of a beam of radiation; two ways in which this can happen for O_2 are shown in Fig. 6.16. Part (a) shows the upper four energy levels of the O_2 molecule, and the way in which they are filled in the ground state (cf. the discussion on p. 222). The ionization process requiring least energy is the removal of an electron from the $2p\pi_g$ level (an antibonding electron) as shown in (b); the resulting O_2^+ is in its lowest possible energy state, so we may refer to it as the ground state of O_2^+. Alternatively, part (c) shows the removal of a bonding $2p\pi_u$ electron, requiring somewhat more energy, and resulting in the formation of an excited O_2^+. Other more highly excited states can be formed by the ejection of more firmly bound electrons (e.g., the $2p\sigma_g$ or the $2s\sigma_u$).

It should, perhaps, be mentioned here that the usual notation for O_2 and its ions describes the ground state molecule as $^3\Sigma_g$ (triplet, since $S = \frac{1}{2} + \frac{1}{2} = 1$, and Σ since the electrons' angular momenta are opposed, so $\Lambda = \lambda_1 + \lambda_2 = 1 - 1 = 0$), while the ground state of O_2^+ is $^2\Pi_g$ (doublet, since there is only one unpaired electron spin), and the excited state is $^4\Pi_g$ (three unpaired spins, hence quartet). We shall not use this notation here, however.

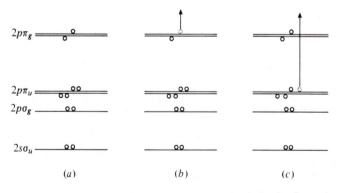

$2p\pi_g$

$2p\pi_u$
$2p\sigma_g$

$2s\sigma_u$

    (a)   (b)   (c)

Figure 6.16 (a) The occupied molecular energy levels in the O_2 molecule: each circle represents an electron. The excitation of an electron (dotted) is shown in the formation of O_2^+ in (b) the ground state, and (c) the first excited state.

The observed photoelectron spectrum of O_2 confirms this pattern of energy levels, since it indicates electrons having an ionization potential of about 1.9×10^{-18} J (12·1 eV) and twice as many electrons with an ionization potential of about 2.6×10^{-18} J (16·2 eV)—clearly these correspond to the two $2p\pi_g$ electrons and the four $2p\pi_u$ electrons respectively. Higher ionization potentials are found for the more firmly held electrons. Thus photoelectron spectroscopy immediately gives us a quantitative estimate of the relative energies of the various molecular orbitals in the oxygen molecule or, of course, of any other diatomic molecule which can be studied by this technique.

The O_2^+ ions formed in this way are stable, but short-lived, molecules. That is to say, in the absence of electron capture (when they revert to O_2) they have no tendency to dissociate into O and O^+, but behave like ordinary diatomic molecules. In particular they vibrate and rotate. We illustrate the situation for O_2 and O_2^+, both in their ground states, in Fig. 6.17(a). Ultra-violet spectroscopy shows that the equilibrium bond lengths of these two molecules are 0·121 nm for O_2 and 0·112 nm for O_2^+ and, if we take the $v = 0$ state of O_2 as the arbitrary zero of energy, then the $v = 0$ state of O_2^+ is some 1.9×10^{-18} J (the ionization potential) above this. Because of

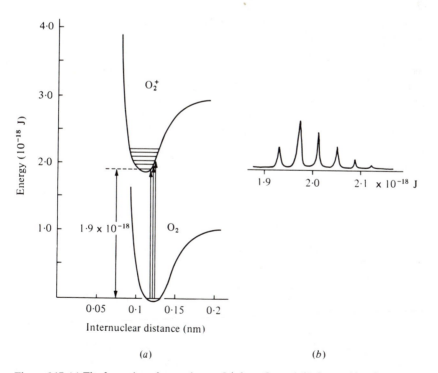

(a)   (b)

Figure 6.17 (a) The formation of ground state O_2^+ from O_2 and (b) the resulting fine structure of the photoelectron spectrum.

the difference in the equilibrium internuclear distances, the Franck–Condon rule (Sec. 6.1.3) indicates that a jump from the ground vibrational state of O_2 will populate several vibrational states of O_2^+; this behaviour is confirmed by the high-resolution photoelectron spectrum, shown diagrammatically in Fig. 6.17(b), which shows a set of discrete ionization potentials, each corresponding to ionization into a particular vibrationally excited state of O_2^+. The spacing between the peaks, about 0.37×10^{-19} J (or 0.23 eV), corresponds to the spacing between the vibrational energy levels of the O_2^+ molecule, which in turn, as we remember from Chapter 3, is equal to the vibrational frequency of the molecule. Since 1000 cm$^{-1} \equiv 0.2 \times 10^{-19}$ J, the spacing gives a vibrational frequency for O_2^+ of about 1850 cm^{-1}. This should be compared with the fundamental frequency of the O_2 molecule itself, determined from Raman spectroscopy, of 1580 cm^{-1}; we see that, very reasonably, removal of an antibonding electron to form O_2^+ in its ground state results in the formation of a stronger, shorter bond.

Similarly observation of the fine structure of the 2.6×10^{-18} J ionization potential reveals a vibrational splitting of some 0.24×10^{-19} J (0.15 eV) or 1200 cm^{-1}—a lower frequency than O_2 itself, consistent with the ionization of a bonding electron.

In general then, observation of a high-resolution photoelectron spectrum can give information on the vibrational frequency of diatomic molecular ions, even though their lifetimes may be of the order of microseconds or less. At present the resolving power of the technique is some 5×10^{-21} J; in other words lines closer than about 250 cm^{-1} cannot be distinguished from each other. Thus *rotational* fine structure cannot yet be studied, but an improvement in resolution by a factor of 10 would allow the rotational levels of ionized hydrides (for example, HF or HCl) to be studied, while a further factor of 10 would open the way to the study of rotational states of the majority of diatomic molecules. Such developments are breathlessly awaited.

6.3 ELECTRONIC SPECTRA OF POLYATOMIC MOLECULES

We have seen in Sec. 3.7 that the vibrational frequencies of a particular atomic grouping within a molecule, for example, CH_3, $C{=}O$, $C{=}C$, etc., are usually fairly insensitive to the nature of the rest of the molecule. Other bond properties, such as length or dissociation energy, are also largely independent of the surrounding atoms in a molecule. Since all these properties depend, in the final analysis, on the electronic structure of the bond, it is plain that we may, at least as an approximation, discuss the structure, and hence the spectrum, of each bond in isolation. Bonds for which this approximation is adequate are usually said to have 'localized' molecular orbitals, i.e., orbitals embracing a pair of nuclei only; other molecules, for which this approximation is invalid, have non-localized orbitals and are often called 'conjugated'. We shall meet some examples of this latter class shortly.

When each bond may be considered in isolation, it is evident that the complete electronic spectrum of a molecule is the sum of the spectra from each bond. The result will plainly be very complex, but a great deal of information about the molecule is contained within it. Thus if some band series can be recognized for a particular bond we immediately know the vibrational frequency of that bond and probably a good estimate of its dissociation energy also. If the rotational structure is resolved, then we have the moment of inertia (from the line spacing) and hence information about the shape and size of the molecule.

Such detailed information is usually obtainable only for molecules studied in the gas phase: in pure liquids, or in solution, molecular rotation is hindered and no rotational structure will be observed. The blurring of the rotational structure often masks the vibrational line series also, and the electronic spectrum of a liquid is usually rather broad and characterless. However, as we shall discuss shortly, it may still be highly characteristic of a particular molecular grouping both in its frequency and its intensity.

Confining our attention, for the moment, to gas-phase spectra, we have already remarked that one of the more important advantages of electronic spectroscopy is that the vibrations, rotations, dissociation energies, and structures of molecules may be investigated in their *excited states*, even though a particular molecule may exist in such a state for not much longer than the time it takes to complete a few rotations. We have not the space to discuss this topic in detail, but one aspect is especially interesting—the fact that electronic excitation often leads to a change in shape of the molecule. That this happens can be seen by studying rotational fine structure in the spectra; here we briefly discuss the theoretical basis for its occurrence.

6.3.1 Change of Shape on Excitation

In Fig. 6.18 we show the orbital picture of a hydride H_2A where A is any polyvalent atom, both in its bent configuration (a), with a bond angle of $90°$, and in a linear form (b), bond angle $180°$. We have seen in Sec. 5.1, that the p orbitals of an atom are at right angles to each other, so we can readily imagine the rectangular molecule to be formed by interaction of two of these p orbitals with hydrogen $1s$ orbitals, leaving the third p orbital unaffected; the latter is called a 'non-bonding' orbital since it plays no part in bonding A and H together, and in general it will have a higher energy than bonding orbitals. In Fig. 6.18(a) we label the non-bonding orbital (which is out of the plane of the paper) N_1, the two bonding molecular orbitals

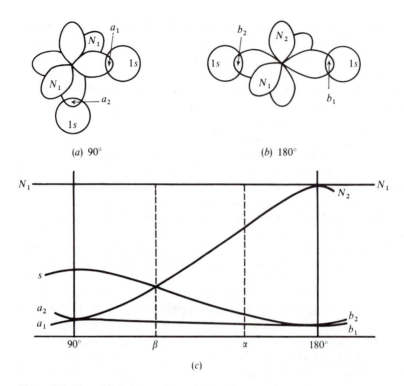

(a) 90°   (b) 180°

(c)

Figure 6.18 The orbital pictures for an AH_2 molecule where the AH bonds are (a) at 90°, and (b) at 180°. In (c) is shown qualitatively the change in energy of the various orbitals as the bond angle changes from 90 to 180°. *(Adapted, with the kind permission of the author, from A. D. Walsh, J. Chem. Soc., 1953, p. 2262.)*

formed from the p and $1s$ atomic orbitals as a_1 and a_2 and the unused (and hence non-bonding) atomic orbital s of A simply as s.

In the linear molecule a new principle must be introduced—that of orbital hybridization. For this, atom A is supposed to mix its s orbital with one of its p orbitals and, from these two orbitals, to form two new orbitals—hybrids—which, it may be shown theoretically, point at 180° to each other. These sp hybrids form rather stronger bonds to other atoms than separate s and p orbitals, so it is energetically favourable, in certain cases, for the atom to 'prepare' hybrid orbitals at the moment of bond formation. In this configuration (Fig. 6.18(b)), there are now two non-bonding p orbitals, labelled N_1 and N_2, and two bonding orbitals formed by overlap between sp hybrids and hydrogen $1s$, called b_1 and b_2.

Now, remembering that a non-bonding orbital is higher in energy than a bonding, and that an sp-bonding orbital is stronger (hence *lower* in energy because the molecule is more stable) than a p-bonding, we can plot the qualitative energy changes for a smooth transition from 90 to 180° bonding.

This we do in Fig. 6.18(c), which is constructed as follows:

1. The non-bonding orbital N_1 remains unchanged throughout, hence its energy is constant;
2. The bonding orbital a_1 passes over into the stronger orbital b_1, hence its energy decreases;
3. The bonding orbital a_2 becomes the non-bonding N_2, thus increasing in energy; N_1 and N_2 are identical in energy at 180°;
4. The bonding orbitals b_1 and b_2 are formed by absorption of the non-bonding s into a_1;
5. If we increase the bond angle beyond 180° (or decrease it below 90°) the reverse changes begin to take place, so 180 and 90° represent maxima and minima on the energy curves as shown.

Now let us see the relevance of this to molecular shapes. Consider first the molecule BeH_2, beryllium hydride. Beryllium, we have seen in Sec. 5.3.1, has the electronic ground state configuration $1s^2 2s^2$, that is, it has two outer electrons with which to form bonds, the two $1s$ electrons being too firmly held by the nucleus to take part in bonding. Each hydrogen atom contributes a further electron, so the BeH_2 molecule must dispose of four electrons into molecular orbitals, with, according to Pauli, a maximum of two electrons per orbital. The most stable state, as can be seen from Fig. 6.18(c), will be for two electrons to go into b_1 and two into b_2, thus producing a *linear* molecule.

When the molecule is excited electronically, the next available orbital to contain the excited electron is N_2 (or N_1), but with a configuration $b_1^2 b_2^1 N_2^1$ it is evident from the figure that the most stable state will be at a bond angle, α, somewhere between 90 and 180°—the increase in the energies of b_1 and b_2 being more than compensated for by the decrease in N_2 until equilibrium occurs at an angle α. Thus we see that the excited state is *bent*. If the electron is so excited as to be ionized completely, leaving the ion BeH_2^+, the three remaining electrons will all be in the b_1 and b_2 orbitals and hence the most stable configuration will again be *linear*.

Now consider the case of water, H_2O. The oxygen atom has an outer electron configuration $2s^2 2p^4$, and so has six electrons to dispose into molecular orbitals. As before each hydrogen contributes one, so water is formed by placing a total of four pairs of electrons into four molecular orbitals. The lowest energy state at which this can be done is shown by the angle β in the figure, which is some angle between 90 and 180° (and is observed experimentally to be about 104°). Thus water is bent in the ground state, with a configuration which may be written $a_1^2 a_2^2 s^2 N_1^2$, since the angle is not far removed from 90°. During excitation one of the N_1 electrons will undergo transitions since these, being of highest energy, are most easily removed. However, since the energy of N_1 does not change with angle, the angle of the remaining $a_1^2 a_2^2 s^2 N_1^1$ state will not change during the transition.

These arguments may be readily extended to other triatomic molecules or to larger polyatomic molecules, although the energy diagram corresponding to Fig. 6.18(c) is more complicated since more orbitals are involved. The results show, however, and experiment confirms, that linear molecules such as CO_2 and $HC{\equiv}CH$ become bent on excitation, the latter taking up a planar zig-zag conformation.

6.3.2 Chemical Analysis by Electronic Spectroscopy

Although rotational and sometimes vibrational fine structure do not appear in the liquid or solid state, both the position and intensity of the rather broad absorption due to an electronic transition is very characteristic of the molecular group involved. In this branch of spectroscopy the position of an absorption is almost invariably given as the *wavelength* at the point of maximum absorption, $\lambda_{max.}$, quoted either in Ångstrom units (1 Å = 10^{-10} m) or in nanometres (1 nm = 10^{-9} m), the latter being more usual. It should be particularly noted that a *large* energy change, corresponding to a high frequency or wavenumber, is represented by a *small* wavelength. For practical reasons the electronic spectrum is divided into three regions: (1) the visible region, between 400 and 750 nm (4000–7500 Å or 25 000–13 300 cm^{-1}), (2) the near ultra-violet region, between 200 and 400 nm (2000–4000 Å or 50 000–25 000 cm^{-1}), and (3) the far (or vacuum) ultra-violet, below 200 nm (below 2000 Å or above 50 000 cm^{-1}). The latter is so called because absorption by atmospheric oxygen is considerable in this region and spectra can only be obtained if the whole spectrometer is carefully evacuated. Thus commercial instruments extend only down to about 185 nm and absorptions below this range are little used for routine chemical purposes.

The *intensity* of an electronic absorption is given by the simple equation:

$$\varepsilon = \frac{1}{cl} \log_{10} \frac{I_0}{I} \ l \ \text{mol}^{-1} \ \text{cm}^{-1} \tag{6.37}$$

where c and l are the concentration and path length of the sample (in mol l^{-1} and in cm^{-1}, respectively), I_0 is the intensity of light of wavelength $\lambda_{max.}$ falling on the sample, and I is the intensity transmitted by the sample. ε is the molar extinction coefficient and ranges from some 5×10^5 for the strongest bands to 1 or less for very weak absorptions.

Electrons in the vast majority of molecules fall into one of the three classes: σ electrons, π electrons, and non-bonding electrons (called n electrons). The first two classes were discussed in Sec. 6.2.2 and the third, which plays no part in the bonding of atoms into molecules, was mentioned briefly in Sec. 6.3.1. In chemical terms a single bond between atoms, such as C—C, C—H, O—H, etc., contains only σ electrons, a multiple bond, C=C,

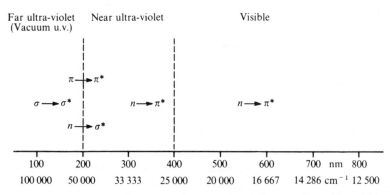

Figure 6.19 The regions of the electronic spectrum and the type of transition which occurs in each.

$C\equiv C$, $C\!=\!N$, etc., contains π electrons in addition, while atoms to the right of carbon in the periodic table, notably nitrogen, oxygen, and the halogens, possess n electrons. In general the σ electrons are most firmly bound to the nuclei and hence require a great deal of energy to undergo transitions, while the π and n electrons require less energy, the n electrons usually (but not invariably) requiring less than the π. Thus, in an obvious notation, $\sigma \rightarrow \sigma^*$ transitions fall into the vacuum ultra-violet, $\pi \rightarrow \pi^*$ and $n \rightarrow \sigma^*$ appear near the borderline of the near and far ultra-violet, and $n \rightarrow \pi^*$ come well into the near ultra-violet and visible regions. These generalizations are indicated schematically on Fig. 6.19, which also shows the relationship between the nanometer and wavenumber scales.

Saturated hydrocarbon molecules, then, which can only undergo $\sigma \rightarrow \sigma^*$ transitions, do not give rise to spectra with any analytic interest since they fall outside the generally available range; examples are the $\sigma \rightarrow \sigma^*$ transitions of methane CH_4, and ethane C_2H_6 which are at 122 and 135 nm respectively.

The insertion of a group containing n electrons, e.g., the NH_2 group, allows the possibility of $n \rightarrow \sigma^*$ transitions in addition and also tends to increase the wavelength of the $\sigma \rightarrow \sigma^*$ absorption; for example, CH_3NH_2: $\sigma \rightarrow \sigma^*$ 170 nm, $n \rightarrow \sigma^*$ 213 nm. It is unsaturated molecules, i.e., molecules containing multiple bonds, which give rise to the most varied and interesting spectra, however. We cannot here discuss the large mass of data in any detail but must be content to indicate a few of the more important generalizations. More detail is to be found in the books by Scott, and by Williams and Fleming, listed in the bibliography at the end of this chapter.

Consider, first, *isolated* multiple bonds within a molecule; the most important factor determining the position of the absorption maxima is, of course, the nature of the atoms which are multiply-bonded. From the following table we see that the $\pi \rightarrow \pi^*$ transitions are relatively insensitive to those atoms while the $n \rightarrow \pi^*$ transitions vary widely:

	$\pi \to \pi^*$ (strong) (nm)	$n \to \pi^*$ (weak) (nm)
$>C=C<$	170	—
$—C\equiv C—$	170	—
$>C=O$	166	280
$>C=N\diagdown$	190	300
$>N=N\diagup$	?	350
$>C=S$	?	500

This behaviour is very reasonable since the n electrons play no part in the bonding and control of them is retained by the atom (O, N, or S) contributing them. The above data is approximate only since different substituents on the A=B group produce slight variations in the wavelength of the $n \to \pi^*$ transition. Thus, considering ketones alone, $\lambda_{max.}$ varies from 272 nm for CH_3COCH_3 to 290 nm for cyclohexanone, and even higher if halogen substituents are included. From the mass of empirical data already assembled, a great deal of information about the substituents to a particular group is obtainable from the electronic spectrum.

More pronounced changes occur, however, when two or more multiple bonds are *conjugated* in the molecule, i.e., when structures having alternate single and multiple bonds arise, for example $—C=C—C=C—$ or $—C=C—C=O$. In this case the $\pi \to \pi^*$ and $n \to \pi^*$ transitions both increase considerably in wavelength and intensity, the increase being greater the more conjugate linkages there are. As a simple example we have the following approximate data for $\pi \to \pi^*$ transitions in carbon–carbon bonds:

	$\lambda_{max.}$ (nm)	ε
$—C=C—$	170	16 000
$—C=C—C=C—$	220	21 000
$—C=C—C=C—C=C—$	260	35 000

while for oxygen-containing molecules we have both $\pi \to \pi^*$ and $n \to \pi^*$ transitions:

	$\pi \to \pi^*$ (strong) (nm)	$n \to \pi^*$ (weak) (nm)
$—C=O$	166	280
$—C=C—C=O$	240	320
$—C=C=C—C=O$	270	350
$O=\langle\bigcirc\rangle=O$	245	435

Thus we see that conjugation immediately brings the very intense $\pi \to \pi^*$ transition into the easily available region of ultra-violet spectrometers. For this reason these techniques are particularly well adapted to the study of conjugated and aromatic systems.

In the last example given above we see that the $n \to \pi^*$ absorption of p-benzoquinone, at 435 nm, has shifted into the blue region of the visible spectrum. When the substance is seen in the ordinary way the complementary colour—yellow—is observed. Colour in large organic molecules is invariably due to the existence of considerable conjugation raising the transition wavelength into the visible region—a fact on which the chemistry of dyestuffs is based.

Substituents on conjugated systems also perturb the ultra-violet transitions in a systematic way; a great deal of empirical data has led to the formulation of rules to predict the effects. These are inown as Woodward's rules, but they have undergone considerable modification and extension since they first formulated by Woodward in 1941. As a simple example, consider butadiene, $CH_2{=}CH.CH{=}CH_2$, which has a strong absorption at 217 nm due to the $\pi \to \pi^*$ transition. This molecule may be considered to be the 'parent' of a whole series of molecules containing the trans conjugated $\diagdown\!\diagup\!\diagdown$ group, whether in hydrocarbon chains or in ring systems. Any substituents or modifications to this parent fragment have each been assigned a positive or negative value which must be added to the basic absorption at 217 nm in order to arrive at the expected absorption frequency of the substituted molecule. Thus the increment for a chlorine atom is 5 nm and for an $-OCH_3$ group 6 nm; if both these substituents occur together, the molecule would absorb at $217 + 5 + 6 = 228$ nm. Excellent accounts of these rules, and tabulations of the values for various substituents, etc., are to be found in the books by Scott, and by Williams and Fleming, mentioned in the bibliography.

6.3.3 The Re-Emission of Energy by an Excited Molecule

After a molecule has undergone an electronic transition into an excited state there are several processes by which its excess energy may be lost; we discuss some of these briefly below.

1. *Dissociation.* The excited molecule breaks into two fragments. This was discussed in some detail for the particular case of a diatomic molecule dissociating into atoms in Sec. 6.1.4. No spectroscopic phenomena, beyond the initial absorption spectrum, are observed unless the fragments radiate energy by one of the processes mentioned below.
2. *Re-emission.* If the absorption process takes place as shown schematically in Fig. 6.20(a), then the re-emission is just the reverse of this, as in (b) of the figure. The radiation emitted, which may be collected and displayed as an emission spectrum, is identical in frequency with that absorbed.

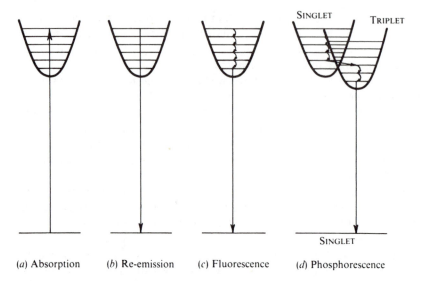

(a) Absorption   (b) Re-emission   (c) Fluorescence   (d) Phosphorescence

Figure 6.20 Showing the various ways in which an electronically excited molecule can lose energy.

3. *Fluorescence.* If, as in Fig. 6.20(a), the molecule is in a high *vibrational* state after electronic excitation, then excess vibrational energy may be lost by intermolecular collisions; this is illustrated in (c) of the figure. The vibrational energy is converted to kinetic energy and appears as heat in the sample; such transfer between energy levels is referred to as 'radiationless'. When the excited molecule has reached a lower vibrational state (for example, $v' = 0$), it may then emit radiation and revert to the ground state; the radiation emitted, called the *fluorescence spectrum*, is normally of lower frequency than that of the initial absorption, but under certain conditions it may be of higher frequency. The time between initial absorption and return to ground state is very small, of the order of 10^{-8} s.

4. *Phosphorescence.* This can occur when two excited states of different total spin have comparable energies. Thus in Fig. 6.20(d), we imagine the ground state and one of the excited states to be singlets (that is, $S = 0$), while the neighbouring excited state is a triplet ($S = 1$). Although the rule $\Delta S = 0$ forbids *spectroscopic* transitions between singlet and triplet states, there is no prohibition if the transfer between the excited states occurs *kinetically*, i.e., through radiationless transitions induced by collisions. Such transfer, however, can only occur close to the cross-over point of the two potential curves (cf. Sec. 6.1.7), and once the molecule has arrived in the triplet state and undergone some loss of vibrational energy in that state, it cannot return to the excited singlet state. It will, therefore, eventually reach the $v' = 0$ level of the *triplet* state. Now al-

though a transition from here to the ground state is spectroscopically forbidden, it *may* take place but much more slowly than an allowed electronic transition. Thus it is that a phosphorescent material will continue to emit radiation seconds, minutes, or even hours after the initial absorption. The phosphorescence spectrum, as a rule, consists of frequencies lower than that absorbed.

Considerable confusion often occurs between the Raman effect, discussed in Chapter 4, and the phenomena of fluorescence or phosphorescence. The main points of difference are as follows:

(*a*) In fluorescence and phosphorescence, radiation must be absorbed by the molecule and an excited electronic state formed; in Raman spectroscopy energy is merely transferred from radiation to molecule, or vice versa, but no excited electronic state is formed.

(*b*) The exciting radiation for fluorescence or phosphorescence must be just that equivalent to the energy difference between electronic states; the exciting radiation for Raman spectroscopy can be of any frequency *except* that which would induce electronic transitions; in the latter case absorption would occur, rather than scattering.

5. *Stimulated emission.* This increasingly important mechanism for removal of excess energy can lead to the production of laser radiation, as discussed in Sec. 1.10. The helium–neon laser, which emits light in the visible region of the spectrum, was considered in Sec. 5.4.6.

6.4 TECHNIQUES AND INSTRUMENTATION

The simple techniques of electronic spectroscopy are familiar to every schoolboy studying physics—a glass prism, some sort of telescope, a bunsen burner, and a pinch of common salt are sufficient apparatus for observing part of the emission spectrum of sodium. And in fact a great deal of rapid and precise analytical work, both qualitative and quantitative, is carried out using flame spectrophotometers not very much more sophisticated in construction than this, except that a photomultiplier or photographic plate is used instead of the rather inaccurate human eye. However, for high-resolution work or for absorption studies, the practical requirements are more stringent.

The choice of a suitable source was formerly one of the main difficulties. The prime requirements of a source are that it should be *continuous* over the region of interest (i.e., there must be no wavelengths at which it does not emit) and it should be as *even* as possible (i.e., there must be no intense emission lines). In the visible region and just into the near ultraviolet—say between 350 and 800 nm—an ordinary tungsten filament lamp is quite suitable. Below this a hydrogen discharge lamp proves adequate, down to about 190 nm, while below this again discharge lamps containing rare gases, such as xenon, must be used. Thus we see that, in contrast to the other forms of spectroscopy discussed in previous chapters, no one source is suitable throughout the region.

Transparent materials for windows and sample cells present no problem, at least in the visible and near ultra-violet regions, since good quality glass or quartz transmit down to 200 nm or better. Below this region alkali fluorides, such as lithium fluoride or calcium fluoride, must be used; these are transparent down to about 100 nm. Prisms, if used, can be made of the same materials. Modern high-resolution instruments, however, employ a reflection grating rather than a prism since the former gives better dispersion and so allows more precise wavelength selection and measurement.

The detector for visible and ultra-violet studies is either a photographic plate or a photomultiplier tube. The chief disadvantage of the photographic method is that the resolving power is limited by the graininess of the image; on the other hand there is no other detector which can record the complete spectrum simultaneously in a small fraction of a second. When studying short-lived species, such as free radicals, it would be quite impossible to scan the complete spectrum using a photomultiplier. Also, at the other end of the time-scale the photographic plate is an efficient integrator of very weak signals—exposure times can be extended to many hours or even days to record a weak emission or absorption. For most routine purposes, however, where the spectrum of a stable material is to be recorded in a time of several minutes, a photomultiplier detector coupled to an amplifier and paper chart recorder is the most flexible and useful combination.

BIBLIOGRAPHY

Bingel, W. A.: *Theory of Molecular Spectra*, Wiley, 1969.
Carlson, T. A.: *Photoelectron and Auger Spectroscopy*, Plenum Press, 1975.
Herzberg, G.: *Molecular Spectra and Molecular Structure, vol. 1, Spectra of Diatomic Molecules*, 2nd ed., Van Nostrand, 1950.
Herzberg, G.: *Molecular Spectra and Molecular Structure, vol. 3, Electronic Spectra and Electronic Structure of Polyatomic Molecules*, Van Nostrand, 1967.
Scott, A. I.: *Interpretation of the Ultraviolet Spectra of Natural Products*, Pergamon Press, 1964.
Sutton, D.: *Electronic Spectra of Transition Metal Complexes*, McGraw-Hill, 1968.
Turner, D. W., C. Baker, A. D. Baker, and C. R. Brundle: *Molecular Photoelectron Spectroscopy*, Wiley-Interscience, 1970.
Williams, D. H. and I. Fleming: *Spectroscopic Methods in Organic Chemistry*, 3rd ed., McGraw-Hill, 1980.

PROBLEMS

(Useful constant: $1 \text{ cm}^{-1} \equiv 11{\cdot}958 \text{ J mol}^{-1}$.)

6.1 Using the data of Fig. 6.5 (some of which is tabulated below), estimate the dissociation energy of the I_2 molecule:

v:	0	5	10	30	50	70	75	80
$\Delta\varepsilon$:	213·3	207·2	200·7	172·1	134·7	82·3	67·0	52·3 cm^{-1}

6.2 The absorption spectrum of O_2 shows vibrational structure which becomes a continuum at $56\,876 \text{ cm}^{-1}$; the upper electronic state dissociates into one ground state atom and one excited atom (the excitation energy of which, measured from the atomic spectrum, is $15\,875 \text{ cm}^{-1}$). Estimate the dissociation energy of ground state O_2 in kJ mol^{-1}.

6.3 The values of $\bar{\omega}_e$ and x_e in the ground state ($^3\Pi_u$) and a particular excited state ($^3\Pi_g$) of C_2 are:

	$\bar{\omega}_e$	x_e
Ground state	1641·4 cm^{-1}	$7{\cdot}11 \times 10^{-3}$
Excited state	1788·2 cm^{-1}	$9{\cdot}19 \times 10^{-3}$

Use Eq. (6.10) to find the number of vibrational energy levels below the dissociation limit and hence the dissociation energy of C_2 in both states.

6.4 The spectrum arising from transitions between the two states of C_2 in Prob. 6.3 shows the v_{00} line at $19\,378 \text{ cm}^{-1}$ and a convergence limit at $39\,231 \text{ cm}^{-1}$. The dissociation is into one ground state atom and one excited atom, the excitation energy of the latter being $10\,308 \text{ cm}^{-1}$; calculate the exact dissociation energies of the two states and compare your answers with those of Prob. 6.3. Explain any discrepancy.

6.5 The band origin of a transition in C_2 is observed at $19\,378 \text{ cm}^{-1}$, while the rotational fine structure indicates that the rotational constants in excited and ground states are, respectively, $B' = 1{\cdot}7527 \text{ cm}^{-1}$ and $B'' = 1{\cdot}6326 \text{ cm}^{-1}$. Estimate the position of the band head. Which state has the larger internuclear distance?

SEVEN

SPIN RESONANCE SPECTROSCOPY

We have seen in earlier chapters that all electrons and some nuclei possess a property conveniently called 'spin'. Electronic spin was introduced in Chapter 5 to account for the way in which electrons group themselves about a nucleus to form atoms and we found that the spin also accounted for some fine structure, such as the doublet nature of the sodium D line, in atomic spectra. Equally, in Sec. 5.6 it was necessary to invoke a nuclear spin to account for very tiny effects, called hyperfine structure, observed in the spectra of some atoms.

In this chapter we shall consider these spins in rather more detail and discuss the sort of spectra they can given rise to directly rather than their influence on other types of spectra. After an introduction discussing the interaction of spin with an external magnetic field we shall consider in some detail the spectra of particles with a spin of $\frac{1}{2}$ (that is, electrons, and some nuclei such as hydrogen, fluorine, or phosphorus), then a brief discussion of some other nuclei whose spins are greater than $\frac{1}{2}$, and finally a few words on the techniques involved in producing electronic and nuclear spin spectra.

7.1 SPIN AND AN APPLIED FIELD

7.1.1 The Nature of Spinning Particles

We have seen that all electrons have a spin of $\frac{1}{2}$, that is, they have an angular momentum of $\sqrt{\frac{1}{2}(\frac{1}{2}+1)}(h/2\pi) = \sqrt{3}/2$ units. Many nuclei also possess spin although the angular momentum concerned varies from nucleus to nucleus.

The simplest nucleus is that of the hydrogen atom, which consists of one particle only, the *proton*. The protonic mass (1.67×10^{-27} kg) and charge ($+1.60 \times 10^{-19}$ C) are taken as the units of atomic mass and charge respectively; the charge is, of course, equal in magnitude but opposite in sign to the electronic charge. The proton also has a spin of $\frac{1}{2}$.

Another particle which is a constituent of all nuclei (apart from the hydrogen nucleus) is the *neutron*; this has unit mass (i.e., a mass equal to that of the proton), no charge and, again, a spin of $\frac{1}{2}$.

Thus if a particular nucleus is composed of p protons and n neutrons its total mass is $p + n$ (ignoring the small mass defects associated with nuclear binding energy), its total charge is $+p$ and its total spin will be a vector combination of $p + n$ spins each of magnitude $\frac{1}{2}$. The atomic mass is usually specified for each nucleus by writing it as a prefix to the nuclear symbol; for example, ^{12}C indicates the nucleus of carbon having a mass of 12. Since the atomic charge is six for this nucleus we know immediately that the nucleus must contain six protons and six neutrons to make up a mass of 12. The nucleus ^{13}C (an *isotope* of carbon) has six protons and seven neutrons.

Each nuclear isotope, being composed of a different number of protons and neutrons, will have its own total spin value. Unfortunately, the laws governing the vector addition of nuclear spins are not yet known so the spin of a particular nucleus cannot be predicted in general. However, observed spins can be rationalized and some empirical rules have been formulated.

Thus the spin of the hydrogen nucleus (^{1}H) is $\frac{1}{2}$ since it consists of one proton only; deuterium, an isotope of hydrogen containing one proton and one neutron (that is, ^{2}H) might have a spin of 1 or 0 depending on whether the proton and neutron spins are parallel or opposed: it is observed to be 1. The helium nucleus, containing two protons and two neutrons (^{4}He) has zero spin, and from these and other observations stem the following rules:

1. Nuclei with both p and n even (hence charge *and* mass even) have zero spin (for example, ^{4}He, ^{12}C, ^{16}O, etc.).
2. Nuclei with both p and n odd (hence charge *odd* but mass = $p + n$, *even*), have integral spin (for example, ^{2}H, ^{14}N (spin = 1), ^{10}B (spin = 3), etc.).
3. Nuclei with odd mass have half-integral spins (for example, ^{1}H, ^{15}N (spin = $\frac{1}{2}$), ^{17}O (spin = $\frac{5}{2}$), etc.).

The spin of a nucleus is usually given the symbol I, called the *spin quantum number*. Quantum mechanics shows that the angular momentum of a nucleus is given by the expression:

$$\text{Angular momentum } \mathbf{I} = \sqrt{I(I + 1)}(h/2\pi) = \sqrt{I(I + 1)} \text{ units} \quad (7.1)$$

where I takes, for each nucleus, *one* of the values 0, $\frac{1}{2}$, 1, $\frac{3}{2}$, ... We can conveniently include the spin quantum number and angular momentum of an *electron* in Eq. (7.1) if we agree to label its spin quantum number I (instead of s as in Chapter 5), and remember that I can be $\frac{1}{2}$ only for an electron. Thus Eq. (7.1) represents the angular momentum of nuclei and of electrons once the appropriate value of I is inserted.

We may note here that many texts use a simpler form of Eq. (7.1), viz.

$$\mathbf{I} = I\,\frac{h}{2\pi} = I \text{ units}$$

This equation is not, however, strictly correct from a quantum mechanical point of view and we shall use the more rigorous equation (7.1) throughout this chapter.

By now the reader will be familiar with the idea that the angular momentum vector $\mathbf{I}$ cannot point in any arbitrary direction, but can point only so that its components along a particular reference direction are either all integral (if I is integral) or all half-integral (if I is half-integral). Thus we can have components along a particular direction z, of:

$$I_z = I, I - 1, \ldots, 0, \ldots, -(I - 1), -I \qquad \text{(for } I \text{ integral)}$$

or

$$I_z = I, I - 1, \ldots, \tfrac{1}{2}, -\tfrac{1}{2}, \ldots, -I \qquad \text{(for } I \text{ half-integral)} \qquad (7.2)$$

giving $2I + 1$ components in each case. These components are normally degenerate—i.e., they all have the same energy—but the degeneracy may be lifted and $2I + 1$ different energy levels result if an external magnetic field is applied to define the reference direction. We shall now consider the effect of such a field.

7.1.2 Interaction between Spin and a Magnetic Field

In general, a charged particle spinning about an axis constitutes a circular electric current which in turn produces a magnetic dipole. In other words the spinning particle behaves as a tiny bar magnet placed along the spin axis. The size of the dipole (i.e., the strength of the magnet) for a point charge can be shown to be:

$$\mu = \frac{q}{2m}\,\mathbf{I} = \frac{q\sqrt{I(I + 1)}}{2m}\,\frac{h}{2\pi} = \frac{qh}{4\pi m}\,\sqrt{I(I + 1)} \quad \text{A m}^2$$

where q and m are the charge and mass of the particle. The magnetic moment is here expressed in the appropriate fundamental SI units, ampere square metre (A m^2); it is useful for later arguments, however, to express the magnetic moment in terms of the magnetic flux density (colloquially 'magnetic field strength'), the SI unit of which is the tesla (symbol T, units $kg \, s^{-2} \, A^{-1}$), where $1 \, T \equiv 10\,000$ gauss. The conversion is:

$$A \, m^2 \equiv (kg \, s^{-2} \, T^{-1}) \, m^2 \equiv J \, T^{-1} \text{ (joules per tesla)}$$

So we may write:

$$\mu = \frac{qh}{4\pi m} \sqrt{I(I+1)} \, J \, T^{-1} \qquad (7.3)$$

and it is this form we shall use in our subsequent discussion. (It should be pointed out that, when the magnetic moment is expressed in c.g.s. units, as in older textbooks, the right-hand side of Eq. (7.3) will be divided by c, the velocity of light.)

When we remove the fiction that electrons and nuclei are *point* charges, Eq. (7.3) becomes modified by the inclusion of a numerical factor G:

$$\mu = \frac{Gqh}{4\pi m} \sqrt{I(I+1)} \, J \, T^{-1} \qquad (7.4)$$

For electrons we have seen (cf. Sec. 5.6) that G is given the symbol g and called the Landé splitting factor; its value depends on the quantum state of the electron and may be calculated from the L, S, and J quantum numbers (cf. Eq. (5.28)). Nuclear G factors, on the other hand, cannot be calculated in advance and are obtainable only experimentally.

For electrons, Eq. (7.4) is usually written

$$\mu = -g\beta\sqrt{I(I+1)} \, J \, T^{-1} \qquad (7.5)$$

where we have expressed the set of constants $eh/4\pi m$ as a (positive) constant β, called the *Bohr magneton*; replacing the electronic charge ($1 \cdot 60 \times 10^{-19}$ C) and mass ($9 \cdot 11 \times 10^{-31}$ kg) in this expression, we can calculate $\beta = 9 \cdot 273 \times 10^{-24} \, J \, T^{-1}$.

Nuclear dipoles, on the other hand, are conveniently expressed in terms of a *nuclear magneton* β_N, which is defined in terms of the mass and charge of the *proton*:

$$\beta_N = \frac{eh}{4m_p \pi} = 5 \cdot 050 \times 10^{-27} \, J \, T^{-1}$$

Thus for a nucleus of mass M and charge pe (where p is the number of protons) we would write:

$$\mu = \frac{Gpe}{2M} \sqrt{I(I+1)} \, \frac{h}{2\pi} = \frac{Gm_p p}{M} \beta_N \sqrt{I(I+1)}$$

$$= g\beta_N \sqrt{I(I+1)} \, J \, T^{-1} \qquad (7.6)$$

where we have collected the parameters $Gm_p p/M$, in which m_p is the protonic mass, into a factor g which is characteristic of each nucleus. This

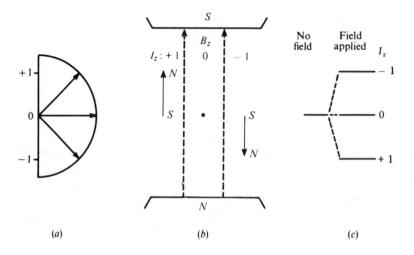

Figure 7.1 Showing (a) the three orientations of the spin of a nucleus with spin quantum number $I = 1$; (b) the resulting magnetic dipole, μ_z, oriented in an applied magnetic field B_z; and (c) the three energy levels allowed to the nucleus.

factor has values up to about six and is positive for nearly all known nuclei (see Table 7.1 later).

Thus the analogous equations (7.5) and (7.6) define the equivalent spin dipole for any spinning particle. The dipole will plainly have components along a reference direction governed by the I_z values:

$$\begin{array}{ll} \mu_z = -g\beta I_z & \text{(for electrons)} \\ \mu_z = g\beta_N I_z & \text{(for nuclei)} \end{array} \right\} \qquad (7.7)$$

where the I_z are given by Eq. (7.2) for a particular particle, and the dipole will interact to different extents, depending on its magnitude, with a magnetic field. The situation for a *nucleus* with $I = 1$ is shown in Fig. 7.1. The angular momentum of the particle (Eq. (7.1)) is:

$$\mathbf{I} = \sqrt{1 \times 2} = \sqrt{2} \text{ units}$$

and if we consider a semicircle with this radius, it is plain that the vector arrow corresponding to $\mathbf{I}$ can point so as to have z components of $+1$, 0, or -1 (the z direction is counted positive towards the top of the paper). This is shown in part (a) of the figure.

Equation (7.7) shows that μ_z and I_z have the same sign (i.e., point in the same direction) for the many nuclei which have positive g values. The three μ_z values for this system are represented by arrows in part (b) of the figure where, conventionally, the lines of force inside the magnet are drawn with an arrow pointing to the N pole. If we imagine the applied magnetic field to be produced by a horseshoe magnet and apply the same convention, it is clear that the lines of force *external* to the magnet will be shown pointing from the N to the S pole and, if we require these to be in the positive z direction, we arrive at the configuration given in Fig. 7.1(b). Thus we see that the state $I_z = -1$ represents a nuclear dipole *opposed* to the magnetic field (i.e., of high energy) while $I_z = +1$ is in the same direction as the applied field and is, therefore, of low energy. The state $I_z = 0$ has no net dipole along the field direction and is therefore unchanged in energy whether the field is applied or not. This is shown in Fig. 7.1(c).

Of course, if the nuclear g factor is negative, μ_z has a sign opposite from I_z, and the order of labelling the energy levels of Fig. 7.1 will be reversed. Similarly, of the two energy levels allowed to an electron in a magnetic field, the lower will be associated with $I_z = -\frac{1}{2}$, the upper with $I_z = +\frac{1}{2}$.

The extent of interaction between a magnetic dipole and a field of strength B_z applied along the z axis is equal to the product of the two:

$$\text{Interaction} = \mu_z B_z$$

Thus the separation between neighbouring energy levels (where I_z differs by unity) is:

$$\Delta E = [E_{I_z} - E_{(I_z - 1)}] = |g\beta_N I_z B_z - g\beta_N (I_z - 1)B_z|$$

$$= |g\beta_N B_z| \quad \text{J} \qquad \text{(when } B_z \text{ is expressed in tesla)} \tag{7.8}$$

Thus in hertz:

$$\frac{\Delta E}{h} = \left| \frac{g\beta_N B_z}{h} \right| \quad \text{Hz} \tag{7.9}$$

(where the modulus sign, $|\ldots|$, indicates that *positive* differences only should be considered). Here, then, is the basis for a spectroscopic technique: a transition of electron or nuclear spins between energy levels (loosely referred to as 'a change of spin') may be associated with the emission or absorption of energy in the form of radiation at the frequency of Eq. (7.9). Further, since the frequency is proportional to the applied field we can arrange, in principle, to study spin spectra in any region of the electromagnetic spectrum, merely by choosing an appropriate field. However, for practical reasons, the fields used are normally of the order of 1–5 tesla for nuclei and 0·3 tesla for electrons. Let us calculate the approximate frequency to be expected under these circumstances.

For nuclei: We have already $\beta_N = 5\cdot05 \times 10^{-27}$ J T^{-1}, and if we choose the rather specific value of $B_z = 2\cdot3487$ T, and the g factor of hydrogen, $g = 5\cdot585$, we calculate

$$\frac{\Delta E}{h} = \frac{5\cdot585 \times 5\cdot05 \times 10^{-27} \times 2\cdot3487}{6\cdot63 \times 10^{-34}} = 100 \times 10^6 \text{ Hz} \tag{7.10}$$

and we see that the appropriate frequency for protons, 100 MHz, falls in the short-wave radiofrequency region; in fact a great many nuclear magnetic

Table 7.1 Properties of some nuclei with non-zero spin

Nucleus	Spin	Resonance frequency (MHz) in field of 2·3487 T	g value
^{1}H	$\frac{1}{2}$	100·00	5·585
^{10}B	3	10·75	0·6002
^{11}B	$\frac{3}{2}$	32·08	1·792
^{13}C	$\frac{1}{2}$	25·14	1·404
^{14}N	1	7·22	0·4036
^{15}N	$\frac{1}{2}$	10·13	−0·5660
^{17}O	$\frac{5}{2}$	13·56	−0·7572
^{19}F	$\frac{1}{2}$	94·07	5·255
^{29}Si	$\frac{1}{2}$	19·87	−1·110
^{31}P	$\frac{1}{2}$	40·48	2·261
^{35}Cl	$\frac{3}{2}$	9·80	0·5472
^{37}Cl	$\frac{3}{2}$	8·16	0·4555
^{107}Ag	$\frac{1}{2}$	4·05	−0·2260
^{119}Sn	$\frac{1}{2}$	37·27	−2·082
^{127}I	$\frac{5}{2}$	20·00	1·118
^{199}Hg	$\frac{1}{2}$	17·83	0·996

resonance spectrometers operate at just this frequency, which is why we chose such a precise value of B_z for the calculation. All other nuclei (except tritium) have smaller g factors, and their spectra fall between 1 and 100 MHz for the same applied field. Table 7.1 collects some data for a few of the more important nuclei.

For electrons: Here $\beta = 9·273 \times 10^{-24}$ J T^{-1}, and let us assume $g = 2$ and $B_z = 0·33$ T. Then

$$\frac{\Delta E}{h} = \frac{2 \times 9·273 \times 10^{-24} \times 0·33}{6·63 \times 10^{-34}} \approx 9000 \times 10^{6} \text{ Hz} \qquad (7.11)$$

Thus electron spin spectra fall at a considerably higher frequency, which is on the long wavelength edge of the microwave region. Because of this difference, techniques of nuclear and electronic spin spectroscopy differ considerably, as we shall see later, although in principle they are concerned with very similar phenomena.

7.1.3 Population of Energy Levels

When first confronted with nuclear and electron spin spectroscopy the student (who has experimented earlier with bar magnets in the earth's field) usually asks: why don't the nuclear (or electronic) magnetic moments immediately line themselves up in an applied field so that they all occupy the lowest energy state?

There are several facets to this question and its answer. Firstly, if we take 'immediately' to refer to a period of some seconds, then spin magnetic moments *do* immediately orientate themselves in a magnetic field, although they do *not* all occupy the lowest available energy state. This is a simple consequence of thermal motion and the Boltzmann distribution. We have seen that spin energy levels are split in an applied field, and their energy separation (Eq. (7.8)) is ΔE joules. Let us confine our attention to particles with spin $\frac{1}{2}$ (and hence just two energy levels) for simplicity—our remarks, however, are easily extended to cover the general case. Classical theory states that at a temperature T K the ratio of the populations of such levels will be given by

$$\frac{N_{\text{upper}}}{N_{\text{lower}}} = \exp\left(-\frac{\Delta E}{kT}\right) \tag{7.12}$$

where k is the Boltzmann constant. Thus at all temperatures above absolute zero the upper level will always be populated to some extent, although for large ΔE the population may be insignificant. In the case of nuclear and electron spins, however, ΔE is extremely small:

$$\Delta E \text{ nuclei} \quad \approx 7 \times 10^{-26} \text{ J} \quad \text{in a } 2.3487 \text{ T field}$$

$$\Delta E \text{ electrons} \approx 6 \times 10^{-24} \text{ J} \quad \text{in a } 0.33 \text{ T field}$$

and since $k = 1.38 \times 10^{-23}$ J K^{-1}, we have, at room temperature $(T = 300 \text{ K})$

$$\frac{N_{\text{upper}}}{N_{\text{lower}}} \approx \exp\left(-\frac{7 \times 10^{-26}}{4.2 \times 10^{-21}}\right) \approx \exp\left(-1 \times 10^{-5}\right)$$

$$\approx 1 - (1 \times 10^{-5}) \quad \text{for nuclei}$$

and

$$\frac{N_{\text{upper}}}{N_{\text{lower}}} \approx \exp\left(-\frac{6 \times 10^{-24}}{4.2 \times 10^{-21}}\right) \approx \exp\left(-1 \times 10^{-3}\right)$$

$$\approx 1 - (1 \times 10^{-3}) \quad \text{for electrons}$$

In both cases the ratio is very nearly equal to unity and we see that the spins are almost equally distributed between the two (or, in general, $2I + 1$) energy levels.

We need to discuss now the nature of the interaction between radiation and the particle spins which can give rise to transitions between these levels.

7.1.4 The Larmor Precession

We have seen (Eq. (7.6)) that the dipole moment of a spinning nucleus is

$$\mu = g\beta_N\sqrt{I(I + 1)} \text{ J T}^{-1}$$

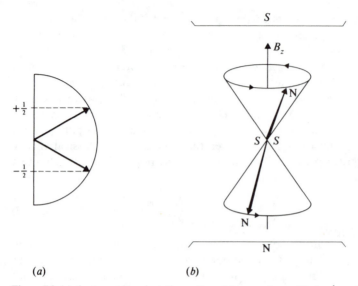

Figure 7.2 (a) the two spin orientations allowed to a nucleus with $I = \frac{1}{2}$, and (b) the Larmor precession of such a nucleus.

and that, according to quantal laws, the vector represented by **μ** can be oriented only so that its components are integral or half-integral in a reference direction. The corollary to this is that, since $\sqrt{I(I + 1)}$ cannot be integral or half-integral if I is integral or half-integral, the vector arrow can *never* be exactly in the field direction. We see one example of this in Fig. 7.1(a) and another for a nucleus with spin $\frac{1}{2}$, in Fig. 7.2(a). For such a particle:

$$\mu = g\beta_N \sqrt{3/2} \quad \text{and} \quad \mu_z = \pm\tfrac{1}{2}g\beta_N \quad \text{only}$$

Thus, whichever energy state a spinning nucleus or electron is in, it will always lie more or less across the field and will therefore be under the influence of a couple tending to turn it into the field direction.

Now the behaviour of a spinning nucleus or electron can be considered analogous to that of a gyroscope running in friction-free bearings. Experiments convince us that the application of a couple to a gyroscope does not cause its axis to tilt but merely induces a *precession* of the axis about the direction of the couple. Essentially the same occurs with a spinning particle and the precession, known as the Larmor precession, is sketched in Fig. 7.2(b). The precessional frequency (or Larmor frequency) is given by:

$$\omega = \frac{\text{magnetic moment}}{\text{angular momentum}} \times B_z \quad \text{rad s}^{-1}$$

$$= \frac{\mu B_z}{2\pi I} \quad \text{Hz}$$

Replacing **μ** and **I** by their expressions in Eqs (7.6) and (7.1)

$$\omega = \frac{g\beta_N \sqrt{I(I+1)}}{\sqrt{I(I+1)}\,\dfrac{h}{2\pi}}\,\frac{B_z}{2\pi} = \frac{g\beta_N B_z}{h} \quad \text{Hz}$$

and, comparing with Eq. (7.9), we see that the Larmor precessional frequency is just the frequency separation between energy levels.

This, then, is a mechanism by which particle spins can interact with a beam of electromagnetic radiation. If the beam has the same frequency as that of the precessing particle, it can interact coherently with the particle and energy can be exchanged; if of any other frequency, there will be no interaction. The phenomenon, then, is one of *resonance*. For nuclei it is referred to as *nuclear magnetic resonance* (n.m.r.), while for electrons it is called *electron spin resonance* (e.s.r.) or, sometimes, *electron paramagnetic resonance* (e.p.r.).

Experimentally there is a choice of two arrangements. We might either apply a fixed magnetic field to a set of identical nuclei so that their Larmor frequencies are all, say, 100 MHz; if the frequency of the radiation beam is then swept over a range including 100 MHz, resonance absorption will occur at precisely that frequency. On the other hand, we could bathe the nuclei in radiation at a fixed frequency of 100 MHz and sweep the applied field over a range until absorption occurs.

The probability of transitions occurring from one spin state to another is directly proportional to the population of the state from which the transition takes place. We have seen in the previous section that these populations are very nearly equal, and so during resonance, upward and downward transitions are induced to almost the same extent. However, while the lower state is more populated than the upper (e.g., at equilibrium in the absence of radiation), upward transitions predominate slightly, and a net (but very small) absorption of energy occurs from the radiation beam. When the populations become equal, upward and downward transitions are equally likely, no further absorption can take place, and the system is said to be *saturated*. The equilibrium populations can be re-established if the system loses its absorbed energy; this it cannot do spontaneously, but only as a result of interaction with radiation or with fluctuations in surrounding magnetic fields of the appropriate frequency. The emitted energy can be collected and displayed as an emission spectrum which, since the emission is induced and not spontaneous, is called the *nuclear induction spectrum*.

7.1.5 Relaxation Times

Let us return to the question posed at the beginning of Sec. 7.1.3, and in particular consider the word 'immediately'. If an external field were suddenly applied to a set of nuclei in bulk material, and if such nuclei can be considered as completely frictionless gyroscopes, then they could not change their orientation in order to produce the correct statistical population of the energy levels unless radiation of the appropriate frequency were present. Without such radiation there would be no mechanism by which the excess energy of the nuclei could be removed from the system (and one would speak of it as having a high 'spin temperature'). However, it is a fact that such nuclei do orient themselves to give the appropriate populations of states for a given temperature without the presence of radiation. The mechanism by which excess spin energy is shared either with the surroundings or with other nuclei is referred to generally as a *relaxation process*; the time taken for a fraction $1/e = 0.37$ of the excess energy to be dissipated is called the *relaxation time*.

Two different relaxation processes can occur for nuclei. In the first, the excess spin energy equilibrates with the surroundings (the *lattice*) by spin-lattice relaxation having a *spin-lattice relaxation time* (or *longitudinal relaxation time*) T_1. Such relaxation comes about by lattice motions (e.g., atomic vibrations in a solid lattice, or molecular tumbling in liquids and gases) having approximately the right frequency to interact coherently with nuclear spins. T_1 varies greatly, being some 10^{-2}–10^4 s for solids, and 10^{-4}–10 s for liquids, the overall shorter times for liquids being due to the greater freedom of molecular movement leading to larger fluctuations of magnetic field in the vicinity of the nuclei. We discuss the measurement of T_1 in Sec. 7.1.8.

Secondly, there is a sharing of excess spin energy directly between nuclei via *spin-spin* (or *transverse*) *relaxation*, the symbol for the time of which is T_2. For solids T_2 is usually very short, of the order 10^{-4} s, while for liquids $T_2 \approx T_1$. The measurement of T_2 is described in Sec. 7.1.7.

T_1 and T_2 have a marked effect on the widths of n.m.r. spectral lines since they reflect the lifetime of a particular spin state. Thus a long relaxation time (both T_1 and T_2 large) means that an excited nuclear spin reverts rather slowly to a lower state (i.e., has a long lifetime), and we know from Heisenberg's uncertainty principle (Eq. (1.10)) that this results in only a small uncertainty in the excited state energy level. Thus taking a typical value of one second for the relaxation time of a nucleus in a liquid, we would calculate:

$$\delta E \approx h/2\pi\delta t \approx 10^{-34} \text{ J}$$

(which uncertainty is to be compared with the spacing between adjacent energy levels for nuclei in a 2.3487 T field, shown in Sec. 7.1.3 to be about 10^{-26} J). The corresponding uncertainty in the radiation frequency associated with the transition would be:

$$\delta v = \delta E/h \approx 0.1 \text{ Hz}$$

In general n.m.r. spectrometers are not capable of resolving lines closer than about 0·5 Hz apart, so that a line width of only 0·1 Hz represents a narrow spectral line.

When either T_1 or T_2 is short the uncertainty is correspondingly larger; for example taking $\delta t = 10^{-4}$ s, as for T_2 in a typical solid, the calculation yields $\delta v \approx 1000$ Hz, which is clearly a very broad line compared with the resolving power. Thus n.m.r. experiments are divided into two main classes: broad-line, usually comprising solid samples, and high-resolution, usually of liquids or gases. But it should be noted that while solids rarely, if ever, give high-resolution spectra, some liquids may give spectra containing broad lines if both T_1 and T_2 are small; such liquids are generally very viscous or contain paramagnetic ions which increase the efficiency of relaxation processes.

Line-width measurement is not a wholly satisfactory method of obtaining relaxation times, for several reasons: line widths themselves cannot always be accurately measured, particularly in a spectrum with overlapping lines; other factors, like magnetic field inhomogeneity or chemical exchange processes, also affect the line width; and a line width can only give an estimate of the efficiency of the whole relaxation process—it does not distinguish between spin-spin and spin-lattice effects. These drawbacks can largely be overcome by using Fourier transform techniques which, in n.m.r., are invariably carried out by applying one or more pulses of exciting radiation to the sample and observing the signal emitted by excited nuclei. The following sections will be devoted to discussing how these measurements are made; we shall discuss some of the applications of relaxation times in Sec. 7.3.2.

7.1.6 Fourier Transform Spectroscopy in N.M.R.

Consider again Fig. 7.2(b), which shows the precession of a nuclear spin in an applied field B_z, where the spin may be either with or against the field. In what follows we shall implicitly consider hydrogen nuclei in a field of 2·5 T, precessing at a frequency of 100 MHz,† that is, 10^{-8} times per second. Remember that there is really a very large number of nuclear spins in the sample and, on average, they will be spread evenly around the precessional 'cone' as in Fig. 7.3(a). Because of this, the total resultant magnetic effect of these spins will be aligned along the cone axis and, since there are slightly more 'up' spins than 'down', we represent the net *bulk magnetic moment vector* of the spins by an upward arrow, as in Fig. 7.3(b). It is sometimes helpful to think of the behaviour of the collection of nuclear spins in terms of this bulk magnetic moment, which we call **M**.

---

† To be exact, 100 MHz precession needs a field of 2·3487 T for hydrogen, as we saw earlier. Since we shall need only approximate calculations in this section, however, we prefer to use the simpler figure of 2·5 T.

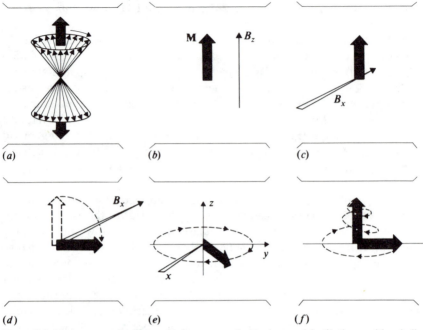

Figure 7.3 (*a*) the precessional 'cone' of a group of spinning nuclei; (*b*) the resulting bulk magnetic moment vector, **M**; (*c*) the application of a 90° pulse along the *x* axis; (*d*) the precession of **M** through 90° about the *x* axis; (*e*) the additional precession of **M** about the main field axis, and (*f*) the overall spiral motion of **M**.

We are now ready to consider the Fourier transform experiment in n.m.r. Imagine that we apply a magnetic field of strength B_x along the x axis (i.e., vertical to the plane of the paper), as in Fig. 7.3(*c*); experimentally this would be achieved by passing a radiofrequency current (of 100 MHz to maximize the resonance interaction between field and nuclei) through a coil whose axis is aligned along the x direction. Such a field will try to turn the bulk magnetic moment, **M**, into the x direction. But **M**, being composed of spinning magnets, behaves like a gyroscope and cannot be turned in this way; instead it precesses about the field direction, i.e., about the x axis, as shown in Fig. 7.3(*d*). The *rate* at which it precesses will be directly proportional to the magnitude of B_x. For simplicity, let us imagine B_x to be $2 \cdot 5 \times 10^{-4}$ T, that is, just 10^{-4} of the main field B_z, so **M** will precess at 10^{-4} times the precessional frequency induced by the main field, i.e., at $10^8 \times 10^{-4} = 10^4$ Hz; thus we see that **M** will take just 10^{-4} seconds (100 microseconds) to complete a revolution. But if we apply B_x for a very short period of time, **M** will not be able to complete a whole revolution; for example, we could apply B_x for 50 microseconds, in which case **M** will complete only half a revolution, and then stop precessing—this is referred to as a 180° pulse of radiation. Similarly a 25 microsecond pulse at an intensity of $2 \cdot 5 \times 10^{-4}$ T is a 90° pulse, the effect of which is shown in Fig. 7.3(*d*). Remember that the actual length, in microseconds, of a 90° pulse will depend on the strength of the pulse field. Increasing B_x by a factor of 10 will increase the rate of **M**'s precession by 10, and so a 90° pulse at $2 \cdot 5 \times 10^{-3}$ T would be $2 \cdot 5$ microseconds long.

In an n.m.r. spectrometer the transmitter and receiver coils are carefully aligned with their axes at right angles to each other so that there is no direct coupling between them. Conventionally we take the transmitter to be aligned along the x axis and the receiver along the y (horizontally in the paper), with the main field in the z axis. Now while the bulk magnetic spin vector is along the z axis there is no fluctuation of magnetic field along y so the receiver coil sees no signal. When the 90° pulse is applied, however, **M** is 'flipped' into the xy plane, and three things immediately start to happen. Firstly **M** experiences a force from the main applied field trying to turn it back into the z direction; as a gyroscope, however, its only response to this force is precession about the force axis, so **M** will begin to precess in the xy plane, as in Fig. 7.3(*e*), at 100 MHz. Secondly, and because of this precession, there *is* now a fluctuating magnetic effect along the x axis—the precessing **M**—so the detector receives a signal, also at 100 MHz. Finally, **M** is in an unstable state—it would 'prefer' to be aligned with the main field axis— and so normal relaxation processes occur to dissipate its excess energy, and **M** gradually (i.e., in a period depending on the relaxation time) returns to the vertical position. The overall motion of **M** after the 90° pulse can be visualized as a spiral, shown in Fig. 7.3(*f*), during which the detector will receive a decreasing signal until **M** is vertical once more. (Strictly, with a precessional frequency of 10^8 Hz and a relaxation time of, perhaps, some 10^{-2} seconds, **M** will precess many thousands or even millions of times during the decay; Fig. 7.3(*f*) should be treated as purely diagrammatic.) This signal can be Fourier transformed (cf. Sec. 1.8), and displayed as the frequency spectrum of the nuclei. Note that the decreasing signal at the detector arises here because the pulse along x excites nuclei to the upper state from which they revert to the lower. Once equilibrium is re-established (vertical **M**) no more signal is emitted unless a further energizing pulse is applied. In fact, of course, all the sample nuclei do *not* emit at precisely 100 MHz—a package of frequencies is emitted (again cf. Chapter 1) which will also, by interference, contribute towards the decay of the signal.

The important point to note is that FT spectroscopy in n.m.r. requires the application of a 90° pulse, observation of the decaying emission signal (often called the *free induction decay*, or FID), and then transformation of the signal for display. Although relaxation processes decide the rate of signal decay, this single experiment is no more suitable than conventional 'frequency domain' methods for measuring relaxation times, and suffers from the deficiencies mentioned at the end of the preceding section. In order to measure relaxation times we need to consider *multiple-pulse* techniques.

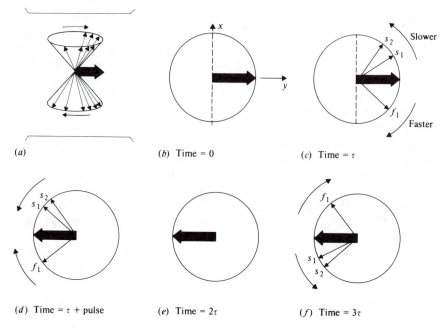

(a)

(b) Time = 0

(c) Time = τ

(d) Time = τ + pulse

(e) Time = 2τ

(f) Time = 3τ

Figure 7.4 The bunching of nuclear spins after a 90° pulse and the effect of a subsequent 180° pulse.

7.1.7 Multiple-Pulse FT: Spin–Spin Relaxation

Let us consider in rather more detail the effect of the 90° pulse described above. Essentially for **M** to be tilted into the horizontal (xy) plane, individual nuclear magnets must have been disturbed from their even distribution round the precessional cone, and have bunched to one side, as shown in Fig. 7.4(a). Their precessional motion still exists, of course, so one should think of the bunch precessing as a whole, thus causing **M** to precess in the xy plane, as indicated earlier. Now for **M** to relax back to equilibrium implies that the energy of the spins must be shared evenly again, and for this to happen we require an exchange of energy *between* the spins. In other words the relaxation process which allows horizontal **M** to become vertical **M** again is *spin-spin relaxation*. Since **M** decays *across* the main field direction, this process is also referred to as *transverse* relaxation.

In order to measure the spin-spin relaxation time (T_2) we need to disentangle this relaxation process from other processes which can cause the spins to precess at different frequencies (mainly field inhomogeneities) and so can also unscramble the bunching. To understand the method, let us imagine looking along the main field direction from the top (i.e., along the axis of the precessional cone). We show the total magnetic vector, seen in this way, in Fig. 7.4(b). Now **M** actually rotates at 100 MHz due to its precession; but to simplify the pictures which follow, imagine that we have 'killed' this rotation by, for example, observing the system with stroboscopic light, flashing at 100 MHz. Thus we shall essentially see **M** once in each rotation at the same place. Field inhomogeneities will mean that some nuclei precess slightly faster, some slightly slower, than average. If we observe the system for a time τ the slower nuclei will appear to lag behind the majority, the faster to drift ahead, as shown in Fig. 7.4(c), where f_1 is a faster nucleus, s_1 and s_2 are slower.

At this point another pulse is applied along the x axis via the transmitter coil, but this time a 180° pulse. Its effect is to rotate all spins by 180° about the x axis (*not* about the centre of the circle), and the net result will be as shown in Fig. 7.4(d): noting carefully where the slower and faster nuclei arrive, we see that, since we have done nothing to change their precessional velocities, the faster nuclei will now catch up with the majority, the slower drift back towards them. Since it took a time τ for the bunch to fan out to Fig. 7.4(c), it will take a further time τ (total time 2τ) for the fanning out to be reversed. The effect of all this on the detector is as follows: immediately after the original 90° pulse the detector receives a maximum signal, which decays due to the fanning out of nuclear spins from field inhomogeneities. The 180° pulse is applied at τ, the fanning out reverses, and the signal builds up again until at 2τ, when the spins are bunched again, the signal reaches a maximum once more. The process continues—at 3τ the fanning out will look like Fig. 7.4(f), and a further 180° pulse will then cause rebunching of the spins at 4τ. This is all collected into Fig. 7.5, where (a) shows the pulse timing, and (b) the detector output. Not surprisingly the technique is often referred to as 'spin-echo', or as the 90°-τ-180° sequence.

(a)

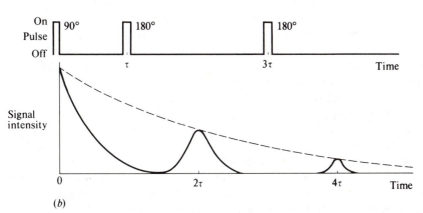

(b)

Figure 7.5 (a) the pulse timing of a 90°-τ-180° sequence and (b) the resultant spin echoes.

Now although the 180° pulses effectively cancel out those *external* effects, such as field inhomogeneity, which produce different precessional frequencies, they cannot cancel the internal decay of **M** from spin-spin interactions which also equalize the distribution of spins round the precessional cone. Thus each spin-echo maximum is smaller than the preceding one as **M** shrinks. The exponential decay of these maxima, drawn in on Fig. 7.5(b), can thus be used to measure T_2. Note that we are here dealing with the decay of the 'raw' detector signal—there is no need to Fourier transform the signal for display as a frequency spectrum, since we are only interested in the change in its overall intensity.

In fact the T_2 measurement described above is by no means simple to carry out—very precise timing of the microsecond pulses is essential otherwise the spin system will rapidly become hopelessly scrambled, and the presence of nuclei in the sample with different precessional frequencies (see the discussion of chemical shifts later) will mask the echo pattern. Thus T_2 measurements are of little practical use, as yet, in structure determination. However, the principle of their measurement is simple to understand, and should be helpful when we now consider spin-lattice relaxation, giving T_1 values which are becoming of increasing practical importance.

7.1.8 Multiple-Pulse FT: Spin-Lattice Relaxation

Instead of starting with a 90° pulse, consider applying a 180° pulse initially to the spin system. Fig. 7.6(a) and (b) show that this will *reverse* the direction of **M**, and will leave the spin system in an unstable state since more spins point against the field than with it. As excited nuclei give up their energy (i.e., relax) **M** will become smaller, pass through zero, then grow upwards until it reaches its equilibrium value once more—an exponential change which is illustrated in Fig. 7.6(c). Remembering that the relaxation time is the time required for $1/e = 0.37$ of the original excitation energy to be dissipated, we can indicate this time on the curve.

Two points are important here. Firstly the excess spin energy is given up by the nuclei to their surroundings, and so the time involved is the spin-*lattice* relaxation time, and we can see that the alternative name for this, the *longitudinal* relaxation time, is because **M** relaxes *along* the main field axis. Secondly all the changes in **M** sketched in Fig. 7.6(c) are confined to the z axis, and so will produce no signal in the receiver coil along y.

We know that we can induce signal emission, however, by applying a subsequent 90° pulse to the system after some delay time τ, as shown in Fig. 7.6(d), thus reaching the situation discussed earlier in Fig. 7.3(c) to (f). **M** will spiral up to equilibrium, emitting a decaying signal, but the *initial* intensity of the signal, immediately after the 90° pulse, will be directly proportional to the magnitude of **M** at the time of that pulse. If we lengthen τ, **M** will have grown more before being tipped into the xy plane, and a

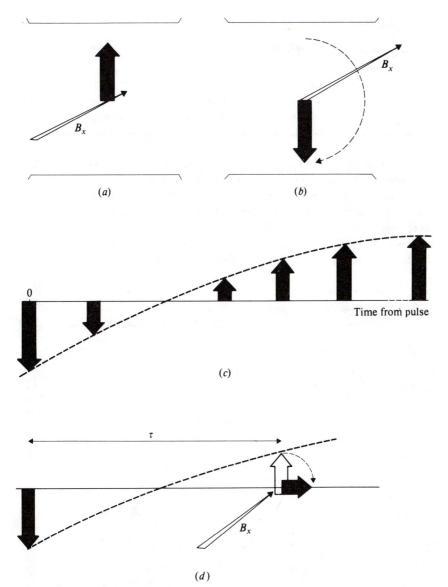

Figure 7.6 A 180°-τ-90° pulse sequence and its effect on **M**.

more intense signal will result. Shorten τ a little, and the signal will be weaker because **M** has not grown so much. Indeed, make τ much shorter, so that **M** is still pointing in the 'wrong' direction when it is tipped into the *xy* plane, and the signal will be *negative*, i.e., it will be an out-of-phase signal which, on Fourier transformation, will give a negative spectrum peak rather than a positive one. In order to measure spin-lattice relaxation, then, we must carry out several 180°-τ-90° experiments with different values of the delay time τ between the pulses, and then plot the intensity of the signal, observed immediately after the 90° pulse, against τ. Typically τ will be of the order of seconds.

Of course the relaxation of **M** into its final equilibrium position and magnitude continues after we have observed the initial signal. We are not interested in this, however—indeed we saw in the previous section that subsequent relaxation is also affected by field inhomogeneities and spin-spin relaxation. We can best think of the 90° pulse as 'freezing' the spin-lattice relaxation of **M** at various times after reversal so that we can measure how far relaxation has progressed at each time.

Naturally the computer attached to the spectrometer is used to control the whole process of measurement. It times the original 180° pulse, selects increasing τ values, times the 90° pulse, and collects and measures the signal intensity. From these data it can readily calculate and report a value for T_1 directly, and there is again strictly no reason to Fourier transform the signal and produce frequency spectra. It has become conventional, however, to continue with this final step, and to display the resulting spectra in a particularly graphic form giving a pseudo three-dimensional picture showing the frequency of lines in the spectrum, and the relationship of their

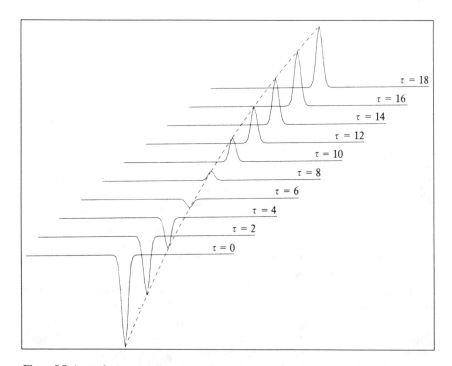

Figure 7.7 A set of spectra produced by 180°-τ-90° sequences with different delay times τ.

intensities to the delay time τ. Typically the computer plots a set of 'stacked' spectra in the form of Fig. 7.7, where each spectrum is offset vertically and to the right by an amount proportional to its delay time. Such a picture shows very pleasingly the initial negative signal from the peak, its diminishment through zero, and subsequent growth to a maximum. On the figure we have joined the peak maxima with a dashed line to show the exponential change in height. Comparison of this figure, viewed sideways, with Fig. 7.6(c), shows the very precise correspondence between the magnitude and direction of **M** and its n.m.r. signal. An approximate value for the spin-lattice relaxation time, T_1 can be found very simply from such a display—in fact T_1 is about $1\frac{1}{2}$ times the time taken for the peak to decay to zero intensity. In this particular (hypothetical) example, the signal becomes zero at about 7 seconds, so the relaxation time is $1\cdot5 \times 7$ or about 10 seconds.

In summary, then, T_1 measurements are relatively easy and usually fairly fast. Some half a dozen 180°-τ-90° experiments, each lasting a few seconds or a minute, can establish T_1 with good accuracy. Small errors in pulse timings are not significant, nor are field irregularities. And complicated spectra, containing several resonances, are no more difficult to tackle than simple ones. Small wonder, then, that T_1 measurements are increasingly being used to assist in spectral assignments and structure determination by n.m.r. We shall return briefly to this topic in Sec. 7.3.2.

7.2 NUCLEAR MAGNETIC RESONANCE SPECTROSCOPY: HYDROGEN NUCLEI

We have seen that a particular chemical nucleus placed in a magnetic field gives rise to a resonance absorption of energy from a beam of radiation, the resonance frequency being characteristic of the nucleus and the strength of the applied field (some data have already been referred to in Table 7.1). Thus n.m.r. techniques may be used to detect the presence of particular nuclei in a compound and, since for a given nuclear species, the strength of the n.m.r. signal is directly proportional to the number of resonating nuclei, to estimate them quantitatively. However an n.m.r. spectrometer is an expensive instrument, and there are many simpler and cheaper methods available to detect the presence or absence of a particular atom in a molecule. Two other characteristics of n.m.r. spectra which have not so far been mentioned make the technique far more powerful and useful; these are the *chemical shift* and the *coupling constant*, which we shall discuss in the following sections.

The vast majority of substances of interest to chemists contain hydrogen atoms and, as this nucleus has one of the strongest resonances, it is not surprising that n.m.r. has found its widest application to these substances.

When discussing chemical shifts and nuclear coupling it is convenient to use one type of nucleus as an example, although all spinning nuclei show these phenomena, and in what follows we shall consider the spectra of hydrogen-containing substances only, leaving discussion of other nuclei to Sec. 7.3.

7.2.1 The Chemical Shift

Up to now we have considered the behaviour of an isolated nucleus in an applied field. Such a situation is not, of course, realizable in practice since all nuclei are associated with electrons in atoms and molecules. When placed in a magnetic field the surrounding electron cloud tends to circulate in such a direction as to produce a field *opposing* that applied (so-called diamagnetic circulation), as shown for a single atom in Fig. 7.8. Plainly the total field experienced by the nucleus is:

$$B_{\text{effective}} = B_{\text{applied}} - B_{\text{induced}}$$

and, since the induced field is directly proportional to the applied field,

$$B_{\text{induced}} = \sigma B_{\text{applied}}$$

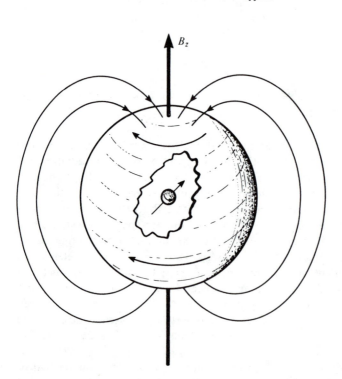

Figure 7.8 Showing the field, produced by diamagnetic circulation of the electron cloud about a nucleus, which opposes the applied field B_z.

where σ is a constant; we have:

$$B_{\text{effective}} = B_z(1 - \sigma) \qquad (7.13)$$

Thus the nucleus can be said to be *shielded* from the applied field by diamagnetic electronic circulation. The extent of the shielding will be constant for a given atom in isolation, but will vary with the electron density about an atom in a molecule; thus we may generalize Eq. (7.13) by writing:

$$B_i = B_z(1 - \sigma_i) \qquad (7.14)$$

where B_i is the field experienced by a particular nucleus i whose shielding constant is σ_i. As an example, since we know that oxygen is a much better electron acceptor than carbon (since oxygen has the greater electronegativity) then the electron density about the hydrogen atom in C—H bonds should be considerably higher than in O—H bonds. We would thus expect $\sigma_{\text{CH}} > \sigma_{\text{OH}}$ and hence

$$B_{\text{CH}} = B_z(1 - \sigma_{\text{CH}}) < B_{\text{OH}} = B_z(1 - \sigma_{\text{OH}})$$

Thus the field experienced by the hydrogen nucleus in O—H bonds is greater than that at the same nucleus in C—H bonds and, for a given applied field, the CH hydrogen nucleus will precess with a smaller Larmor frequency than that of OH. Put conversely, in order to come to resonance with radiation of a particular frequency (for example, 100 MHz), a CH hydrogen requires a greater applied field than OH.

The effect of a steadily increasing field on the energy levels of the CH_3 and OH hydrogens in CH_3OH is shown in Fig. 7.9. The OH hydrogen nucleus, having a smaller shielding constant, experiences a greater field, hence its energy levels are more widely spaced than those of the more shielded CH_3 nuclei at any given applied field. If the system is irradiated with a beam of radiation at, say, 100 MHz while the applied field is increased from zero, the OH nucleus will come into resonance first and absorb energy from the beam, the CH_3 nuclei absorbing at a higher field. This is shown in the spectrum of methyl alcohol, CH_3OH, at the foot of the figure. The fact that the ratio of the absorption intensities (strictly the ratio of the areas under the peaks) is 1 : 3 immediately allows us to identify the smaller peak with the single hydrogen nucleus in the OH group, the larger with the CH_3 group. Since neither carbon nor oxygen have nuclei with spin, they do not contribute to the spectrum.

Two very important facets of n.m.r. spectroscopy appear in Fig. 7.9: (1) identical nuclei (that is, H nuclei) give rise to different absorption *positions* when in different *chemical* surroundings (for this reason the separation between absorption peaks is usually referred to as their *chemical shift*) and (2) the *area* of an absorption peak is proportional to the number of *equivalent* nuclei (i.e., nuclei with the same *chemical shift position*) giving rise to the

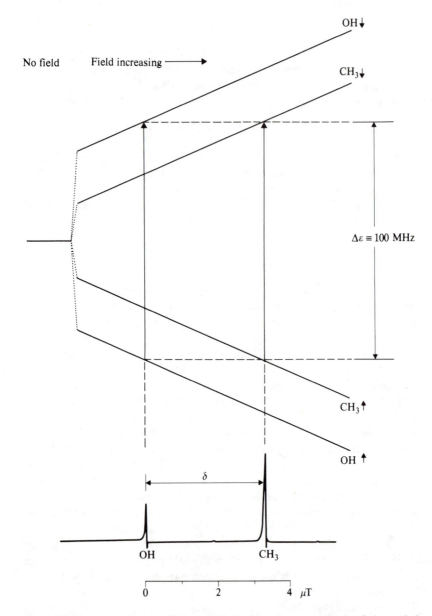

Figure 7.9 The effect of an applied magnetic field on the energy levels of the methyl and hydroxyl hydrogen nuclei of methyl alcohol, CH_3OH. The applied field is increased rapidly initially (dotted portion) until near resonance at $2·3487T$, then the increase is much slower. The n.m.r. spectrum of methyl alcohol is shown at the foot of the figure.

absorption. We see here the basis of a qualitative and quantitative analytical technique.

It should be noted that electron density is not the only factor determining the value of the shielding constant. Another, frequently very important, contribution to shielding arises from the field-induced circulation of electrons in neighbouring parts of a molecule which gives rise to a small magnetic field acting in opposition to the applied field. Two of the many possibilities are shown in Fig. 7.10. In (a) the circulation of the cylindrical charge cloud comprising an acetlylenic triple bond is shown to *reduce* the effect of the applied field at a hydrogen nucleus on the axis of the circulating charge. The nucleus is thus shielded and will resonate to high applied field. In (b), on the other hand, we show the circulation of the annular cloud of π-orbital electrons around a benzene ring (only the top annulus is shown, to simplify the diagram, but the identical annulus underneath the ring circulates in the same direction). The induced field is here in the *same direction* as the applied field in the vicinity of the hydrogen nuclei, and these nuclei are consequently deshielded and resonate to low applied field. Obviously both the effects noted here will be somewhat reduced by molecular tumbling, but they do not average to zero.

Thus both shielding and deshielding may arise from induced electron circulation, and the total of such effects at any particular site, together with an electron density contribution, is included in the shielding constant σ.

There are several ways in which the chemical shift difference, δ, between the OH and CH_3 signals of Fig. 7.9 may be expressed. Firstly, since we have imagined the spectrum to be recorded by varying the applied field, we could attach a tesla scale (or, rather, a microtesla (μT) scale, since the separation is very small) to the spectrum, and observe that $\delta = 3\cdot26\ \mu$T. Or, remembering that the spectrum could equally well have been obtained by varying the applied frequency at constant field, we could quote the chemical shift

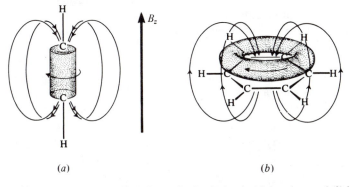

(a)   (b)

Figure 7.10 The field-induced electronic circulation in (a) acetylene and (b) benzene, showing shielding and deshielding respectively at nearby hydrogen nuclei.

difference in hertz as follows:

$$2 \cdot 3487 \text{ T} \equiv 100 \text{ MHz}$$

hence

$$3 \cdot 26 \ \mu\text{T} \equiv 139 \text{ Hz} = \delta$$

In practice, however, neither of the above chemical shift units is entirely satisfactory, although both appear in the older literature. The difficulty can be seen as follows: rewriting Eq. (7.9) to take account of shielding at nucleus i and combining with Eq. (7.14):

$$\frac{\Delta E_i}{h} = \frac{g\beta_N B_i}{h} = \frac{g\beta_N B_z(1 - \sigma_i)}{h} \quad \text{Hz}$$

and so, in hertz:

$$\delta = \frac{g\beta_N B_z}{h} (\sigma_{\text{CH}} - \sigma_{\text{OH}}) \quad \text{Hz} \qquad (7.15)$$

Now the shielding constant, σ_i, is independent of the applied field or frequency, so plainly chemical shift separation measured in hertz or microtesla is directly proportional to the applied field, B_z: it is, at the least, inconvenient to measure a quantity which changes with the operating conditions of the instrument, particularly since different instruments use fields varying between 0·6 and 10 tesla.

The difficulty can be overcome if we quote chemical shifts as a *fraction* of the applied field or frequency. Thus for methyl alcohol, $\delta = 139$ Hz at 100·0 MHz is equivalent to $\delta = 1 \cdot 39 \times 10^{-6}$ or 1·39 p.p.m. (parts per million) of the operating frequency; similarly, $\delta = 3 \cdot 26 \ \mu\text{T}$ at 2·3487 T is also 1·39 p.p.m. of the applied field. If we were now to double the field and frequency, the separation expressed in microtesla or hertz would also double according to Eq. (7.15), but it would still be just 1·39 p.p.m.

Chemical shift measurements are, of course, formally based on the resonance position of the bare hydrogen nucleus (the proton) as the primary standard; for this there is no shielding and hence $\sigma = 0$. Since this is a quite impracticable standard, it is necessary to choose some reference substance as a secondary standard and to measure the resonance positions of other hydrogen nuclei from this in parts per million; the substance now almost universally selected for hydrogen resonances is tetramethyl silane, $Si(CH_3)_4$, or TMS (except for aqueous solutions where, since TMS is immiscible with water, the salt $(CH_3)_3SiCH_2CH_2CH_2SO_3^-Na^+$, referred to as DSS, is preferred, since the CH_3 groups resonate in the same position as those of TMS).

Tetramethyl silane has several advantages over other substances which have been used as standards:

1. Its resonance is sharp and intense since all 12 hydrogen nuclei are equivalent (i.e., have the same chemical environment) and hence absorb at exactly the same position.
2. Its resonance position is to high field of almost all other hydrogen resonances in organic molecules (that is, σ_{TMS} is large) and hence can be easily recognized.
3. It is a low boiling point liquid (b.p. 27°C) so can be readily removed from most samples after use.

Thus if about five per cent of TMS is added to a sample and the complete n.m.r. spectrum produced, the sharp, high-field resonance of the TMS is easily recognized and can be used as a standard from which to calibrate the spectrum and to measure the chemical shift positions of other molecular groupings. Conventionally n.m.r. spectra are displayed with the field increasing from the left, which places the TMS resonance to the extreme right. Two measurement scales have been used for chemical shifts, both in parts per million. One scale, the τ scale, arbitrarily sets the resonance of TMS at a scale value of 10 p.p.m. and numbers p.p.m. downwards to the left, as shown in Fig. 7.11(b). Although nearly all proton spectra fall between 0 and 10 on this scale, it suffers from the slight disadvantage that

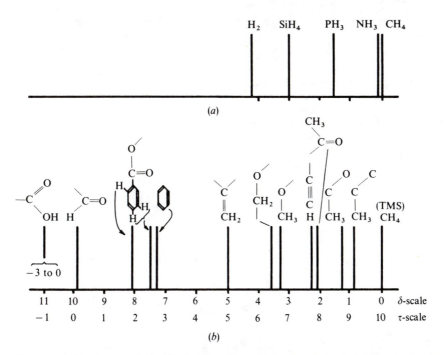

Figure 7.11 Showing the relationship between the τ and δ scales of chemical shift values, and the approximate proton chemical shifts of some simple molecules and groups.

negative numbers are required to represent some resonances. The τ scale was much favoured, however, and many proton spectra found in the literature use it. Its main disadvantage became apparent as n.m.r. spectroscopy of other nuclei, particularly ^{13}C, became more widely available. It happens that TMS is also a good reference material for ^{13}C spectra and that its ^{13}C resonance also occurs to high-field (i.e., on the right) of other ^{13}C resonances. These resonances, however, span some 220 p.p.m. and so, to be consistent, TMS should be given the value 220 in ^{13}C spectra while remaining at 10 in ^1H spectra. This illogicality has been avoided by setting the TMS resonance arbitrarily to zero in both types of spectrum; this scale, the δ scale, is also shown in Fig. 7.11(b), and we see that δ values decrease with increasing field. The δ scale is now almost universally used, but conversion between the two scales, when necessary, is very simple: clearly $\delta = 10 - \tau$.

While showing both scales, Fig. 7.11 also indicates the approximate resonance positions of some molecules and groups. In studying this figure we must not lose sight of the fact that the resonances indicated are due to only the *hydrogen* nucleus in the molecule or group concerned.

As one would expect, we note from Fig. 7.11(a) that the position of the hydrogen resonance depends upon the atom to which it is directly attached—cf. the series CH_4, NH_3, PH_3, SiH_4, and H_2 between $\delta = 0$ and $\delta = 4.2$. Far more important, however, is the fact that, when the hydrogen is attached directly to a particular atom, e.g., carbon, its resonance position depends markedly on the nature of the *other* substituents to that atom—cf.

Table 7.2 Some chemical shift data

Resonance positions are given in δ and are, with the exception of those indicated, accurate to about $\pm\ 0.5\ \delta$; R represents a saturated alkyl group, Ph the phenyl group, C_6H_5.

Group	δ	Group	δ
R.COOH	11.5†	CH_3.O.COR	3.7
R.CHO	10.0†	—CH_2.OR	3.5
Ph.H	7.3	CH_3.OR	3.3
		—CH_2.COR	2.4
		≡CH	2.2
$=C\stackrel{H}{\underset{R}{\diagdown}}$	5.3	CH_3.COOR	2.1
		CH_3.COR	2.1
$=C\stackrel{H}{\underset{H}{\diagdown}}$	5.0	$=C\stackrel{CH_3}{\underset{R}{\diagdown}}$	1.8
—CH_2.O.COR	4.2	C—CH_2—C	1.5
CH_3.O.COPh	4.0	C—CH_3	0.9

† Range $\pm\ 3.0\ \delta$.

the series CH_4, CH_3O—, $C=CH_2$, $O=CH$, etc., in Fig. 7.11(b); a small selection of the available data is also shown in Table 7.2. When we combine this with the statement, made earlier, that the area of each resonance is proportional to the number of hydrogen nuclei contributing to that resonance, we see that n.m.r. techniques provide ready means of both qualitative and quantitative group analysis in organic chemistry. After we have discussed the phenomenon of energy coupling between different nuclei in the following sections, we shall see that the usefulness of n.m.r. extends even further to the determination of the structure and configuration of molecules.

7.2.2 The Coupling Constant

Suppose that two hydrogen nuclei in different parts of a solid, e.g., a crystal lattice, are sufficiently close together in space that they exert an appreciable magnetic effect on each other—in n.m.r. terms 'appreciable' means 0.01 μT or more. We show such a case in Fig. 7.12 for two nuclei labelled A and X. Here we ignore the spin direction of X since this is immaterial for the moment, but we show the two possible directions of the z component of A's spin, either in the field direction (conventionally 'up') or opposed to that direction ('down'). The lines of force originating from A, considered as a simple bar magnet, are seen to *oppose* the applied field at X when A is up, and to reinforce it when A is down. Remembering (Sec. 7.1.3) that the two A directions are virtually equally likely, we see that nucleus X will find itself in an applied field $B_z + B_A$ or in $B_z - B_A$ with equal probability, where B_A is the field at X due to nucleus A. Clearly this will result in the X nuclei of a sample showing two different resonance positions in the spectrum—the X

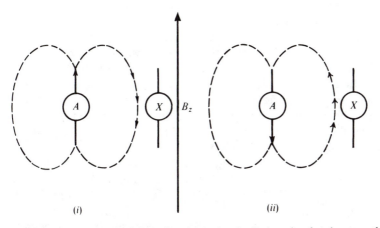

Figure 7.12 The direct coupling of nuclear spins. In (i) the spin of A decreases the net magnetic field at X, while in (ii) the field at X is increased.

signal will be a *doublet*—separated, on a tesla scale, by $2B_A$. If we now include X's spin in the argument, it is obvious that the effect of X on A is precisely the same as that of A on X, so the A signal will be split into an identical doublet.

The extent of this direct spin–spin interaction, or *coupling*, is large for hydrogen nuclei, being of the order 10^{-4} T when the nuclei are 0·1 nm apart, and is of practical use in solid-state n.m.r. spectra in measuring with some precision the interatomic distances in compounds.

When liquid or gaseous samples are considered, however, this direct spin-coupling mechanism is found not to apply because random molecular motions within the sample (molecular tumbling) continually change the orientation of the molecules within the applied field. For example, if we imagine the AX 'molecule' of Fig. 7.12(i) to be rotated through 90° in the plane of the paper (remember that the nuclear spins will *not* rotate with the molecule, since they are governed only by the field direction), nucleus X will sit vertically above or below A and will thus be in a position where the field due to A *reinforces* the applied field B_z—this is just the opposite situation from that previously considered. It can be shown by proper integration of the effect of A on X at all possible orientations that continued rotation of the molecule averages the coupling exactly to zero—all situations in which A reinforces the field at X are just balanced by all those where it diminishes that field.

Nonetheless the n.m.r. spectra of liquids do show the phenomenon of coupling but the effect is very much smaller (by a factor of 10^2–10^4 than that observed for direct coupling in solids); clearly a different mechanism is involved. Consider first Fig. 7.13(i) which shows two hydrogen nuclei joined by a pair of bonding electrons, as in the hydrogen molecule. To a first approximation we can assume that each electron 'belongs' to a particular nucleus, so we associate electron (a) with nucleus A and electron (x) with nucleus X. Plainly the most stable state energetically is that in which the electronic magnetic dipole is opposed to that of its own nucleus; but theories of chemical bonding tell us that the electrons, which occupy the same orbital, will have their spins opposed also, and so we see that the most stable state will be that in which the nucleus-electron-electron-nucleus spins alternate as shown in the figure; consequently the spins of A and X will preferentially be paired. This is similar to the situation found for the 'across-space' effect, but is several orders of magnitude smaller; however, it represents coupling between the nuclei.

Note carefully that the above argument is not intended to be rigorous; we are not saying that the spins of A and X are immutably fixed in the paired state, simply that the paired configurations ↑↓ and ↓↑ are very slightly lower in energy than the parallel configuration ↑↑ and ↓↓; the energy difference is so minute (some 10^{-32} J) that both configurations are equally likely to occur. The effect which this has on the spectrum is shown by Fig. 7.14. At (i) of the figure we show the four energy levels of the AX system,

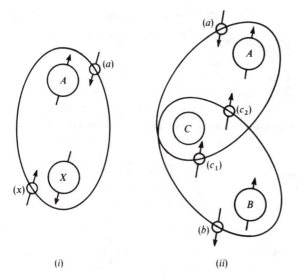

(i)   (ii)

Figure 7.13 The coupling of nuclear spins via bonding electrons for (i) directly bonded atoms and (ii) atoms bound to a third atom having no spin.

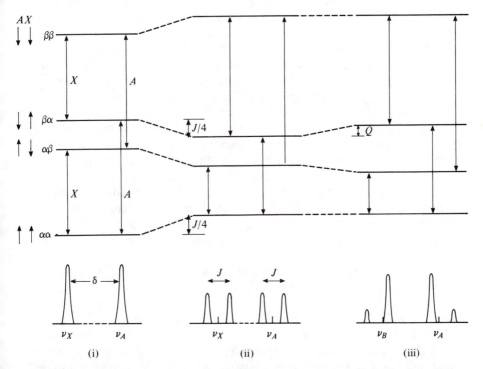

Figure 7.14 To illustrate the effect of coupling between two nuclei. At (i) coupling is ignored, at (ii) the coupling is taken to be small compared with the chemical shift, and at (iii) the coupling is large compared with the chemical shift.

each corresponding to one of the four possible spin combinations, ↑↓, ↑↑, ↓↓ and ↓↑; a convenient notation for these spin states is also indicated on the diagram, whereby each 'up' spin is designated α and each 'down' spin β. In this part of the diagram we imagine there to be no coupling between A and X; hence the separation between levels $\alpha\alpha$ and $\alpha\beta$, corresponding to a spin change of X only, is just the resonance energy of X, and this is clearly identical with the separation between levels $\beta\alpha$ and $\beta\beta$. Similarly the separations $\alpha\alpha$–$\beta\alpha$ and $\alpha\beta$–$\beta\beta$ are identical and give the resonance energy of A. Transitions between the various levels give rise to two spectral lines, one at each of the chemical shift positions of A and of X, designated v_A and v_X.

When coupling occurs we know that the levels $\alpha\alpha$ and $\beta\beta$ are destabilized, and that $\alpha\beta$ and $\beta\alpha$ are stabilized; this is shown in (ii) of the figure, where the amount of the energy change is symbolized $J/4$—the convenience of this will become clear shortly. Now the two possible X transitions have different energies: there is $\alpha\alpha \leftrightarrow \alpha\beta$ which is smaller than in (i) by $\frac{1}{2}J$, and $\beta\alpha \leftrightarrow \beta\beta$, higher by $\frac{1}{2}J$; hence one spectral line appears $\frac{1}{2}J$ above v_X and another $\frac{1}{2}J$ below, the separation between these two lines being just J (hence the convenience of defining the stabilization and destabilization as $J/4$ above). Similarly the A transitions will be split into an identical doublet. The 'legs' of the doublet will be the same intensity since (1) all spin states are virtually equally likely, and (2) detailed calculation shows that the probabilities of transitions occurring between any levels are identical. Thus we arrive at an AX spectrum: a pair of 1 : 1 doublets, each pair separated by the coupling constant J, and the midpoints of the doublets separated by the chemical shift δ. Since J is observed to be a field-independent quantity, it is conveniently expressed in hertz rather than in parts per million like δ.

At this point we should emphasize that Fig. 7.14 is by no means drawn to scale. If A and X are both hydrogen nuclei, then in part (i) the transitions labelled X and A represent the resonance energies of such nuclei in, say, a 2·5 T field, i.e., some 100 MHz. The separation between $\alpha\beta$ and $\beta\alpha$, on the other hand, represents the chemical shift difference, δ, between v_A and v_X, perhaps a few hundred hertz. Thus if the diagram were to scale, the distance between $\alpha\alpha$ and $\alpha\beta$ should be some million times greater than that between $\alpha\beta$ and $\beta\alpha$. Turning to (ii) of the diagram, the quantity $J/4$ is a few hertz only, so the displacements of the various energies have been much exaggerated.

Up to now we have taken the chemical shift difference of A and X (some 100 Hz) to be large compared with the coupling constant J (a few hertz). In (iii) of Fig. 7.14 we show what happens when the chemical shift difference decreases, i.e., when the levels $\alpha\beta$ and $\beta\alpha$ are assumed to be closer together than shown. It is a well-known fact in quantum mechanics that energy levels of the same symmetry, as are $\alpha\beta$ and $\beta\alpha$, tend to repel each other if they become close, while those of different symmetry, like $\alpha\alpha$ and $\beta\beta$, are not affected. If both $\alpha\beta$ and $\beta\alpha$ are repelled by an amount Q, as shown on the diagram, then clearly both X transitions are decreased in energy by Q, while both A transitions are increased by the same amount. The result is to push the spectral lines apart so that the chemical shift positions, v_A and v_X, are no longer at the midpoints of their respective doublets. Actually in the figure we have relabelled the nuclei A and B instead of A and X: this is conventional notation in n.m.r. spectroscopy, where letters close together in the alphabet are used to indicate nuclei whose chemical shift difference is small, and vice versa.

One further effect of a small chemical shift difference is illustrated in Fig. 7.14(iii); this is that, although the populations of the levels are unaffected by the perturbation Q, the relative transition probabilities are very much affected, and detailed calculations show that the result is for the centre lines of the AB spectrum to gain intensity at the expense of the outer. In the limit when the chemical shift difference becomes zero (A_2 system instead of AB, since A and B become identical), the centre lines coalesce into one and the outer lines have vanishing intensity. Thus although coupling between the A_2 nuclei certainly exists, its effect is not observable in a spectrum. A typical AB spectrum is illustrated in Fig. 7.15 where it is evident that δ (measured in hertz) is about five times J.

Let us now return to Fig. 7.13(ii). This shows the coupling of hydrogen nuclei which are each attached to a third non-spinning nucleus, such as carbon; an example of this situation is the methylene fragment

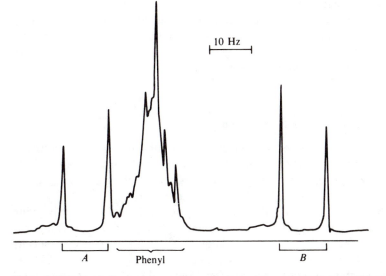

Figure 7.15 Part of the n.m.r. spectrum of ethyl cinnamate, $C_6H_5CH{=}CH\cdot COOC_2H_5$, showing the typical AB pattern of the olefinic hydrogens. The broad resonance in the centre is due to the phenyl group; the ethyl resonance is off the scale to the right.

$>CH_2$. Now the chain of reasoning runs as follows: the spins of A and (a) are paired, as are those of (a) and (c_1) since both the latter occupy the same orbital; (c_1) and (c_2) have parallel spins, however, since they occupy degenerate orbitals in the same atom (cf. Hund's principle, Sec. 5.3.1); finally (b) and (c_2) are paired and nucleus B has its spin paired with electron (b). We see that the lowest energy state, according to this electron path, is that wherein the spins of A and B are *parallel*, not paired. In this situation the coupling constant, J_{AB}, is defined to be negative, whereas it is defined positive for the previous case shown in Fig. 7.13(i).

Similar arguments to the above indicate that J is again positive for coupling via three bonds (for example, H—C—C—H), and negative via four, etc., provided this type of electron path is the predominant contributor to the coupling; in some systems, particularly unsaturated ones (i.e., containing multiple bonds), other electron paths may become important. However, the magnitude of the coupling constant attenuates rapidly with increasing number of bonds; thus for H_2 the coupling (measured indirectly from the spectrum of HD) is some 240 Hz, for the H—C—H fragment (e.g., in methane) it is some 12 Hz, for H—C—C—H (e.g., in ethanol, CH_3CH_2OH) it is 7 Hz, and for H—C—C—C—H it is on the present limit of measurement, 0·5 Hz or less. However in unsaturated molecules, in which electrons occupy orbitals extending over more than two nuclei and are thus more mobile, the couplings are somewhat larger and have been observed over more than four bonds.

It is not usually easy to determine the sign of a coupling constant experimentally, but where such data exist they are often in good agreement with the alternation theory outlined above. Further, calculations can be made predicting the magnitudes of certain couplings by applying quantum mechanical methods to the very diagrammatic approach of Fig. 7.13; these, too, are in generally good agreement with experimental values.

7.2.3 Coupling between Several Nuclei

So far we have considered the effect of coupling between two hydrogen nuclei only; but such nuclei often occur in molecules as groups, particularly CH_3 and CH_2 groups. We turn now to consider coupling between groups, using the ethyl fragment, CH_3CH_2, as our example.

In the ethyl fragment the three hydrogens of the methyl (CH_3) group have the same chemical shift since all the CH bonds are identical and the shielding at each of the nuclei is the same. Such nuclei are called *chemically equivalent*. In the same way the two nuclei of the methylene (—CH_2—) group are chemically equivalent but their chemical shift is, of course, different from that of the methyl nuclei. Further there is some freedom of rotation of the methyl group in this fragment and hence the interaction between the methyl and methylene nuclei is averaged to the same value—the coupling constant is the same between any methyl and any methylene hydrogen. The nuclei in the methyl group (or the methylene group) are said to be *magnetically equivalent* as well as chemically equivalent. This property affords a considerable simplification of the overall spectrum because the *couplings within a group of magnetically and chemically equivalent nuclei do not affect the spectrum* and can be ignored, as in the case of A_2 considered earlier. In the present example, this means we can neglect coupling between the methyl hydrogens themselves, or between the two methylene hydrogens, and need consider only the coupling between a methyl and methylene nucleus. Since all the latter couplings are equal, as explained above, the system has just one J value to be considered. If the chemical shift between methyl and methylene is large, then we have a system A_3X_2 (where A is a CH_3 hydrogen nucleus, and X a CH_2 nucleus) with a coupling constant J_{AX}.

When considering the spectrum resulting from an A_3X_2 system it is quite possible to construct an energy level diagram analogous to that of Fig. 7.14 for the AX case. Such a diagram is complex, however, and it is much simpler to use another approach, which may be called the 'family tree' method; in this, coupling between grouped nuclei is considered stepwise.

Thus in Fig. 7.16(i) we start by imagining the A_3 and X_2 groups to be uncoupled, and hence as giving rise to one line each of intensity 3 and 2 units respectively. When we let the A_3 group couple to just *one* of the X_2 pair, the A_3 line is split into a doublet, separation J_{AX}, each line having intensity $1\frac{1}{2}$ units. If now each leg of this doublet is considered to couple with the second X nucleus, each will be split into a doublet, separation J_{AX}, intensity $\frac{3}{4}$ unit. However, the inner line of each doublet will overlap, because the coupling is identical for both, so the overall spectrum will have the appearance of a *triplet*, intensity of each line $\frac{3}{4}$, $1\frac{1}{2}$, $\frac{3}{4}$ (that is, a $1:2:1$ triplet) with the coupling J_{AX} appearing twice in the spectrum.

The splitting of X_2 by coupling with A_3 proceeds similarly. Coupling to one A yields a $1:1$ doublet, to the second a $\frac{1}{2}:1:\frac{1}{2}$ triplet, and to the third a $\frac{1}{4}:\frac{3}{4}:\frac{3}{4}:\frac{1}{4}$ quartet (i.e., a $1:3:3:1$ quartet) with the coupling constant J_{AX} repeated three times.

This argument is very readily generalized: a group of p equivalent nuclei splits a neighbouring group into $p+1$ lines with intensities given by the pth line of Pascal's triangle:

$$
\begin{array}{ccccccccc}
p = 1 & & & & 1 & & 1 & & \\
p = 2 & & & 1 & & 2 & & 1 & \\
p = 3 & & 1 & & 3 & & 3 & & 1 \\
p = 4 & 1 & & 4 & & 6 & & 4 & & 1 \\
\end{array}
$$

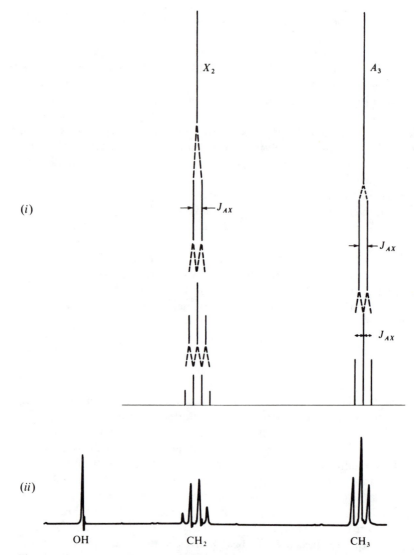

Figure 7.16 (*i*) A theoretical $A_3 X_2$ spectrum and (*ii*) the actual n.m.r. spectrum of ethanol, CH_3CH_2OH.

where each entry is obtained by summing the two numbers to its right and left in the line immediately above. Thus the two nuclei of the X_2 group split the A_3 resonance into $2 + 1 = 3$ lines, while the X_2 resonance is itself split into $3 + 1 = 4$ lines by the A_3 nuclei.

The family tree method thus predicts a quartet and triplet structure for the $-CH_2CH_3$ spectrum, the former having a total intensity of two units, the latter of three. Comparison with the spectrum of CH_3CH_2OH at the foot of Fig. 7.16 shows that this prediction is amply justified. It must be stressed, however, that if the chemical shift separation of the CH_3 and CH_2 resonances were to be decreased (forming an $A_3 B_2$ system), for example, by lowering the operating field of the instrument, or by changing the nature of the substituent on the CH_3CH_2— group, then the family tree method becomes a successively poorer approximation to the observed spectrum. Already in the $A_3 X_2$ spectrum of Fig. 7.16 the inner lines are seen to have a slightly enhanced intensity with respect to the outer (cf. the AB pattern of Fig. 7.14) even though the chemical shift separation is some 20 times the coupling constant. At smaller shifts not only is the intensity distortion increased, but additional lines begin to appear, quite inexplicable with the family tree method. Under these conditions accurate fit between theory and experiment is only possible by detailed calculation of the energy levels and transition probabilities for the system.

Some other typical coupling patterns are shown in Fig. 7.17. In (a) we see an $A_2 X_2$ spectrum given by the substance cyanhydrin, $CNCH_2CH_2OH$. The methylene groups have different chemical shifts and so it is to be expected that coupling between them will split each resonance into a $1:2:1$ triplet; this is to be observed in the spectrum. Other chemical structures giving rise to similar spectral patterns are 1,2-disubstituted benzenes, or five-membered unsaturated heterocyclic systems such as furan, although in such molecules the spectra are rather more complicated owing to additional couplings which arise between the nuclei. In (b) we show an $A_6 X$ spectrum arising from an isopropyl group, $(CH_3)_2CH$—. In this all six methyl hydrogens are equivalent and hence they split the lone methylene hydrogen into a septet, while they themselves are split into a doublet by the single nucleus. Such a pattern is easily recognizable even if the two outer lines of the septet are too weak to be observed, and is very characteristic. Finally in (c) we show the spectrum produced when three different nuclei couple together—an AMX system if all the chemical shifts are large. Within this system there are three different coupling constants: J_{AM}, J_{AX}, and J_{MX} so that, for instance, the A resonance is split into a doublet of spacing J_{AM} and then each line of the doublet is further split into a doublet by J_{AX}. Thus each resonance gives rise to a symmetrical $1:1:1:1$ quartet as shown in the figure. This pattern might arise from the three ring hydrogens of a monosubstituted furan or similar molecule (in this example we show the spectrum of an α-furan) or it might be from a vinyl group $CH_2{=}CH$—, where again all three hydrogen nuclei have a different chemical shift.

These coupling patterns, like that for the CH_3CH_2— fragment, are considerably complicated if the chemical shift between coupled nuclei is small. Usually, however, the additional fine structure produced and the intensity perturbations do not prevent recognition of the overall pattern, particularly when some experience has been gained from studying actual spectra. The tremendous analytical value of such patterns is obvious since

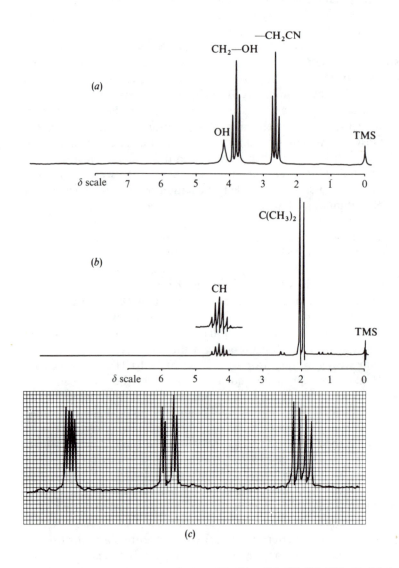

Figure 7.17 The n.m.r. spectra of (*a*) cyanhydrin, $CN.CH_2CH_2OH$, (*b*) 2-iodo-propane, $(CH_3)_2CHI$, and (*c*) 2-furanoic acid,

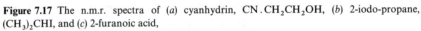

$$OCH{=}CH.CH{=}C.COOH$$

In (*a*) and (*b*) the normal δ scale is shown, while in (*c*) the scale is expanded to 1 Hz per division in order to show clearly the three quartets. The resonance of the COOH group is off the scale to the left.

their recognition immediately gives information about the chemical groupings present in the molecule under examination. We discuss this aspect of
n.m.r. spectra more fully in the next section.

7.2.4 Chemical Analysis by N.M.R. Techniques

In the preceding sections we have built up a picture of the application of
n.m.r. to constitutional and structural studies. Thus the observation of the τ
values of lines in a spectrum (or of the centres of multiplets if coupling is
occurring) immediately indicates, with very little ambiguity, the types of
hydrogen-containing groups within the molecule, while the relative intensities of the lines yield directly the proportions in which these groups occur.

Further, the multiplet structure of each group in the spectrum gives
information on the number of hydrogen nuclei coupled to that group and in
this way shows which groups are near neighbours in the molecule. Thus
groups such as CH_3CH_2, $-CH_2CH_2-$, $(CH_3)_2CH-$, etc., can be instantly
recognized from the n.m.r. spectrum.

As an example of the use and limitations of n.m.r. spectroscopy in
analysis, consider the spectrum shown in Fig. 7.18. Resonances are centred
at $\delta = 8.2$, 7.5, 4.4, and 1.3, the position of the former two and the coupling
pattern of the latter two suggesting that they arise from a phenyl group
(C_6H_5-) and an ethyl group (C_2H_5-) respectively. The integral trace on
the spectrum shows these resonances to have relative intensities of $5:2:3$
(the two phenyl resonances here being summed) so we know the phenyl and
ethyl to be present in $1:1$ ratio. If we also know that the molecular formula
of the substance is $C_9H_{10}O_2$, we can rapidly deduce that the resonances at
$\sigma = 8.2$ and 7.5 are due to hydrogen nuclei respectively ortho and meta/
para to a carbonyl group (cf. Fig. 7.11), while those at 4.4 and 1.3 are

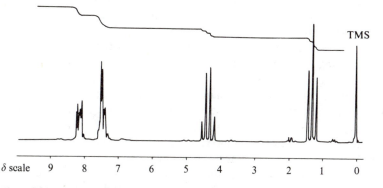

δ scale

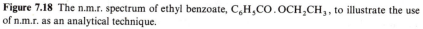

Figure 7.18 The n.m.r. spectrum of ethyl benzoate, $C_6H_5CO.OCH_2CH_3$, to illustrate the use
of n.m.r. as an analytical technique.

consistent with the grouping $CH_3CH_2 . O . CO—$. The molecule is thus ethyl benzoate, $C_6H_5CO . O . CH_2CH_3$.

Note that the n.m.r. spectrum, while allowing us to deduce directly the presence of phenyl and ethyl groups, does not indicate the *presence* of groups not containing magnetic nuclei, in this case O and CO. However, once these groups are known to be present, either by determination of the molecular formula or by observation of the infra-red spectrum, then the n.m.r. spectrum does indicate their *position* in the molecule. Thus the coupling pattern shows clearly that the CH_3 and CH_2 groups are directly bound and not, for example, joined via O or CO ($CH_3 . O . CH_2—$ or $CH_3 . CO . CH_2—$) since in these latter configurations the coupling constant would be immeasurably small.

This very simple example serves to indicate the method of approach when using n.m.r. for analytical purposes. Of course only in the simplest cases is a complete structural determination possible from the n.m.r. spectrum alone, but when taken in conjunction with other techniques, in particular infra-red spectroscopy, a great deal of useful structural information can usually be obtained about an unknown molecule.

7.2.5 Exchange Phenomena

The student may have been puzzled by one aspect of the alcohol spectra shown in Figs 7.9, 7.16, and 7.17; in these three spectra the resonance of the —OH hydrogen is shown as a single line whereas we might now expect it to be coupled with neighbouring CH_3, CH_2, or CH nuclei and hence have multiplet structure—quartet, triplet, or doublet, respectively. The reason it does not so couple is attributable to the fact that the hydroxyl hydrogen is readily exchanged with other hydrogen nuclei or ions in its vicinity; when this happens, the replacement hydrogen does not necessarily have the same spin direction as that being displaced and, if the exchange occurs sufficiently rapidly, the neighbouring CH nucleus experiences a 'coupling field' which is averaged to zero. Thus there is no net coupling between the OH and the neighbouring groups.

If the exchange is prevented by rigorous drying of the alcohol samples, coherent coupling appears and the OH resonance has the expected multiplet structure. It can be shown theoretically that for a coupling constant of J Hz the coupled nucleus must exchange more rapidly than $J/2\pi$ times per second for the multiplet to collapse to a singlet. For the alcohols considered above $J \approx 6$ Hz, and hence an exchange rate of only about once per second is sufficient to destroy coherent coupling. Of course, the transition from coupling to no coupling is not abrupt; at exchange rates rather lower than $J/2\pi$ the lines of the OH multiplet begin to broaden, at $J/2\pi$ they are so broad that all trace of line splitting is obliterated and, at higher rates of exchange, the broad line sharpens until the single very sharp resonance of the illustrated spectra is seen.

An obvious application of this effect is to the study of hydrogen exchange kinetics; exchange rates frequently vary with temperature or concentration and such variations can be followed very precisely by observation of the change in line shape of n.m.r. signals.

Nuclei in different chemical surroundings (i.e., having different chemical shifts) may, in addition, have their chemical shift positions averaged by exchange phenomena. If the exchange is considerably more rapid than the difference between the two chemical shifts (expressed in hertz) then only one sharp resonance signal will appear, midway between the separate chemical shift positions; for a much slower exchange, two sharp resonances will be observed at the proper positions, while for intermediate rates either one or two broad resonances occur.

The exchange giving rise to the averaging may be a physical exchange of the nuclei (as in the exchange of the OH proton in alcohols considered earlier) or merely an internal rearrangement, such as the rotation of a methyl group or the interconversion of two chair forms of cyclohexane. In the latter, interconversion results in all the equatorial hydrogens becoming axial, and vice versa; experiments on substituted cyclohexanes, in which the interconversion is inhibited, show that equatorial and axial protons have chemical shifts differing by up to about 1 p.p.m., but in the unsubstituted compound only one sharp resonance line is observed, showing that the interconversion takes place considerably more rapidly than 100 times per second.

7.2.6 Simplification of Complex Spectra

Frequently when studying all but the simplest compounds, the n.m.r. spectrum is a complicated mass of overlapping lines—groups showing two or more different coupled splittings with their patterns distorted by small chemical shifts, for example. There are several techniques now available to simplify such spectra and to convert them to more readily recognizable and analysable patterns.

Firstly we remember that chemical shift differences vary with the strength of the applied field, whereas coupling constants do not. Thus if faced with a spectrum obtained with an applied field of 2·35 T (a 100 MHz spectrum) which shows a distorted pattern because a chemical shift difference is similar in magnitude to a coupling, it is relatively simple to re-examine the sample in an instrument operating at a higher field. Fields of 5·1 T (220 MHz) are now common, and some instruments have fields of up to 10 T (400 MHz), the latter using powerful 'superconducting' magnets. 220 MHz spectra yield chemical shifts twice as large as those at 100 MHz, and usually offer considerable simplification.

Another method by which chemical shift separations may be increased is by adding various materials to the sample solution. It has long been known that a change of solvent or of concentration could have a small

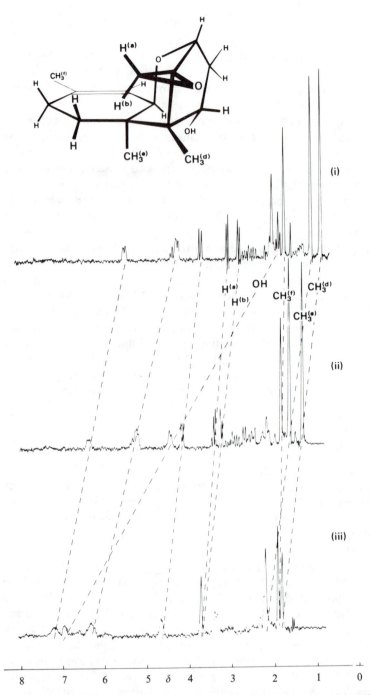

Figure 7.19 100 MHz spectra of 4-epi-trichodermol; (i) a solution containing 10 mg of the pure compound; (ii) solution plus 5 mg Eu(fod)₃; (iii) solution plus 10 mg Eu(fod)₃. *(Thanks are due to Dr. J. R. Hanson and Mr. P. Dew of the University of Sussex for assistance in obtaining this spectrum.)*

effect on chemical shift positions, the magnitude of the effect depending on the extent of solute–solvent interactions, but usually not being more than 1 p.p.m. However, it has been discovered that some complexes of rare-earth metals, particularly of europium and praseodymium, have exceptionally marked effects. For example, Fig. 7.19 shows the 100 MHz spectrum of 4-*epi*-trichodermol, both pure in solution, (i), and with 5 mg and 10 mg of a europium agent added, (ii) and (iii) respectively; dashed lines connect the bases of some resonances to show how they move under the influence of the 'chemical shifting agent'.

The actual reagent used here is tris(1,1,1,2,2,3,3-heptafluoro-7,7-dimethyl-4,6-octanedione) Europium III (or Eu(fod)$_3$ for short); as befits its impressive name, this is a large molecule and, when it is attached to the OH group of 4-*epi*-trichodermol, it not only perturbs the resonance of the OH, but its influence spreads over the whole molecule, decreasing inversely with distance—the further away a particular hydrogen nucleus from Eu(fod)$_3$, the less it is shifted. Thus the OH resonance moves most rapidly, and, of the two hydrogen atoms labelled (*a*) and (*b*) on the molecular picture, (*b*), the nearer to the shift reagent, moves more rapidly than (*a*); this is seen in the spectrum by the collapse of their coupled doublets into one line as the chemical shifts of (*a*) and (*b*) coincide—the *AB* system becomes A_2. Equally, of the three CH$_3$ groups labelled (*d*), (*e*), and (*f*) in the picture, (*d*) and (*e*) show markedly greater shift than (*f*).

In this particular example most of the individual resonances are well resolved and distinct in the spectrum of the pure compound, but the use of the shifting agent helps in the assignment of certain resonances to particular nuclei. In other cases, however, where the original spectrum is badly over-lapped, a shift reagent may offer considerable simplifications. In this context it should be mentioned that, while europium reagents shift resonances to low field (high δ) as seen here, praseodymium shift reagents normally cause shifts in the opposite direction.

A third approach towards simplification concentrates on the coupling constants. If coupling can be destroyed between nuclei an obvious simplification follows; thus in the spectra of alcohols discussed earlier, the OH signal is seen to be a single line rather than a multiplet due to the kinetic exchange of the hydrogen nucleus, and similarly the neighbouring proton group is decoupled from the OH and exhibits a simple spectrum.

This decoupling can be deliberately carried out in other cases by rapid exchange of the *spin* of the nucleus, rather than exchanging the nucleus itself. Consider two coupled nuclei, A and X, giving the typical AX spectrum of Fig. 7.14(ii); bathing the sample in radiation of frequency v_A allows the spectrum of nucleus A to be observed, but it is quite simple *simultaneously* to apply a strong radiofrequency field at frequency v_X. This causes nucleus X to undergo rapid transitions between its two spin states and, although these changes are not directly observed by the spectrometer (since that is tuned to frequency v_A) the reversals do result in a decoupling of the X spin from that of A; consequently the original doublet at v_A collapses into a single line. Since two separate radiofrequencies are applied to the sample, this technique is known as 'double resonance'.

Double resonance has further uses besides simplification of spectra. For instance, the technique may be used to locate a particular resonance. Thus if the chemical shift of nucleus X places its resonance so that it is over-lapped and lost within other spectral lines, it may be found by double irradiation of a nucleus coupled to X; the decoupling results in X's signal collapsing from a multiplet to a single sharp line, a change which is rela-tively easy to observe. Again, in a weak sample where some signals may be lost in the background noise, decoupling may result in a multiplet changing to a single line which shows up well above the noise. This application will be considered further in particular connection with ^{13}C spectra in Sec. 7.3.2. A final application of double-resonance techniques involves the application of a *weak* secondary radiofrequency, rather than the strong radiation re-quired for decoupling. Under these conditions only partial decoupling may occur and, in some circumstances, *extra* multiplet splitting may be ob-served. These experiments, known graphically as 'spin tickling' experiments, lead to measurements of fundamental n.m.r. constants, particularly the rela-tive signs of coupling constants.

7.3 NUCLEAR MAGNETIC RESONANCE SPECTROSCOPY: NUCLEI OTHER THAN HYDROGEN

7.3.1 Nuclei with spin $\frac{1}{2}$

In general, any nucleus with a spin of $\frac{1}{2}$ will give rise to n.m.r. signals provided the appropriate magnetic field and radiofrequency are applied. There are many such nuclei, of which probably ^{13}C is the most important for the general chemist; formerly this nucleus was difficult to study since it gives rise to extremely weak signals and, even after concentration of the isotope considerably above its naturally occurring one per cent level, satis-factory n.m.r. spectra were not easy to obtain. The introduction during the last few years of Fourier transform methods, however (see Sec. 7.4.2), has caused a big upswing in the study of this nucleus. Other spin $\frac{1}{2}$ nuclei, frequently studied in the past, include ^{19}F and ^{31}P, both being the only isotopes of their particular elements, ^{29}Si (4·7 per cent) and various metals such as ^{117}Sn, ^{119}Sn, ^{195}Pt, ^{205}Tl, or ^{207}Pb.

For all these nuclei the phenomena of chemical shift and spin-spin coupling are observed in the spectra and generally both are considerably larger than their hydrogen counterparts. Thus, while a range of some 15 p.p.m. contains virtually all the known hydrogen chemical shifts, values for phosphorus-containing compounds span some 400 p.p.m., fluorine some 600 p.p.m., and a few metals, notably ^{205}Tl and ^{207}Pb, range over 14 000 p.p.m. or more. This behaviour can be traced to the greater number and increased mobility of the extranuclear electrons leading to greater variation in diamagnetic shielding.

The increase in spin-spin coupling constants is probably attributable to the same cause although the increase is not so marked as in the case of chemical shifts. Thus the coupling between two directly bonded phosphorus nuclei is some 600 Hz, while it may be as high as 1400 Hz for phosphorus bonded to fluorine. These figures, however, are only some 3–6 times larger than the corresponding direct H—H coupling of 240 Hz. A further general point is that couplings tend to attenuate less rapidly with increase in the number of bonds separating the coupled nuclei than do those of hydrogen.

Two practical advantages follow from the large chemical shifts of these nuclei. Firstly, less precise instrumentation is required in order to obtain data useful for structural determinations since the tolerances with which chemical shifts must be measured are correspondingly larger. Secondly, the spectra obtained are simpler to analyse in that they have less of the complications which arise in hydrogen spectra when the chemical shifts are comparable in magnitude with coupling constants. The spectra are thus of the $A_a M_m X_x \ldots$ type, rather than $A_a B_b C_c \ldots$.

In the following section we give a very brief discussion of ^{13}C n.m.r., concentrating on areas where there are differences with proton spectroscopy, and in Sec. 7.3.7 an even shorter discussion of some recent studies in ^{31}P spectroscopy. The interested reader is referred to books in the bibliography for more detailed coverage of the n.m.r. of these and other nuclei.

7.3.2 ^{13}C N.M.R. Spectroscopy

The principles governing ^{13}C n.m.r. are exactly similar to those of ^{1}H spectroscopy, although the scale of observed shifts and couplings is, as has been stated, greater for the former. Chemical shifts in ^{13}C work are very conveniently measured (in p.p.m.) from tetramethyl silane (TMS) as reference so that the same sample can be used to study both nuclei; the δ scale, where TMS is arbitrarily assigned the value $\delta = 0.0$, is invariably used. On this scale ^{13}C shifts range from 0 to about 250 p.p.m. and Fig. 7.20 shows the approximate regions of the spectrum in which resonances of carbon nuclei in different chemical surroundings are found. Also, as in proton spectra, the precise chemical shift of a nucleus depends on the atom or atoms attached to it, and there are correlations with the electronegativity of substituents. In ^{13}C considerable success has been achieved in assigning observed resonances to specific atoms within molecules by using additivity rules for the combined shift effect of several substituents.

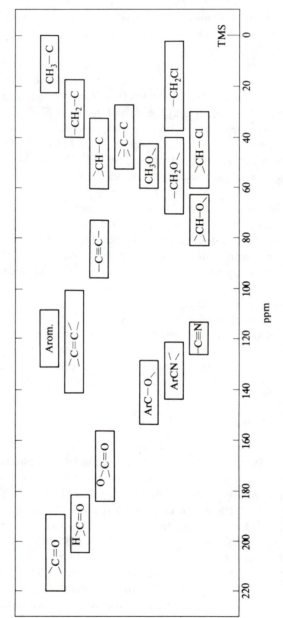

Figure 7.20 The approximate chemical shift ranges of some ^{13}C resonances.

Because of the greater range of chemical shifts ^{13}C spectra nearly always contain a separate and distinct resonance for each chemically shifted nucleus in the molecule—very little overlap of resonances occurs. This obviously simplifies spectra-structure correlation, but we shall see later that the *intensities* of ^{13}C resonances usually cannot be so simply correlated with the number of atoms contributing to each resonance.

A further simplification in ^{13}C spectra is apparent when one considers spin-spin coupling—with normal samples one can completely ignore direct coupling between neighbouring ^{13}C nuclei. This is not to say that such coupling does not exist—it most certainly does, and two ^{13}C nuclei bound directly together in a molecule, or separated from each other by two or three bonds, will give rise to the AX-type spectrum mentioned in Sec. 7.2.2 for ^1H spectra. However, remembering that ^{13}C is present in a normal, non-enriched sample, only to the extent of about one per cent natural abundance, we can see that the probability of two such nuclei being found within a few bonds of each other is very remote. Indeed, if there are fewer than 100 carbon atoms in each molecule of sample, then there will be, on average, only one ^{13}C nucleus per *molecule*.

Coupling between ^{13}C and neighbouring *protons*, on the other hand, shows up clearly in ^{13}C n.m.r. spectra, where it can be both a help and a hindrance. It is helpful in assigning resonances to particular nuclei in a molecule. For example, a $\geq$C—H group in a molecule will show a doublet structure in the ^{13}C spectrum, with a J_{CH} coupling constant of about 125 Hz; a double-bonded carbon, as in $>$C$=$C—H, will show a coupling of some 170 Hz, and a triply bonded one, C$\equiv$C—H, some 250 Hz. These values are large, and thus easily seen and distinguished in a spectrum. Substituents often affect the values considerably, but again useful empirical correlations exist; equally long-distance coupling, over two or three bonds, is well characterized. As in ^1H spectra, the *pattern* of the coupling is also very diagnostic. Thus a C—H fragment yields a doublet in the ^{13}C spectrum, CH$_2$ gives a triplet pattern and CH$_3$ a quartet. Figure 7.21 shows just such a spectrum for CH$_3$C*OOH; here it is the starred carbon which is recorded, and its coupling to the three distant protons by some 10 Hz gives a clear quartet pattern.

^{13}C coupling is a hindrance in that it 'spreads out' the intensity of an already weak resonance, thus making observation more difficult. This can be overcome very elegantly by an extension of the double resonance technique mentioned on p. 283. There we saw that selective radiation at a particular chemical shift position in a ^1H spectrum destroys coupling to neighbouring nuclei and thus simplifies spectra. In ^{13}C work it is common to decouple *all* the protons in the molecule from the carbons. To do this requires irradiation of the sample at the ^1H frequency (100 MHz, for example), while observing the spectrum at the ^{13}C frequency (25·4 MHz in the same field). If the 100 MHz radiation is strong and sufficiently 'wide

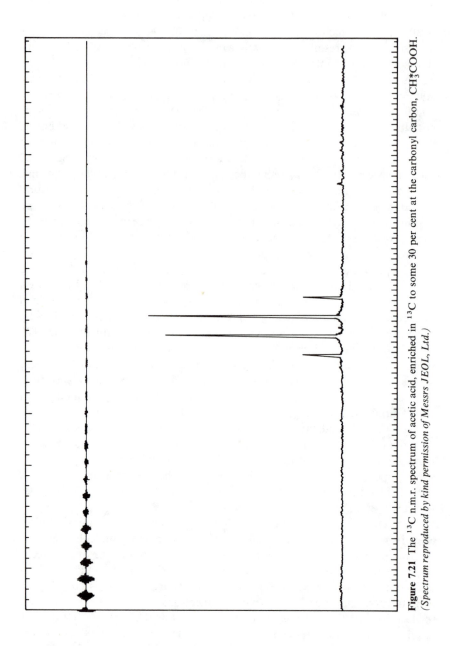

Figure 7.21 The ^{13}C n.m.r. spectrum of acetic acid, enriched in ^{13}C to some 30 per cent at the carbonyl carbon, CH$_3^*$COOH. (*Spectrum reproduced by kind permission of Messrs JEOL, Ltd.*)

band' to cover all the proton frequencies in the sample—i.e., some 100 p.p.m., or 1000 Hz wide—then the protons will change their spins sufficiently often to be effectively decoupled from the carbons. This technique, often called *noise decoupling*, simplifies and intensifies the ^{13}C spectrum by collapsing the spin–spin multiplets to single lines. This results in the loss of the pattern recognition advantages mentioned in the previous paragraph, and it is often highly informative to compare the noise-decoupled spectrum with the non-decoupled one, thus finding how many hydrogens are attached to each ^{13}C.

Noise decoupling also leads to an intensity bonus due to a phenomenon known as the nuclear overhauser effect, or NOE. Briefly, we already know that the intensity of an n.m.r. signal is closely dependent on the (small) difference in population of excited and ground spin states, and that once sufficient nuclei have become excited, no more energy can be absorbed from the exciting radiation until relaxation has, at least partially, restored equilibrium conditions. In the presence of noise-decoupling radiation it turns out that other relaxation processes, which do not themselves absorb or emit radiation, are induced, so that equilibrium conditions are attained more rapidly. This effect, which in ^{13}C spectra typically gives rise to an approximate doubling of the signal intensity, is also known as *spin pumping*.

Relative intensity measurements of ^{13}C peaks often give apparently anomalous results. In Fig. 7.22(a) we sketch the ^{13}C spectrum of ethyl acetate as an example. The assignment of the four resonances to the four carbon nuclei in the molecule is very straightforward; the resonance at

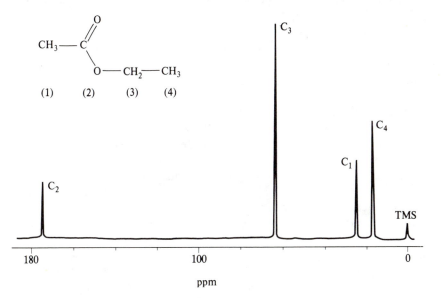

Figure 7.22 Representation of the ^{13}C spectrum of ethyl acetate.

$\delta = 175$ is obviously in the C=O region (cf. Fig. 7.20), and the others are due to the saturated CH's. In general, the resonance of CH_2 groups attached to electronegative atoms (like oxygen) occurs between $\delta = 40$ and $\delta = 70$, which establishes C_3; and finally the proximity of the carboxyl group (O . C=O) to C_1 shifts its resonance to higher δ from C_4. The intensities, however, are all 'wrong', since we would expect four resonances of equal intensity. The reason for this is to be found in the differing relaxation times of the four ^{13}C nuclei, as we discuss below.

When observing n.m.r. spectra of low-intensity samples, like ^{13}C, by Fourier transform methods, the computer averaging technique (see Sec. 1.9) is invariably used. This requires repeated observation and accumulation of the spectrum, the accumulated real signals becoming enhanced, while the noise components essentially cancel and disappear. However, this technique can only be really effective if sufficient time is given between excitation pulses to allow the sample to return to equilibrium, otherwise maximum energy absorption from successive pulses will not occur. Strictly an infinite time should be allowed to elapse, but effectively a period of some four or five relaxation times is sufficient for virtually complete equilibration. If, as often happens in ^{13}C spectra, the relaxation times for nuclei in different surroundings are quite different, then four or five times the *longest* relaxation time should be allowed. For protons relaxation times are usually fairly short, and a delay of two or three seconds between pulses is adequate. For ^{13}C, however, relaxation times are in general rather longer, and often very much longer (1–100 seconds is typical), and a delay of four or five times 100 seconds between pulses would obviously make the collection of data very lengthy. Normally a delay of some 10 seconds is chosen, which is sufficient for nuclei with a short relaxation time to decay and so give rise to their 'proper' intensity in a spectrum, but those with a longer relaxation time will be somewhat smaller than they should be, and those with a very long relaxation period may well be very small indeed. A very rough rule is that the intensity of a particular signal is inversely proportional to the relaxation time of the nucleus concerned.

To explain differences in relaxation times we need to understand a little about the relaxation process itself. In order to give energy to their surroundings, excited nuclei require a magnetic (or electric) fluctuation to occur at approximately their precessional frequency—in this way resonance interaction with the surroundings can be set up. The most common source of local fluctuating magnetic fields is other spinning nuclei and, since ^{13}C nuclei are very isolated from each other, it is essentially only hydrogen nuclei attached to a particular ^{13}C nucleus, or perhaps two or three bonds away, which provide the relaxation mechanism. These ^1H nuclei offer a fluctuating magnetic field by virtue of the random movements of the molecule and, because we require these movements to occur at about 10^7–10^8 Hz (so as to resonate with the ^{13}C), we are restricted to fairly rapid tumbling of molecules in solution. Molecular rotations and vibrations are too fast (10^{10}–10^{13} Hz, approximately) and so do not contribute.

We can explain the intensity 'anomalies' of Fig. 7.22, then, as follows. The C_2 carbon has no attached hydrogens and so has a relatively long relaxation time (typically about 30 seconds), thus giving rise to a line of low intensity. The CH_2 has, of course, two hydrogens and so can relax relatively easily (typical relaxation time of some 2–4 seconds), to give an intense line. The two CH_3 groups, although having three hydrogens, undergo rotation about the C—CH_3 bond, which motion is sufficiently fast to destroy resonance relaxation. Typical relaxation times for 'free' CH_3 groups are 10–20 seconds (although any steric hindrance to this rotation lowers the time), and the resultant low intensity of their ^{13}C peaks is often used to distinguish them from CH or CH_2 groups in a spectrum.

The intensity anomaly can be overcome either by spending *very* much longer collecting a spectrum, or by adding a solution of a paramagnetic ion, such as iron or chromium. The 'free' electrons in such ions can be very efficient relaxers, and a ^{13}C spectrum usually exhibits a near-normal intensity pattern after the addition of such a reagent. At times, however, and particularly if the reagent forms a chemical complex with the molecule of interest, relaxation at some sites becomes too efficient, and the n.m.r. signal becomes too broad to be observed.

Observation of spin-lattice relaxation times, therefore, can sometimes be of assistance in assigning complex spectra—the low intensity of CH_3 group spectra and of carbon atoms not attached to hydrogens, for example. In addition, however, T_1 values are beginning to give insights into various aspects of molecular motion such as hindered rotation or movements of parts of a large molecule, e.g., side chains in polymers. And, although we have illustrated this section with reference to ^{13}C spectra, relaxation is a phenomenon observed with all spinning nuclei, and relaxation times will become just as useful for other nuclei.

7.3.3 Biological N.M.R.: ^{31}P Spectroscopy

^{31}P, a nucleus with spin $\frac{1}{2}$ and 100 per cent natural abundance, has always been of some interest in n.m.r. studies. Although it has a tendency to yield rather broad spectral lines, it has a wide and diagnostic range of chemical shifts, and its coupling patterns to neighbouring hydrogens can give useful structural information. Recently, however, it has become of some importance in the study of biological materials. Of course, all biological substances contain carbon and hydrogen, and so these nuclei could be, and are, studied in this respect. However ^{13}C has only a low natural abundance, and proton spectra are dominated by the large amount of water in biological samples; ^{31}P gives spectra of good intensity which are often relatively simple to interpret.

Recently, too, techniques have been developed whereby *living* systems can be examined by n.m.r. Whole plants, small animals, or parts of larger animals (e.g., a human arm or leg—still attached to its living host, of course) can be placed between the poles of a suitably designed magnet and, with the provision of surface coils to direct the radiofrequency radiation to the area of interest, ^{31}P spectra of relatively small areas can be studied. In order further to limit the site to be examined, one technique makes use of the breadth of phosphorus resonances. Local 'shim' coils are arranged to give a good homogenous field only over the volume of interest while allowing the field outside this volume to change rapidly with distance. In this way ^{31}P resonances from outside the volume are very broad and merge with the background, whereas those from the experimental site are relatively sharp and stand out clearly.

Spectra obtained in this way are already being used for medical diagnosis—some disorders are associated with an alteration in the type or concentration of phosphorus-containing chemicals in body systems. And the same methods have been applied to obtaining ^{13}C spectra of living systems, particularly to trace the destination of ^{13}C atoms from enriched samples.

Another technique which has been used on living systems is the so-called n.m.r. imaging. Here the object is not to produce an n.m.r. *spectrum*, but simply to chart the density of, say, hydrogen nuclei in various regions of the sample. The results are very like an X-ray photograph, particularly in that bone has a considerably lower proportion of hydrogen than have other tissues. They are obtained, of course, without the potential damage which can occur from X-radiation—at least it is not currently thought that radiofrequency radiation is in any way harmful to living tissues.

7.3.4 Nuclei with Spin Greater than $\frac{1}{2}$

We saw at the beginning of this chapter that the application of a magnetic field to any nucleus with spin I causes the spin vector to become oriented in any one of $2I + 1$ possible directions, each associated with a slightly different energy level. Thus Fig. 7.1 shows the situation for a nucleus with a spin of 1, for example, ^{14}N. Since, in a given field, the spacing between the energy levels of a particular nucleus are all identical, transitions induced between any neighbouring levels will result in the emission or absorption of energy at the same frequency. Thus only one resonance line will appear for each nucleus.

In principle, then, *any* spinning nucleus will give rise to a single resonance line when the appropriate field and frequency are applied; the position of the resonance for a given nucleus will vary with its chemical surroundings and it will be split into a multiplet by interaction with other spinning nuclei—in other words, the phenomena of chemical shift and spin-spin coupling will be observed. In practice, however, the spectra of nuclei with spin greater than $\frac{1}{2}$ are usually intrinsically weak and not easy to observe and they are not much used, in themselves, for analytical or structural studies. However, elements such as ^{14}N (spin $I = 1$), and the halogens, chlorine, bromine (both of spin $= \frac{3}{2}$), and iodine (spin $= \frac{5}{2}$) occur widely in chemistry, and we should consider what effect the presence of these nuclei might have, by virtue of spin-spin coupling, on the n.m.r. spectrum of neighbouring *hydrogen* nuclei.

Let us take the ^{14}N—H group as an example; we would expect the coupling here to be reasonably large because the two nuclei are directly bonded (J, in fact, is observed to be some 50 Hz for this group). Now the unit spin of the nitrogen nucleus can take up one of three orientations in an applied field (cf. 7.1), which we may represent as $\uparrow$, $\rightarrow$, and $\downarrow$. Comparing with Fig. 7.12 we see that the spin direction $\uparrow$ will reduce the field at the hydrogen nucleus, $\downarrow$ will reinforce it to the same extent, while it is plain that $\rightarrow$ will leave it unaltered. We expect the hydrogen resonance to be split into a *triplet*, therefore, and since all three spin orientations of the nitrogen are equally likely, each line of the triplet will have the same intensity. The formation of this $1:1:1$ triplet is shown schematically in Fig. 7.23(a). In Fig. 7.23(b) we reproduce the spectrum of acidified methylamine, CH_3NH_2, to show the triplet structure of the hydrogen resonance. The solution is acidified merely to suppress the otherwise rapid exchange of the hydrogens attached to nitrogen, which exchange would destroy the N—H coupling. One result of this is that the molecule becomes converted to the methylammonium *ion*, $CH_3NH_3^+$, and so the methyl resonance is split into a $1:3:3:1$ quartet by spin coupling with the —NH_3^+ hydrogen nuclei; the fact that this quartet is sharply defined implies that the exchange rate is small. In spite of this we note that the NH triplet consists of very broad resonance lines; this broadening is due to quadrupole relaxation which we shall discuss shortly.

We can easily generalize the above discussion to other nuclei. A nucleus with spin I can take up one of $2I + 1$ equally likely spin orientations in an applied field; of these, some will reinforce and some reduce the field experienced by a neighbouring nucleus, so the resonance of the neighbour will be split into a multiplet of $2I + 1$ lines with equal intensity. Thus a single chlorine nucleus would tend to produce quartet structure in its neighbours (spin $= \frac{3}{2}$), while a single iodine nucleus ($I = \frac{5}{2}$) would produce sextets. In practice, however, these splittings are not observed because of quadrupole relaxation which causes rapid transitions between the spin states (see next section).

7.3.5 Quadrupole Effects

In addition to the magnetic moment discussed throughout this chapter, all nuclei with a spin ≥ 1 also possess an *electric quadrupole moment*, which arises because the nuclei are not spherical. Such nuclei, in fact, are shaped

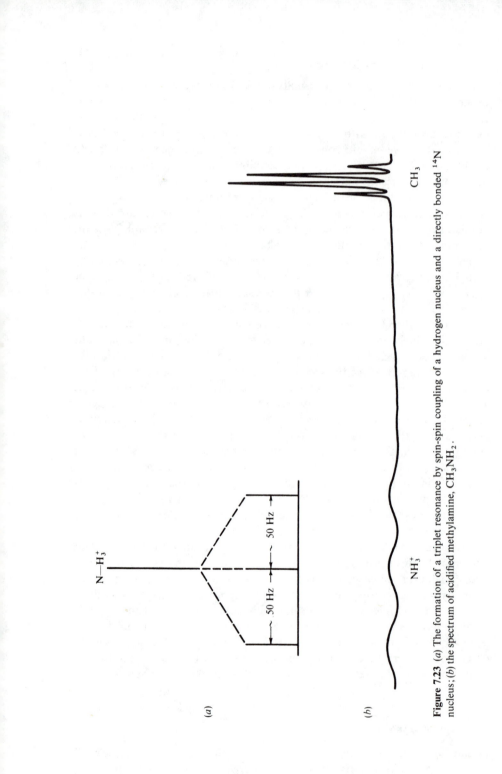

(a)

(b)

Figure 7.23 (a) The formation of a triplet resonance by spin-spin coupling of a hydrogen nucleus and a directly bonded ^{14}N nucleus; (b) the spectrum of acidified methylamine, CH_3NH_2.

either like a symmetrical egg or like a tangerine (elongated, or flattened at the poles, respectively). Even if the charge density within the nucleus is constant, the distorted shape gives rise to a charge distribution which is non-spherical; the electric quadrupole moment is a measure of the departure from sphericity, being positive for egg-shaped and negative for tangerine-shaped nuclei. For spherical nuclei (i.e., spin $\frac{1}{2}$ or 0) the electric quadrupole moment is zero.

The electric moment interacts strongly with an applied electric field and, if such a field has a pronounced gradient at the nucleus, the nuclear moment will tend to lie in the field direction; further if the gradient changes direction, the nuclear moment will try to follow this change. Thus, consider the pyramidal molecule of ammonia, NH_3. This has the three hydrogen nuclei arranged at the corners of the base of a pyramid, the nitrogen nucleus being above and equidistant from them. The presence of three positively charged nuclei on one side of the nitrogen nucleus produces a strong electric field gradient at that nucleus which tends to orient its quadrupole moment axis perpendicularly to the plane containing the hydrogen nuclei. If, due to molecular tumbling, the molecule rotates as a whole, the nitrogen nucleus tends to follow this rotation. So strong is the coupling between nuclear electric moment and field gradient that it is sufficient to prevent the coherent alignment of the nuclear *magnetic* moment in an applied magnetic field—in other words the quadrupole electric moment supplies a mechanism by which the spin orientation may be relaxed. Here, then, we have another—and usually very efficient—relaxation mechanism (cf. Sec. 7.15) called *quadrupole relaxation*.

If, on the other hand, we consider the ammonium ion, NH_4^+, which consists of a nitrogen nucleus at the centre of a regular tetrahedron of hydrogen nuclei, there is no field gradient at the nitrogen nucleus. Thus its spin can be oriented by an applied magnetic field quite independently of the orientation of the hydrogen nuclei—the quadrupole relaxation is extremely weak in this case. We see, then, that the efficiency of quadrupole relaxation will depend upon the symmetry of the surroundings. Univalent atoms, such as the halogens, will always be situated in a field gradient which will experience efficient relaxation; the relaxation of polyvalent atoms, such as nitrogen, will vary from molecule to molecule.

Quadrupole relaxation has two effects in n.m.r. spectroscopy. The first of these, the broadening of the n.m.r. signal from the nucleus possessing a quadrupole moment, is very similar to the broadening caused by fast relaxation (cf. Sec. 7.1.5). When quadrupole relaxation occurs the lifetime of a particular spin state may be as short as 10^{-4} s or less. This, as we saw in Sec. 7.1.5, is comparable to the relaxation time of nuclei in solids, for which we calculated a line width of some 1000 Hz. Such a 'line' would be so broad that it would be impossible to detect with a normal high-resolution n.m.r. spectrometer, since it would be indistinguishable from a very gentle wandering of the background noise.

For the second effect of quadrupole relaxation we recall the discussion of double resonance in Sec. 7.2.6, wherein an experimental method for 'stirring' or relaxing coupled spins was described. In quadrupole relaxation we have another method for destroying coherent coupling between nuclei. If the relaxation is highly efficient as, for example, in halogen nuclei, the 'stirring' is so rapid that coupling is completely destroyed and the resonances of neighbouring nuclei remain sharp; for less efficient relaxation, as in many nitrogen-containing molecules, the coupling is only partially destroyed and multiplet but broad resonances result from neighbouring nuclei. An example of this has already been seen in the very broad triplet formed by the $-NH_3^+$ group of the methylammonium ion of Fig. 7.23. In routine n.m.r. spectroscopy, the only commonly occurring quadrupolar nucleus whose effect is noticeable in spectra is, in fact, that of ^{14}N: here the effect is usually to broaden the resonance of neighbouring nuclei, sometimes with the appearance of multiplet structure. Often the broadening is so great that the hydrogen resonances disappear completely into the background noise.

A single crystal of a solid substance has a well-ordered and regular structure, apart from possible lattice defects, and any quadrupolar nucleus contained in a crystal will find itself in exactly identical surroundings as similar nuclei in other regions of the lattice. Thus all the quadrupole moments will tend to lie in the same direction, insofar as this is consistent with the Boltzmann energy distribution. Even without the application of an external magnetic field such nuclei will be able to absorb energy coherently from a beam of radiation at an appropriate frequency. These absorptions, which occur in the region 1–1000 MHz, are known as the *nuclear quadrupole resonance* (n.q.r.) spectrum or sometimes the *pure quadrupole spectrum* to emphasize the fact that an external field is not applied. N.q.r. spectroscopy essentially uses the quadrupolar nucleus as a probe to detect and estimate electric field gradients in the crystal and the data obtained is invaluable in applications of crystal field theory. The spectra are not easy to interpret, however, and will not be discussed here; the interested reader is referred to the book by Lucken listed in the bibliography.

7.4 TECHNIQUES AND INSTRUMENTATION

Basically an n.m.r. spectrometer requires components with functions similar to those already described for other spectroscopic techniques (radiofrequency source, sample holder, radiofrequency detector, recorder, etc.) but with the addition of a powerful magnet. We have seen that a resolving power of 0·5 Hz requires that the magnetic field be stabilized to within one part in 10^8 or better and, in addition, it must be the same over all parts of the sample under test. These requirements are very stringent.

Many earlier spectrometers used permanent magnets which, when carefully thermostatted, gave very precise and constant fields. They had two disadvantages, however. Firstly, the field could not be varied, so a separate crystal oscillator was required for each nucleus studied; and secondly the field strength was limited to about 2 T (some 90 MHz for protons) and, as we saw earlier, there are advantages in using higher fields.

Electromagnets can achieve fields up to a little over 2·5 T, although very sophisticated electronics is needed to stabilize the current to the 1 in 10^8 needed. Higher fields (the current limit is about 10 T) require the use of 'superconducting' magnets. These consist essentially of a few turns of heavy gauge wire cooled in liquid helium; at such a low temperature the electrical resistance of the wire is negligible and so a large, steady current, once started, can pass without loss or change in the field. The obvious disadvantage of such a system is the expense and difficulty of working at low temperatures. The sample, of course, remains at room temperature (or whatever other temperature is required), since it is placed in a thermally insulated cavity in the magnet.

Homogeneity of the field over the sample is ensured, in an electromagnet, by very careful machining of the pole faces, and by the use of secondary 'shim' coils to adjust the field gradients. Also the sample is often spun rapidly to give effective averaging of the field 'seen' by all the nuclei.

Basically the same apparatus is used to observe spectra in the continuous wave mode (frequency domain) and in the Fourier transform mode. We deal with the former first.

7.4.1 Continuous Wave N.M.R. Spectroscopy

The usual arrangement of the components for this method is shown in Fig. 7.24 which, it should be noted, is not drawn to scale. The sample, held in a glass tube some 15 cm long and 0·5 cm or more in diameter, consists of about $\frac{1}{2}$ ml of liquid, either pure or a solution, with a trace of a reference compound added (for example, TMS for hydrogen spectra). The magnet poles are 20–30 cm in diameter, and the gap between them is only some 2–3 cm. When recording a spectrum in the field-sweep mode, the radiofrequency oscillator (the source) bathes the sample in radiation of, say, 100 MHz by means of the coil placed near the sample in a vertical plane. The magnetic field is set to 2·5 T and this is smoothly raised (swept) by means of a current produced in the sweep generator fed to auxiliary coils round the magnet poles. As each nucleus is brought to resonance it absorbs energy from the oscillator and then, when it reverts to the ground state, the emitted energy is collected by the detector coil wound round the sample, amplified, and passed to the recorder.

This method is by far the simplest technically, since it is easy to change a magnetic field smoothly, and fairly easy to maintain a precise frequency

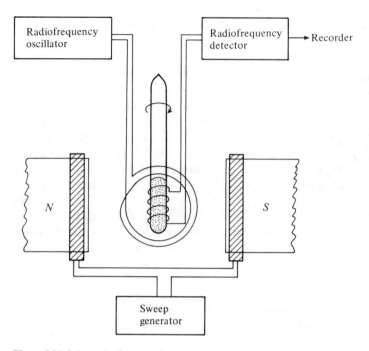

Figure 7.24 Schematic diagram of a continuous-wave n.m.r. spectrometer.

by using a carefully thermostatted crystal oscillator. The alternative frequency-sweep mode, wherein the magnetic field is maintained constant and the radiofrequency oscillator swept through a frequency range, is more difficult, but has advantages, particularly for some double-resonance experiments, and is sometimes employed.

Both of these sweep techniques have proved successful for an adequate quantity of a sample containing either hydrogen, fluorine, or phosphorus nuclei. For instance, some 20 mg of sample dissolved in $\frac{1}{2}$ ml of a solvent enable a good spectrum to be obtained in about 5 min; 10 mg samples take longer, perhaps 20 min. However, it is scarcely feasible to operate with samples containing less than 1 mg of material, or with even quite concentrated samples of nuclei whose signals are inherently weak, such as those from ^{13}C nuclei.

7.4.2 Fourier Transform N.M.R. Spectroscopy

When applied to n.m.r., the FT technique is invariably used in the emission mode; we saw in Sec. 1.8 that this entails collecting the radiation emitted by excited nuclei in the sample as a function of time, storing the collected information in a computer, and mathematically transforming the result into the conventional frequency-domain spectrum. The apparatus required for this is thus very similar to that for continuous sweep measurements, except that the magnet sweep coils are not needed, and of course a computer must be added.

The sample nuclei must initially be excited by irradiation with a beam of the appropriate frequency (i.e., about 100 MHz for protons in a field of 2·5 T); the radiation must contain frequencies spanning the whole spectrum of interest—i.e., 'white' radiation—and must have sufficient power to excite all the nuclei. This immediately raises problems of 'saturation' and, because of this, FT n.m.r. spectroscopy is always carried out by using one or more short pulses of exciting radiation. We discussed the application of a 90° pulse and the subsequent decay of the emission signal in Sec. 7.1.6, and in fact Fig. 7.3 summarizes the essentials of a single FT n.m.r. experiment.

A 90° pulse is ideal, since it produces the maximum signal intensity— the bulk magnetic vector, **M**, of Fig. 7.3 will be tipped completely into the xy plane. One disadvantage of using this size of pulse, however, is that we need to allow a period of some four or five relaxation times between pulses in order to allow spin equilibrium to be reached. For nuclei with a long relaxation time, this delay period is inconveniently long. In practice, therefore, a shorter energizing pulse is often used; this produces a smaller initial signal (because the projection of **M** in the xy plane is less) but, since we are effectively starting part way along the decay curve, needs a shorter time to reach equilibrium. As always a compromise is sought, and a pulse length of some 30° to 50° is often used.

7.5 ELECTRON SPIN RESONANCE SPECTROSCOPY

7.5.1 Introduction

In Chapter 6 we saw that the majority of stable molecules are held together by bonds in which electron spins are opposed; in this situation there is no net electron spin, no electronic magnetic moment, and hence no interaction between the electron spins and an applied magnetic field. On the other hand some atoms and molecules contain one or more electrons with unpaired spins and these are the substances which are expected to show *electron spin resonance* (e.s.r.) spectra; since such substances show bulk paramagnetism, this type of spectroscopy is often referred to as *electron paramagnetic resonance* (e.p.r.).

Substances with unpaired electrons may either arise naturally or be produced artificially. In the first class come the three simple molecules NO, O_2, and NO_2, and the ions of transition metals and their complexes, for example, Fe^{3+}, $[Fe(CN)_6]^{3-}$, etc. These substances are stable and easily studied by e.s.r. Unstable paramagnetic materials, usually called *free radicals* or *radical ions*, may be formed either as intermediates in a chemical reaction or by irradiation of a 'normal' molecule with ultra-violet or X-ray radiation or with a beam of nuclear particles. Provided the lifetimes of such radicals are greater than about 10^{-6} s they may be studied by e.s.r. methods; shorter-lived species may also be studied if they are produced at low temperature in the solid state—so-called *matrix techniques*—since this increases their lifetimes.

Virtually all the theory which we shall need in the discussion of e.s.r. spectra has been dealt with in preceding sections of this chapter; Sec. 7.1 is particularly relevant here, since it covers such matters as the electron's magnetic moment, its interaction with an applied magnetic field to give two energy levels, the populations of those levels, the Larmor precession, and relaxation processes. In fact, with the exception of the magnitudes of some quantities, the whole of Sec. 7.1 can be carried straight over into the discussion of e.s.r. spectroscopy. Parts of Sec. 7.2, such as the remarks regarding spin-spin coupling, are also relevant, as we shall see.

As in all forms of spectroscopy, four properties of the spectral lines are of importance, viz., their intensity, width, position, and multiplet structure. The first two we can deal with quite briefly, the final pair merit separate sections.

The *intensity* of an e.s.r. absorption is proportional to the concentration of the free radical or paramagnetic material present. Thus we have immediately a technique for estimating the amount of free radical present; the method is extraordinarily sensitive, in favourable cases some 10^{-13} mol of free radical being detectable.

The *width* of an e.s.r. resonance depends, as in the case of n.m.r., on the relaxation time of the spin state under study. Of the two possible relaxation processes, the spin-spin interaction is usually very efficient, unless the sample is extremely dilute, and gives a relaxation time of 10^{-6}–10^{-8} s; the spin-lattice relaxation is efficient at room temperature (some 10^{-6} s) but becomes progressively less so at reduced temperatures, often becoming several minutes at the temperature of liquid nitrogen. For most samples, then, we could choose 10^{-7} s as a typical relaxation time and, using this in the Heisenberg uncertainty relation of Eq. (1.11) we calculate a frequency uncertainty (line width) of $(2\pi\ \delta t)^{-1} \approx 1$ MHz. A shorter relaxation time will increase this width, and 10 MHz is not uncommon. Clearly this is a *much* wider spectral line than in the case of n.m.r., where we found a normal line width for a liquid to be some 0·1 Hz.

The wider e.s.r. lines have advantages and disadvantages. On the credit side, the homogeneity of the applied magnetic field is far less critical and where, for n.m.r., it is essential to use a magnetic field homogeneous to 1 in 10^8 over the sample, for e.s.r. a figure of 1 in 10^5 is adequate; this represents a considerable easing of manufacturing tolerances. On the debit side, however, a broad line is more difficult to observe and measure than a sharp one (and it is for this reason that e.s.r. spectrometers nearly always operate in the derivative mode, as described in Sec. 1.4), and any effects equivalent to the chemical shift in n.m.r. will be masked by the overlapping of broad lines.

7.5.2 The Position of E.S.R. Absorptions; the g Factor

We know from Sec. 7.1.2 that the spin energy levels of an electron are separated in an applied magnetic field, B_z, by an amount:

$$\frac{\Delta E}{h} = \frac{g\beta B_z}{h} \quad \text{Hz} \tag{7.16}$$

where β is the Bohr magneton ($9\cdot273 \times 10^{-24}$ J T^{-1}) and g the Landé splitting factor. A resonance absorption will thus occur at a frequency $v = \Delta E/h$ Hz. From Eq. (7.16) we see that the position of absorption varies directly with the applied field and, since different e.s.r. spectrometers operate at different fields, it is far more convenient to refer to the absorption in terms of its observed g value. Thus, rearranging Eq. (7.16) we have:

$$g = \frac{\Delta E}{\beta B_z} = \frac{hv}{\beta B_z} \tag{7.17}$$

and if, for example, resonance were observed at 8388·255 MHz in a field of 0·30 T, it would be reported as resonance at a g value of 2·0023. This very precise figure is the g factor for a free electron (rather than the slightly approximate value of two given by putting $L = 0$ in Eq. (5.28)), but it is a remarkable fact that virtually all free radicals and some ionic crystals have a g factor which varies only, some $\pm 0\cdot003$ from this value. The reason for this is essentially that in free radicals the electron can move about more or less freely over an orbital encompassing the whole molecule (as we shall see in the next section) and it is not confined to a localized orbital between just two of the atoms in the molecule. In this sense it behaves in very much the same way as an electron in free space, having $L = 0$.

Some ionic crystals, on the other hand, have very different g factors, values between about 0·2 and 8·0 having been reported. The difference here is that the unpaired electron is contributed by, and 'belongs' to, a particular atom in the lattice, usually a transition metal ion. Thus the electron is localized in a particular orbital about the atom, and the orbital angular momentum (L value) couples coherently with the spin angular momentum giving rise to a g value consistent with Eq. (5.28).

Nonetheless, many ionic crystals show a g factor very close to the free electron value of 2; this may come about in two ways:

1. The ion contributing the electron may exist in an S state (that is, $L = 0$). For example, the ground state of Fe^{3+}, in which five d electrons are unpaired (that is, $S = \frac{5}{2}$, $2S + 1 = 6$), has zero orbital momentum. Thus $L = 0$, $J = S + L = S$, and the term symbol is $^6S_{5/2}$ (cf. Chapter 5). Since $J = S$, $g = 2$ (Eq. (5.28)).

2. The electric fields set up by all the ions in a crystal may be sufficiently strong to *uncouple* the electron's orbital momentum from its spin momentum—i.e., coherent Russell–Saunders coupling breaks down and, on the application of a magnetic field, the electron spin vector precesses *independently* about the field direction. Thus the value of L is immaterial and the g factor reverts to two. On the other hand, if the internal crystal field is weak, or if the paramagnetic electron is well shielded from the field (e.g., as in a rare-earth metal, where the relevant electron orbit is buried deep within outer electron shells), L and S couple to produce a resultant J which itself precesses about the applied magnetic field, and g is given by Eq. (5.28). Intermediate cases also occur where L and S are only partly uncoupled, the residual orbital contribution to the energy giving rise to a g value not easily predictable theoretically.

7.5.3 The Fine Structure of E.S.R. Absorptions

In e.s.r. spectroscopy we must distinguish between two kinds of multiplet structure; there is the *fine structure* which occurs only in crystals containing more than one unpaired electronic spin, and the smaller *hyperfine structure*. We shall deal with the fine structure first.

Consider the case of a crystal which contains molecules or ions with two parallel, rather than paired, electron spins, resulting in a total spin of 1; this is, of course, a *triplet* state, since $S = 1$ and $2S + 1 = 3$. Molecular triplet states are often unstable, reverting to the singlet state with paired spins. For example, in the case of naphthalene irradiated with ultra-violet light, individual molecules undergo excitation to the triplet state which decays quite rapidly to the ground state. Here, however, by cooling the crystal to low temperatures, or by diluting the naphthalene in a solid, inert lattice, the triplet state can be maintained and examined by e.s.r. techniques. On the other hand transition metal ions often exist quite stably in a triplet state and are easily studied at room temperature by e.s.r.

The two electrons per molecule or per ion forming the triplet state can be treated for most e.s.r. purposes as a single particle of spin 1. Thus the angular momentum vector corresponding to $S = 1$ is given by:

$$\mathbf{S} = \sqrt{S(S + 1)}\, \frac{h}{2\pi} = \sqrt{2} \text{ units}$$

and, in the absence of a field, the vectors orient themselves randomly. When a magnetic field is applied, or if an internal field exists within the crystal, the

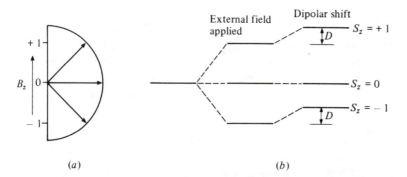

Figure 7.25 (a) The allowed orientations of two parallel electron spins in a magnetic field, B_z. (b) The splitting of the energy into three levels, and the dipolar shift, D, raising the $S_z = \pm 1$ states.

vectors can take up one of three directions only—essentially with, across, or against the field direction, as shown in Fig. 7.25(a); the components of the angular momentum in the field direction are, of course, $S_z = +1$, 0, or -1. In the field the $S_z = +1$ state is raised in energy, the $S_z = -1$ lowered, and the $S_z = 0$ unaffected, as shown in the centre of Fig. 7.25(b); (note that the $+1$ state is *raised* in energy, whereas (cf. Fig. 7.1(c)) when dealing with nuclei of spin 1 the $+1$ state is lowered; this reflects, of course, the opposite sign of nuclear and electron charges). Under the selection rule, $\Delta S_z = \pm 1$ only, two transitions are allowed, but both would have identical energy, and hence give rise to just one spectral line. In fact *two* lines of different energy are invariably observed, and we must now consider why it is that the energy levels are split unsymmetrically; there are several reasons.

First, remembering that each $S = 1$ state is, in fact, made up of two electrons with parallel spins, we know that each spin produces a small magnetic field in the vicinity of its partner. This effect is identical with that of dipolar spin-spin coupling described in Fig. 7.12 for two *nuclei*, except that it is in the opposite direction, since nuclei and electrons have opposite charges—thus here an up spin increases its neighbour's field, while a down spin decreases it. Hence for our two electrons, *both* spins in the $S_z = +1$ state feel an applied field rather greater than the external field, B_z, and this state is *raised* in energy; for $S_z = -1$ both feel a smaller field, which again *raises* their energy (remember that increasing the applied field *raises* the $S_z = +1$ energy, but *lowers* the $S_z = -1$). In the case of $S_z = 0$, the dipolar field is *across* the main applied field so neither adds nor subtracts from it. The $S_z = 0$ energy is, therefore, quite unaffected by dipolar coupling.

The net effect of dipolar interaction is, then, to raise *both* the $S_z = +1$ and $S_z = -1$ states with respect to $S_z = 0$, as shown on the right of Fig. 7.25(b). Here we show the dipolar raising, D, to be smaller than the splitting caused by the main applied field; for small fields, however, it is clear that both $S_z = +1$ and $S_z = -1$ may lie above $S_z = 0$ in energy, since the magnitude of the dipolar shift is quite independent of the applied field.

Secondly, there may be spin-orbit coupling, whereby the electrons' orbital angular momenta and spin momenta are combined. Since the net spins of $S_z = +1$ and $S_z = -1$ are oppositely aligned, such coupling may again disturb the energy level pattern, but the $S_z = 0$ state will be unaffected. The spin-orbit effect may be in the same sense or opposed to the dipolar interaction.

Finally, if there is a strong internal electric field within the crystal (and there will be, unless the substance is highly symmetrical), this will result in further perturbations to the energy levels of $S_z = +1$ and $S_z = -1$, but not of $S_z = 0$.

The net result of these three energy perturbations is usually considered under the one heading of the *zero-field* (or *crystal-field*) effect, since they produce an energy level shift of the $S_z = +1$ and -1 states with respect to $S_z = 0$ even in the absence of an external field. This situation is shown at Fig. 7.26(a), where the $S_z = \pm 1$ states are assumed to lie above the $S_z = 0$; the opposite shift may also arise, but this does not alter the following discussion of the spectrum. When a steadily increasing magnetic field is applied to the crystal, as in (b) of the figure, the $S_z = \pm 1$ levels diverge and, for a given radiation frequency, it is clear that there will be transitions at two different applied fields; thus the spectrum will consist of two fine-structure lines.

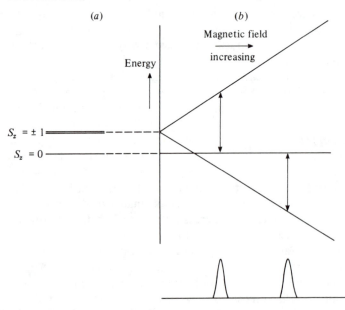

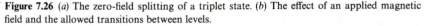

Figure 7.26 (a) The zero-field splitting of a triplet state. (b) The effect of an applied magnetic field and the allowed transitions between levels.

The zero-field splitting is usually a large effect; for example the observed separation between the $S_z = 0$ and the $S_z = \pm 1$ levels in the naphthalene triplet state is equivalent to about 3000 MHz, which is to be compared with the energy separation of some 8000 MHz for an externally applied field of 0·3 T; some transition metal ions have much larger zero-field splittings. Thus in a field-sweep spectrum, as indicated at the foot of Fig. 7.26(b), the fine structure separation is large, some 0·1–0·2 T or more, compared with the line width of 10^{-3}–10^{-4} T.

Extension of the above arguments shows that fine structure-splitting will occur similarly for crystals containing molecules or ions with more than two parallel spins. In general, if there are n parallel spins, there will be n equally spaced resonances in the e.s.r. spectrum.

7.5.4 Hyperfine Structure

This arises through coupling of the unpaired electron with neighbouring nuclear spins in much the same way as the coupling between nuclear spins discussed in Sec. 7.2.2. In general a nucleus with spin I will split the resonance line of an electron into a multiplet with $2I + 1$ lines of equal intensity. The separation between the lines is usually of the order of 10^{-3}–10^{-4} T (i.e., some 50 MHz) which is larger by a factor of approximately 10^6 than nucleus–nucleus coupling. The reason is that an electron can approach a nucleus much more closely than can another nucleus, and so will interact more strongly with it.

The biggest factor influencing the magnitude of electron–nucleus coupling is the amount of time which the electron spends in the vicinity of the coupled nucleus or, in other words, the *electron density* at the nucleus. We can express this as:

$$A = R\rho \tag{7.18}$$

where A is the observed coupling, ρ the electron density, and R the intrinsic coupling for unit density. Thus for the hydrogen atom in the ground state the observed hyperfine splitting is some 50 mT (incidentally the largest known) and, since the electron density in the $1s$ orbital must be unity (the electron is nowhere else), R is also 50 mT. Now in the methyl radical, CH_3, the electron resonance is observed to have a quartet structure with lines of intensity ratio $1:3:3:1$ and a separation of 2·3 mT. The quartet pattern is consistent with the interaction of the electron *equally* with all three hydrogens (cf. the similar pattern for the CH_2 group coupled to CH_3 in the n.m.r. spectrum of Fig. 7.16), and we calculate from Eq. (7.18) that $\rho = 2\cdot3/50 = 0\cdot046$. This implies that the electron spends some five per cent of its time in the $1s$ orbital of each hydrogen, the remaining 85 per cent in the neighbourhood of the carbon atom.

We can take this argument further. The coupling between each ring hydrogen and the unpaired electron in the free radical p-benzosemiquinone:

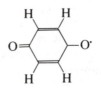

is about 0·24 mT. This tells us, firstly, that the electron density at each hydrogen is 0·24/50 ≈ 0·005. Now in the methyl radical considered above, the coupling between electron and hydrogen when the electron density at the carbon is 0·85, is 2·3 mT; it follows that the electron density at each of the ring carbon atoms in p-benzosemiquinone must be just one-tenth of this, 0·085, in order to give a coupling one-tenth as large. Thus we know the electron density at each atom of the molecule: each H, 0·005; each C, 0·085; each O (by difference), 0·24; or $\frac{1}{2}$, 8·5, and 24 per cent of the electron's time is spent at each nucleus, respectively.

This use of e.s.r. techniques allows us to build up a qualitative picture of the electron distribution within a molecule which may help, for instance, in understanding chemical reactions—a positively charged reactant will plainly tend to attack that part of a molecule where the electron density is greatest, and vice versa. However, the observed results are not always in good quantitative agreement with the predictions of the molecular orbital theory, particularly for situations where simple molecular orbital theory predicts zero electron density at some points within a molecule—often coupling is observed to occur with nuclei at these points. For instance, the methyl radical already mentioned is a planar molecule, with the carbon and three hydrogen nuclei in the same plane (Fig. 7.27(a)); the unpaired electron is considered to be held in a p orbital with lobes above and below this plane and, since the plane is a region of zero electron density for the orbital, there can be no electron density at the hydrogen nuclei arising directly from the

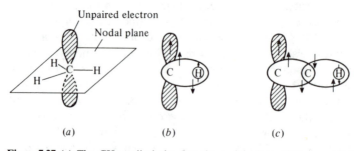

(a)   (b)   (c)

Figure 7.27 (a) The CH_3 radical showing the nodal plane of the p orbital containing the unpaired electron; (b) positive and (c) negative coupling between the unpaired electron and a neighbouring nuclear spin.

unpaired spin. Thus $\rho = 0$ and so, since A is observed to be non-zero, it would seem that Eq. (7.18) does not apply.

A better account of the observed coupling can be given by following an argument very similar to that given in Sec. 7.22 and Fig. 7.13(*i*) for indirect nucleus–nucleus coupling. Thus in Fig. 7.27(*b*) the unpaired electron, with spin up, say, can be considered to polarize the other carbon electrons preferentially into the spin up situation (Hund's rule), so that the other electron in each C—H bond will have a spin down orientation (Pauli's principle), and this will tend to align the hydrogen nuclear spins in the up direction. Hence the unpaired spin and the nuclear spin will tend to lie in the same direction and coupling exists. We can still apply Eq. (7.18) provided we agree to take ρ as a measure of the *spin* density of the unpaired electron (i.e., its influence at a remote point) rather than its *physical* density.

By the same argument it is clear that negative coupling—*negative spin density*—can occur. If we add a further carbon atom into the chain, the situation is as shown in Fig. 7.27(*c*); here the two electrons on the new carbon atom will tend to have parallel spins (Hund's rule again), and so the unpaired electron–nucleus configuration will be in the opposite sense to that in (*b*).

These 'configuration interaction' effects can be allowed for quantitatively in a refined version of the molecular orbital theory, and then excellent agreement is observed with experimental results. It seems that there is now ample justification for using e.s.r. spectrocopy in the accurate measurement of electron spin densities at various points within a molecule.

7.5.5 Double Resonance in E.S.R.

The concept of double resonance in e.s.r. spectroscopy exactly parallels that in n.m.r.—observation of a spectrum at one frequency while simultaneously irradiating at another. In e.s.r. there are two possibilities—the second frequency may be either at *nuclear* or at *electron* resonance frequencies—and these are called *endor* (electron nuclear double resonance) and *eldor* (electron–electron double resonance) respectively. We shall deal only with the former; it is technically the simpler and is used more often than *eldor*.

Consider the simple case of a single unpaired electron interacting with a nucleus of spin $\frac{1}{2}$—an organic radical where the electron couples with a hydrogen nucleus is an obvious example. In an applied field both the electron and the nucleus will occupy one of two different energy states (spin 'up' or 'down'), and so there is a combined total of four energy states available. We may build up the energy level diagram for the system in the following way (Fig. 7.28). The electron energy is split by the applied field into two widely separated states—if the field is 0·3 T then the separation is some 9000 MHz—as shown on the extreme left. Remember (and cf. Sec. 7.1.2) that the 'down' spin of an electron is more stable than the 'up' spin, because

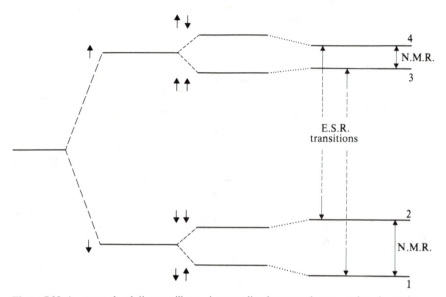

Figure 7.28 An energy level diagram illustrating coupling between electron and nuclear spins.

of the electron's negative charge. In the same field the hydrogen energy is split by some 13 MHz only, so we can initially simply add or subtract the nuclear energy, depending upon whether the nuclear spin is down or up, arriving at the four levels labelled ↓↑, ↓↓, ↑↑, and ↑↓ in the figure. Here we show the electron spin first, and remember again that the nuclear spin 'up' is stabilized in a field. This gives the energies in increasing order as shown.

The arguments so far, however, have ignored electron–nucleus coupling—we have simply superimposed the nuclear energy values on to those of the electron and have not included the possibility of mutual interaction of the energy states. When we do include this, we remember that states with paired spins (↓↑ and ↑↓) tend to be stabilized, or lowered in energy, while parallel spins (↑↑ and ↓↓) are raised. The dotted lines of Fig. 7.28 indicate the effects of these interactions, yielding the set of energy levels on the right of the figure. The reader should note the similarity between this figure and Fig. 7.14, which describes the energy levels for two coupled *nuclei*. The only difference is that of scale—in Fig. 7.14 the nuclear frequencies are all the same order of magnitude; here the electron and nuclear energy-level spacings differ by a factor of 1000.

Transitions in the e.s.r. region (those which change the electron's spin only) are shown in Fig. 7.28 and, as expected, give rise to a doublet due to coupling with H. Similarly the n.m.r. transitions, involving the hydrogen spin only, also show two resonances. It is not particularly easy to observe the n.m.r. resonance of a radical directly—many factors are against this, such as the small concentration of radical giving a weak signal; broadening due to the efficient relaxation by the 'free' electron; stability problems with the radical, etc.). We can, however, see the *effect* of applying n.m.r. frequency radiation while looking at the e.s.r. spectrum.

Imagine bathing the sample in fairly intense e.s.r. radiation so as to saturate, say, the $1 \rightarrow 3$ transition. This means that there will be no net e.s.r. signal observed, because the populations of levels 1 and 3 are equal. If we now also apply radiofrequency (n.m.r.) radiation to the sample, while still observing at the e.s.r. frequency, we can imagine sweeping the n.m.r. radiation slowly upwards. At some point we shall induce the $3 \rightarrow 4$ transition, thus raising some molecules to level 4 and so decreasing the population of level 3. As soon as this happens, e.s.r. radiation will be absorbed once more, because there is now 'room' in level 3 for the transition $1 \rightarrow 3$ to occur. So sweeping through the *n.m.r.* resonance causes a signal in the *e.s.r.* spectrum. If we continue to raise the n.m.r. frequency until resonance with the $1 \rightarrow 2$ transition occurs, we shall again disturb the relative populations, and another e.s.r. signal will be seen.

Initially, then, the endor spectrum of this system does not look very exciting—two lines separated by the coupling constant, which is exactly the same information as can be gained from the e.s.r. spectrum alone. There are, however, two reasons why endor techniques are useful.

Electron spin resonances are often very broad and it is not always possible to resolve spectra sufficiently well to see splitting due to nuclear coupling. Endor spectra, on the other hand, depend on the width of the *nuclear* resonance, because it is only while passing through the latter that an e.s.r. signal is observed. Although the nuclear signal is often considerably broadened by relaxation *via* the free electron, it is nonetheless much sharper than e.s.r., so nuclear coupling is often very much easier to observe, and the coupling constants themselves are more accurately evaluated, from the sharper spectral lines given by endor.

Additionally, if the e.s.r. coupling is to a nucleus with a spin greater than $\frac{1}{2}$, perhaps to nitrogen ($I = 1$) in an organic molecule, or to a metal ion in a crystal, the e.s.r. signal is split into $I + 1$ lines. If resolved, this splitting gives a complex spectrum; if unresolved, it gives a very broad signal. The endor spectrum is much simpler; whatever the spin of the coupled nucleus, its resonance is split only into a doublet by interaction with the single electron.

We should point out that the above description of endor does not represent the way in which such experiments are actually carried out. We have discussed fixed energy levels and varying applied frequencies. In fact it is technically simpler to keep the applied e.s.r. frequency constant and to vary the energy levels by varying the magnetic field until resonance occurs. It is simpler to think of the process in the way we have done above, however, and none of the principles are altered.

7.5.6 Techniques of E.S.R. Spectroscopy

We have already seen that e.s.r. spectroscopy, as usually carried out, falls into the microwave region of the spectrum. The brief description of micro-wave apparatus already given in Sec. 2.5.1 applies equally to e.s.r., therefore, with the additional requirement for a magnet operating at some 0·3 T. The magnet is invariably an electromagnet, and is equipped with subsidiary coils and a sweep generator to give the wide field scan necessary for e.s.r. work.

Additionally, coils are usually fitted to supply an oscillating magnetic field of strength a few microtesla and frequency some hundreds of cycles per second; this has two purposes. Firstly, it provides a modulator (or 'chopper') frequency to which the detector-amplifier system may be tuned in order to improve the signal-to-noise ratio. Secondly this arrangement can be used to display the spectrum in the derivative mode, as is usual for e.s.r. In Fig. 7.29(a) we show the (constant) amplitude of the modulation field as

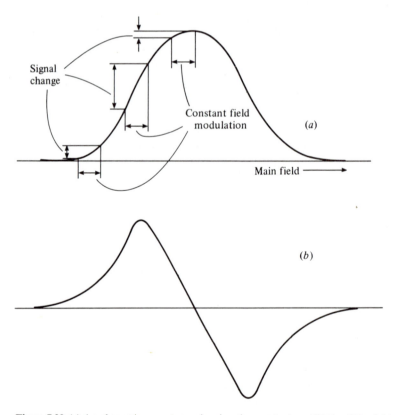

Figure 7.29 (a) An absorption spectrum showing the constant amplitude of the field modulation and consequent signal change leading to (b), the corresponding dispersion spectrum.

the main field is swept over the absorption spectrum. In regions where the absorption spectrum is varying slowly the signal change monitored over each cycle of the modulation is small, whereas when the spectrum changes rapidly, the signal change is large. Thus the signal change measures the *slope* of the spectrum and this may be recorded directly as the derivative curve, as in Fig. 7.29(*b*). The latter has two main advantages for e.s.r. spectroscopy: (1) the point of maximum absorption is difficult to measure accurately with a broad absorption curve, but is shown with much greater precision as the intersection of two lines in the derivative mode; (2) it is also more accurate to estimate the intensity of a derivative signal than that of the corresponding broad absorption.

BIBLIOGRAPHY

Abraham, R. J., and P. Loftus: *Proton and Carbon-13 N.M.R. Spectroscopy: An Integrated Approach*, Heyden & Son, 1978.

Akitt, J. W.: *N.M.R. and Chemistry: An Introduction to N.M.R. Spectroscopy*, Chapman & Hall, 1973.

Carrington, A., and A. D. McLachlan: *Introduction to Magnetic Resonance*, Chapman & Hall, 1979.

Casy, A. F.: *PMR Spectroscopy in Medicinal and Biological Chemistry*, Academic Press, 1971.

Dixon, W. T.: *Theory and Interpretation of Magnetic Resonance Spectra*, Plenum Press, 1972.

Farrar, T. C., and E. D. Becker: *Pulse and Fourier Transform N.M.R.: Introduction to Theory and Methods*, Academic Press, 1971.

Gerson, F.: *High Resolution E.S.R. Spectroscopy*, John Wiley, 1971.

Griffiths, P. R.: *Transform Techniques in Chemistry*, Plenum Press, 1978.

Hecht, H. C.: *Magnetic Resonance Spectroscopy*, John Wiley, 1967.

Kevan, L., and R. N. Schwartz (eds): *Time Domain Electron Spin Resonance*, Wiley, 1979.

Lucken, E. A. C.: *Nuclear Quadrupole Coupling Constants*, Academic Press, 1969.

Martin, M. L., G. J. Martin, and J. J. Delpuech: *Practical N.M.R. Spectroscopy*, Heyden & Son, 1980.

Mooney, E. F.: *An Introduction to ^{19}F N.M.R. Spectroscopy*, Heyden & Son, 1970.

Mullen, K., and P. S. Pregosin: *Fourier Transform in N.M.R. Techniques: A Practical Approach*, Academic Press, 1974.

Shaw, D.: *Fourier Transform N.M.R. Spectroscopy*, Elsevier, 1976.

Wehrli, F. W., and T. Wirthlin: *Interpretation of Carbon-13 N.M.R. Spectra*, Heyden & Son, 1976.

Wertz, J. E., and J. R. Boulton: *Electron Spin Resonance: Elementary Theory and Practical Applications*, McGraw-Hill, 1972.

Williams, D. H., and I. Fleming: *Spectroscopic Methods in Organic Chemistry*, 3rd ed., McGraw-Hill, 1980.

PROBLEMS

(Useful constants: $h = 6.626 \times 10^{-34}$ J s; $\beta_N = 5.051 \times 10^{-27}$ J T^{-1}.)

7.1 A particular n.m.r. instrument operates at 30·256 MHz; what magnetic fields are required to bring a hydrogen nucleus and a ^{13}C nucleus to resonance at this frequency? (Use data from Table 7.1.)

7.2 The four lines from an AX spectrum are observed at $\delta = 5\cdot8$, $5\cdot7$, $1\cdot1$, $1\cdot0$ (measured from TMS with an instrument operating at 100 MHz). What are the chemical shift positions (in δ) of the A and X nuclei, and the coupling constant (in hertz) between them?

7.3 Vinyl fluoride exhibits the following approximate coupling constants:

$$J_{H'H''} = 5 \text{ Hz} \qquad J_{H'F} = 85 \text{ Hz}$$
$$J_{H'H'''} = 13 \text{ Hz} \qquad J_{H''F} = 50 \text{ Hz}$$
$$J_{H''H'''} = -3 \text{ Hz} \qquad J_{H'''F} = 20 \text{ Hz}$$

Sketch the n.m.r. spectrum of the fluorine nucleus, assuming that the chemical shift differences between the hydrogen nuclei are all large compared with their couplings.

7.4 A fictitious non-spinning nucleus B forms pentavalent compounds; one such is H_4B—CH_3, where the BH_4 group forms a square pyramid with B at the apex, and the methyl group is free to rotate about B—C bond. Sketch the n.m.r. spectrum, assuming that all chemical shifts are large compared with the couplings.

7.5 Predict the number of lines in the e.s.r. spectrum of each of the following radicals:

    (a) $[CF_2H]^{\bullet}$ (b) $[^{13}CF_2H]^{\bullet}$
    (c) $[CF_2D]^{\bullet}$ (d) $[CClH_2]^{\bullet}$

(The spin of D is 1; other spins are given in Table 7.1.)

7.6 CD_3COCD_3 (deuterated acetone) is often used as a solvent for n.m.r. spectra since it should exhibit no proton resonance. In fact samples usually contain some residual CD_3COCD_2H. Predict the spectrum of this molecule, assuming no coupling between the CD_3 and CD_2H groups, and remembering that the deuterium nucleus has a spin of 1.

7.7 Although the low-temperature n.m.r. spectra of solids usually show broad resonance, raising the temperature often causes a sudden narrowing, even though the sample is still well below its melting point. Explain this observation.

EIGHT

MÖSSBAUER SPECTROSCOPY

Mössbauer spectroscopy, named after its discoverer who received a Nobel prize in 1961 for his work, is concerned with transitions between energy levels within the nuclei of atoms. About a third of the known elements, principally the heavier ones, when formed by the radioactive decay of an isotope of the same or a different element, are initially produced in an excited nuclear state; after a very short delay, of the order of microseconds, the excited nucleus reverts to the ground state and emits energy of a very high frequency, usually in the γ-ray region of the spectrum. It is the study of this γ-ray emission and subsequent reabsorption which constitutes Mössbauer or γ-ray spectroscopy.

8.1 PRINCIPLES OF MÖSSBAUER SPECTROSCOPY

In this section we outline the essentials of γ-ray spectroscopy using as our example the iron nucleus—one of the most thoroughly studied in this respect. The isotope ^{57}Fe is conveniently produced by the decay of radioactive ^{57}Co, which is a relatively long-lived species, having a half-life of some 270 days. A simplified energy level diagram for the process is shown in Fig. 8.1 where, following electron capture, the cobalt nucleus is seen to produce ^{57}Fe in an excited energy state, designated Fe*; Fe* very rapidly

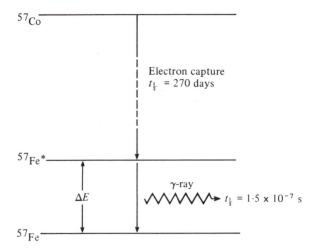

Figure 8.1 Simplified energy level scheme showing the decay of radioactive ^{57}Co to excited state and then ground state ^{57}Fe.

drops to the ground state Fe, the energy change involved being $\Delta E = 2 \cdot 30 \times 10^{-15}$ J (per nucleus). The frequency of the emitted γ-ray is thus $v = \Delta E/h = 3 \cdot 5 \times 10^{18}$ Hz. Now if a second Fe nucleus, initially in the ground state, were to be put in the emitted radiation beam, it would be expected to absorb energy from the beam and to become excited into the Fe* state. A γ-ray detector (a scintillator or a geiger counter) placed in the beam behind the absorber should show that this has happened. In fact, it was found possible to observe γ-ray absorption only if the source material were moved relative to the sample; the two main reasons for this are of fundamental importance, and we discuss them now.

Firstly, consider the width of the γ-ray emission; the half-life of the excited state Fe* is about $1 \cdot 5 \times 10^{-7}$ s and, replacing this in Eqs (1.10) and (1.11), we see that the Heisenberg uncertainty principle gives the energy uncertainty as $\delta E \approx 10^{-34}/\delta t \approx 10^{-27}$ J, and the frequency uncertainty as $\delta v \approx 10^{6}$ Hz. These uncertainties are very small when compared with the energy change of 10^{-15} J, and the associated frequency of $3 \cdot 5 \times 10^{18}$ Hz— the excited state energy level is very precisely defined. In fact, the relative line width, $\delta v/v \approx 10^{-12}$, is very much smaller than that found in any other spectroscopic technique. For instance in n.m.r. spectra we have encountered a line width of some $0 \cdot 1$ Hz in 100 MHz, for which $\delta v/v \approx 10^{-8}$; in the infra-red region, $\delta v/v$ is about 10^{-5} for gases, rather larger for liquids.

The second factor is that, with such very sharp emission frequency, any effect which produces a change in the nuclear energy levels or in the radiation frequency itself will prevent resonance reabsorption. One very important effect exists in the motion of the emitting nucleus. A photon of frequency 10^{18} Hz has a relatively large momentum (given, according to the de Broglie relationship, by h/λ where λ is the wavelength) and, when this is ejected by the nucleus, the nucleus will recoil considerably in order to conserve the total momentum. A simple calculation shows that the nuclear recoil velocity is of the order 10^2 m s^{-1}. Now it is well known that, when a moving body emits radiation (or sound) a stationary observer sees (or hears) a shifted frequency; this, of course, is called the Doppler effect. Specifically the frequency shift Δv is given by:

$$\Delta v = \frac{vv}{c} \quad \text{Hz} \tag{8.1}$$

where v and c are the frequency and velocity of the emitted radiation, respectively, and v is the relative velocity of source and observer. For our Fe* nucleus, recoiling at 10^2 m s^{-1} and emitting radiation at $3 \cdot 5 \times 10^{18}$ Hz:

$$\Delta v = \frac{3 \cdot 5 \times 10^{18} \times 10^2}{3 \times 10^8} \approx 10^{12} \text{ Hz}$$

This shift, although small relative to the emission frequency of 10^{18} Hz, is very large indeed compared with the line width of 10^6 Hz.

It is these factors then—nuclear recoil moving the emitted frequency some millions of line widths away from the position of the absorption frequency—which earlier prevented the study of γ-ray spectroscopy. Mössbauer's main contributions—elegantly simple when seen in retrospect—were, firstly, to use solid crystal lattices as emitters, in which the emitting nucleus is firmly fixed to surrounding nuclei, and hence has a very large apparent mass within which the recoil energy can be dissipated, and further to cool both source and sample to low temperatures so that thermal motions of the lattice atoms are reduced to a minimum.

In this way it was shown, for instance, that the γ-ray emission from a piece of radioactive cobalt metal (that is, Fe* nuclei in a metal lattice) could be absorbed by metallic iron. However, it was also found that if the iron sample were in any other chemical state, the different chemical surroundings of the iron nucleus produce a sufficient effect on the nuclear energy levels for absorption no longer to occur. The final requirement in our spectroscopy—we already have source, sample, and detector—is some scanning device so that we may search for the precise absorption frequency of a particular sample. The Doppler effect here becomes useful rather than a nuisance.

We have seen that a velocity of 10^2 m s^{-1} produces a huge Doppler shift; a similar calculation from Eq. (8.1) shows that a velocity of 1 cm s^{-1} (10^{-2} m s^{-1}) produces a shift of 10^8 Hz—which, being about 100 line widths, represents a reasonably wide scan of the spectrum. Experimentally,

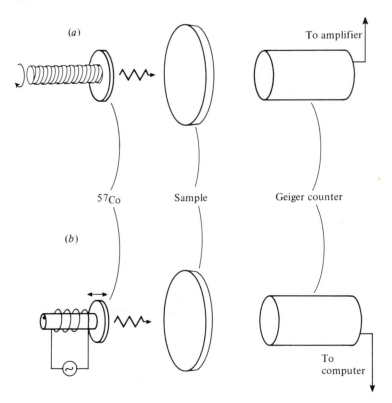

Figure 8.2 Two experimental arrangements for Mössbauer spectroscopy; (*a*) shows a screw-thread drive, and (*b*) an oscillating drive for the ^{57}Co source.

then, we could proceed as in Fig. 8.2(*a*). Here a piece of ^{57}Co is mounted on a screw thread, rotation of which gives a steady velocity drive to the source. A geiger counter mounted behind the sample will show a sudden fall in the count rate when the sample starts to absorb the γ-rays emitted by the source. A complete spectrum will have to be examined point by point, since for any one source velocity the Doppler shift is constant, and so one would need to set up perhaps a hundred different relative velocities of source and sample, varying from $+1$ cm s^{-1} through 0 to -1 cm s^{-1}, to cover a reasonable range of frequencies.

A much more convenient arrangement is shown in Fig. 8.2(*b*) where the source metal is seen mounted on what is essentially a loud-speaker coil. An alternating current of a few cycles per second is applied to the coil so that the source oscillates back and forth. At the extremes of its motion it will have zero velocity relative to the sample, whereas at the centre it will have its maximum velocity either towards or away from the sample. The geiger counter output now needs to be fed to a multi-channel computer wherein the results from each point of the source's movement are collected and summed over each cycle. A time of several minutes to a few hours is usually sufficient to record a good spectrum. The final spectrum is displayed on counts per second versus centimetres per second scales, a fall in counts per second indicating γ-ray absorption by the sample.

In essence then, Mössbauer spectroscopy is simple. Excited nuclei must be available from some source, which should be fairly stable (i.e., have a half-life of at least several weeks) so that its frequent replacement is unnecessary, and so that the rate of decay, and hence the intensity of γ-emission, stays sensibly constant during an experiment of several hours. The excited nuclei must decay to the ground state rapidly, emitting γ-rays in the 10^{17}–10^{20} Hz range. A Doppler shifting device, a geiger counter, and a small computer complete the apparatus. Against this apparent simplicity, however, it must be remembered that the relative velocity of source and sample must be *very* precisely controlled (an error of 0.01 cm s^{-1} in the velocity shifts the frequency by more than one line width and could easily render an absorption undetectable), and that the source and sample should ideally be maintained at the temperature of liquid helium. However, as stated earlier, many nuclei fulfil the required conditions and have been studied by γ-ray techniques. We turn now to consider some applications.

8.2 APPLICATIONS OF MÖSSBAUER SPECTROSCOPY

8.2.1 The Chemical Shift

It has already been mentioned that a nucleus in chemical surroundings different from those of the source does not absorb at the same frequency; this effect is referred to as the chemical shift, or sometimes as the isomer shift, and an observed shift is usually reported in cm s^{-1}, conversion to hertz or megahertz being trivial but unnecessary. Such a shift is illustrated in the Mössbauer spectrum of the ferrocyanide ion, $[Fe(CN)_6]^{4-}$, in Fig. 8.3(a), where we see that the single sharp absorption peak occurs at a velocity of about $-\frac{1}{2}$ mm s^{-1} with reference to the *Fe in a ^{57}Co source as zero.

The main factor affecting the magnitude of chemical shifts is the electron density at the nucleus concerned and, since p, d, etc., orbitals have zero density at their nuclei, we may be quite specific and say that it is the s orbital density which is important. Observation of shifts, then, allows measurement of relative s electron density which, in turn, gives an estimate of the bond character of atoms or ions chemically attached to the Mössbauer nucleus. Thus in a series of tin compounds, using the Mössbauer isotope ^{119}Sn, the chemical shifts given in Table 8.1 are observed; the outer electron structure of the tin atom is $5s^2 5p^2$, so in the 4+ state tin has no

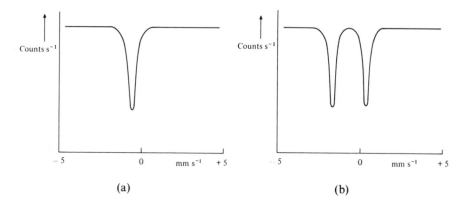

Figure 8.3 The Mössbauer spectra of (a) the $[Fe(CN)_6]^{4-}$ ion and (b) the $[Fe(CN)_5NO]^=$ ion, showing quadrupole splitting.

outer s electrons, whereas in the $2+$ state it is assumed that the (higher) $5p^2$ electrons are removed, leaving two s electrons. In its four-covalent state, where the compounds are tetrahedral and the configuration is $5(sp^3)$, there is essentially only one s electron. The chemical shifts reflect these structures almost linearly. Clearly it is a relatively simple matter to discover the valence state of an unknown tin compound from its Mössbauer spectrum.

Another important factor affects chemical shifts; nucleic in excited states usually have a different radius from those in the ground state—they may be either smaller or larger. Theoretically it may be shown that the chemical shift is given by:

$$\text{Chemical shift} = \text{const } (\rho_{ex.}^2 - \rho_{gd.}^2)\left(\frac{R_{ex.} - R_{gd.}}{R_{gd.}}\right) \qquad (8.2)$$

where ρ^2 represents s electron density at the excited and ground state nuclei, and R is the radius of each nucleus; the constant is simply a universal proportionality factor. For any given nucleus, of course $(R_{ex.} - R_{gd.})/R_{gd.}$ is constant, and so relative electron densities can be determined as stated above. However, the *sign* of the chemical shift will depend on whether $R_{ex.}$

Table 8.1 Chemical shift of some ^{119}Sn compounds

Valence state	Electron configuration	Chemical shift (mm s^{-1})
Sn^{4+}	$5s^0 5p^0$	0
Sn (4-covalent)	$5(sp^3)$	2·1
Sn^{2+}	$5s^2 5p^0$	3·7

is larger or smaller than $R_{gd.}$. One way to discover which is the larger is to examine a sample under high pressure; here the assumption is that the compression increases the electron density near the sample nucleus, so $(\rho_{ex.}^2 - \rho_{gd.}^2)$ decreases, and then the sign of the observed chemical shift change gives the sign of $(R_{ex.} - R_{gd.})$.

Measurement of change of nuclear size on excitation clearly has great importance for theories of nuclear structure. For instance, it is an observed (but as yet unexplained) fact that in ^{129}I the excited nucleus is larger, whereas in ^{127}I the ground state nucleus has the larger radius—it is rather surprising that the addition of two neutrons should have such a profound effect on the nuclear structure. These nuclei are formed in the excited state by decay from the radioactive tellurium nuclei, ^{129}Te and ^{127}Te.

8.2.2 Quadrupole Effects

It happens that the majority of Mössbauer nuclei have non-zero spin and, further, that most of them have half-integral rather than integral spins. The spin of the excited state is invariably different from that in the ground state (and, indeed, there is a selection rule which says that this must be so), and so it follows that either or both of the nuclear states must involve a spin greater than $\frac{1}{2}$, that is, one or both will have a quadrupole moment (cf. Sec. 7.3.5), and this will interact with electric field gradients in the vicinity.

A fairly common situation is for the excited state nucleus to have $I = \frac{3}{2}$ (and thus a quadrupole moment), and the ground state $I = \frac{1}{2}$; this is found, for example, in ^{57}Fe, ^{119}Sn, and ^{129}Xe. For this case the four (that is, $2I + 1$) possible orientations of the excited nucleus in an electric field along the vertical (z) direction are shown in Fig. 8.4(a). Now since the angle which the cigar-shaped nucleus (assuming the nucleus to have a *positive* quadrupole moment) makes with the field gradient is the same in the $I_z = +\frac{1}{2}$ and the $I_z = -\frac{1}{2}$ states, these two states have the same energy in the field. Similarly the $I_z = \pm\frac{3}{2}$ states have the same energy, although obviously this energy will be different from that of the $\pm\frac{1}{2}$ states. Thus the excited nuclear energy, shown in (b) of the figure, splits into two levels, part (c), when an electric field gradient exists at the nucleus. With a positive quadrupole moment the $\pm\frac{3}{2}$ states are raised in energy and the $\pm\frac{1}{2}$ states lowered, whereas if the quadrupole moment is negative (tangerine-shaped nucleus) the reverse is true. The splitting is small, with normal electric fields but, as the spectrum of the nitroprusside ion, $[Fe(CN)_5NO]^=$, in Fig. 8.3(b) shows, it gives quite an observable effect. It should be noted that the electric field causing the splitting here is *not* externally applied, but is inherent in the structure of the ion. Thus comparing Fig. 8.3(a) with (b), we see that the $Fe(CN)_6^{4-}$ ion is sufficiently symmetrical for there to be no net field gradient at the iron nucleus, while replacement of one CN group by NO produces an internal field.

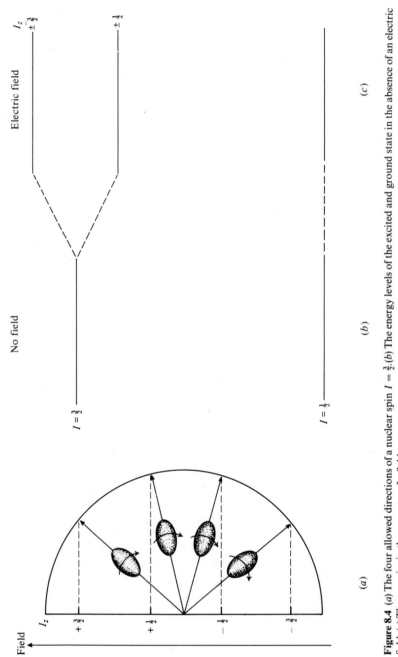

Figure 8.4 (*a*) The four allowed directions of a nuclear spin $I = \frac{3}{2}$. (*b*) The energy levels of the excited and ground state in the absence of an electric field. (*c*) The energies in the presence of a field.

While useful data on the magnitudes of internal crystal fields can be obtained from such spectra, it should be pointed out that a simple explanation is not available for every spectrum so far observed. Thus some compounds which might be expected to possess internal electric fields do not show splitting, and some which would be expected to show a single spectrum, show unexplained multiplet structure.

When other spin quantum numbers are involved, or when both ground and excited state nuclei have a quadrupole moment, no new principles are involved and an extrapolation of the above arguments usually yield a convincing explanation of observed spectral structure.

8.2.3 The Effect of a Magnetic Field

The non-zero spin which we have seen to be associated with either the excited or the ground state nucleus—and usually with both—will interact with a magnetic field in the manner already described in Chapter 7. Each energy state will split into $2I + 1$ separate energy levels, the spacing between them being $\Delta E = g\beta_N B_z/h$, where B_z is the magnetic field at the nucleus. In general the g values of the excited and ground states will differ, so the splitting in each state will be different; and in addition, the g values may have opposite signs (as in the case of ^{57}Fe, where the excited state g is negative).

Figure 8.5(a) and (b) show the situation for this nucleus; at (a) the energy levels are shown in the absence of a magnetic field, whereas in (b) the field is shown splitting the ground state ($I = \frac{1}{2}$, g positive) into two sublevels, and the excited state ($I = \frac{3}{2}$, g negative) into four. Following the conventions of Sec. 7.1.2, we show the $I_z = +\frac{1}{2}$ sub-level of the ground state nucleus to be lower than that of $I_z = -\frac{1}{2}$, while the sub-levels of the excited state increase in energy in the order $I_z = -\frac{3}{2}, -\frac{1}{2}, +\frac{1}{2}, +\frac{3}{2}$.

The selection rules for Mössbauer spectroscopy are complicated; essentially all transitions for which $\Delta I_z = 0$ or ± 1 are allowed, as shown on the diagram, but the relative transition probabilities are found to differ according to the states involved. Thus if we consider the six transitions in pairs, and write the Fe* state first, we have (i) $\frac{3}{2} \leftrightarrow \frac{1}{2}$, $-\frac{3}{2} \leftrightarrow -\frac{1}{2}$; (ii) $\frac{1}{2} \leftrightarrow \frac{1}{2}$, $-\frac{1}{2} \leftrightarrow -\frac{1}{2}$; (iii) $-\frac{1}{2} \leftrightarrow +\frac{1}{2}$, $+\frac{1}{2} \leftrightarrow -\frac{1}{2}$. Detailed calculation shows that, while the two members in each pair have the same probability, the relative probabilities of the three pairs are $3 : 2 : 1$ for (i) : (ii) : (iii). The overall spectrum will, therefore, be as shown at the foot of Fig. 8.5(b), and it is precisely this type of spectrum which has been observed, for instance, in the case of metallic iron.

We should note that the magnetic field necessary to cause the energy-level splitting may be applied externally, but it also often happens that

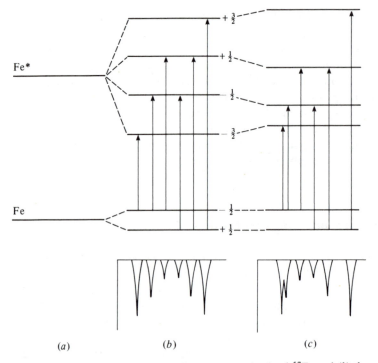

Figure 8.5 (*a*) The excited and ground state energy levels of ^{57}Fe and (*b*) the splitting produced by a magnetic field. In (*c*) the effect of a simultaneous magnetic and electric field is shown.

internal effects within the sample produce sufficient field to cause observable splitting. By using an external field to 'calibrate' the spectrum, it is quite possible to estimate the magnitude of the internal fields, and in this way fields 20–50 T have been found for various compounds of ^{57}Fe. Such fields are large compared with the 5–10 T fields created by superconducting magnets, but one should not suppose that the internal field extends uniformly through the bulk sample—the field is formed by interaction of the nucleus with its surrounding electrons and this is probably an extremely localized effect.

If an *electric* field exists within the sample as well as an internal or external magnetic field, then the quadrupolar shifts are superimposed on the magnetic splitting. In Fig. 8.4 we saw that, for a positive quadrupolar moment, the $\pm\frac{3}{2}$ states are moved upwards in energy, while the $\pm\frac{1}{2}$ states move down. This effect is shown in part (c) of Fig. 8.5 and we see that, since the selection rules and transition probabilities are unchanged, a six-line spectrum is again produced, but that the lines are no longer equally spaced. A spectrum of this type is observed for iron difluoride, FeF_2.

BIBLIOGRAPHY

Greenwood, N. N.: 'The Mössbauer Spectra of Chemical Compounds': *Chemistry in Britain*, **3**, 56, 1967.
Greenwood, N. N. and T. C. Gibb: *Mössbauer Spectroscopy*, Chapman & Hall, 1971.
Goldanskii, V. I., and R. H. Herber: *Chemical Applications of Mössbauer Spectroscopy*, Academic Press, 1968.
Wertheim, G. K.: *The Mössbauer Effect*, Academic Press, 1964.

PROBLEMS

(Useful constants: $h = 6.626 \times 10^{-34}$ J s; $c = 2.998 \times 10^8$ m s^{-1}.)

8.1 A photon of wavelength λ has an equivalent momentum of h/λ. Calculate the recoil velocity of a free Mössbauer nucleus of mass 1.67×10^{-25} kg (that is, atomic weight 100) when emitting a γ-ray of 0.1 nm wavelength. What is the Doppler shift of the γ-ray frequency to an outside observer?

8.2 A particular Mössbauer nucleus has spins of $\frac{5}{2}$ and $\frac{3}{2}$ in its excited and ground states, respectively. Into how many lines will the γ-ray spectrum split if: (*a*) the nucleus is under the influence of an internal electric field gradient, but no magnetic field is applied; (*b*) there is no electric field gradient at the nucleus but an external magnetic field is applied; (*c*) both an internal electric field gradient and an external magnetic field are present.

GENERAL BIBLIOGRAPHY

Some of the general texts on spectroscopy, each covering a wide range of techniques, are listed below:

Bingel, W. A.: *Theory of Molecular Spectra*, Wiley, 1969.

Brand, J. C. D., and J. C. Speakman: *Molecular Structure: The Physical Approach*, 2nd ed., Edward Arnold, 1975.

Brittain, E. F., W. O. George, and C. H. Wells: *Introduction to Molecular Spectroscopy, Theory and Experiment*, Academic Press, 1970.

Browning, D. R. (ed.): *Spectroscopy*, McGraw-Hill, 1969.

Chang, R.: *Basic Principles of Spectroscopy*, Krieger, 1978.

Cottrell, T. L.: *Dynamic Aspects of Molecular Energy States*, Oliver and Boyd, 1965.

Crooks, J. E.: *The Spectrum in Chemistry*, Academic Press, 1978.

Dunford, H. B.: *Elements of Diatomic Molecular Spectroscopy*, Addison-Wesley, 1968.

Steinfeld, J. I.: *Molecules and Radiation: An Introduction to Modern Molecular Spectroscopy*, Harper & Row, 1974.

Straughan, B. P., and S. Walker: *Spectroscopy*, Vols. 1–3, Chapman and Hall, 1976.

Wheatley, P. J.: *The Chemical Consequences of Nuclear Spin*, North-Holland Pub. Co., 1970.

Williams, D. H., and I. Fleming: *Spectroscopic Methods in Organic Chemistry*, 3rd ed., McGraw-Hill, 1980.

Chapter 1

1.1 $v = 2.998 \times 10^{13}$ Hz; $\bar{v} = 10^3$ cm^{-1}; $\Delta E = 1.987 \times 10^{-20}$ J mol^{-1} or $\Delta E = 1.196 \times 10^4$ J mol^{-1}; $\lambda = 5$ μm (that is, half as large).

1.2 (a) HBr only; (b) HBr and CS$_2$ (for the asymmetric stretch and the bend only).

1.3 (a) $\delta E \approx 10^{-33}$ J; $\delta v \approx 10^{-2}$ Hz. (b) $\delta E \approx 10^{-31}$ J; $\delta v \approx 1$ Hz.

1.4 For 4.005×10^{-22} J: (a) 368; (b) 818; (c) 905; (d) 990. For 4.005×10^{-21} J: (a) 0; (b) 135; (c) 368; (d) 905.

Chapter 2

2.1 $B = 0.357\,17$ cm^{-1}; $I = 7.837 \times 10^{-42}$ kg m^2; $r = 1.756 \times 10^{-10}$ m. For $J = 9 \rightarrow J = 10$ the value of $\Delta \varepsilon$ is 7.1434 cm^{-1}; the maximum population is at $J = 17$.

2.2 In general, $\omega/2\pi = 2Bc\sqrt{J(J + 1)}$ r/s, hence (a) zero; (b) 3.02×10^{10}; (c) 11.24×10^{10} r/s.

2.3 H^{37}Cl: $B = 10.5739$ cm^{-1}; D^{35}Cl: $B = 5.446$ cm^{-1}.

2.4 The quickest way to evaluate J, B, and D is to note that the separations between the given lines are about 16.8 cm^{-1}; this suggests that $2B$ is about 17 cm^{-1}, from which the first line, at 84.544 cm^{-1}, obviously corresponds to the $J = 4 \rightarrow J = 5$ transition. Using this, B and D can be calculated from two of the lines and checked using the third. Hence: $B = 8.473$ cm^{-1}; $D = 3.71 \times 10^4$ cm^{-1}; $r = 1.414 \times 10^{-10}$ m; $\bar{\omega} = 2560$ cm^{-1}, approx.

2.5 Maximum population is at $J = 10$; some other points on the graph are:

J	Approx. relative population	J	Approx. relative population
0	1·0	12	12·4
2	4·9	14	11·3
4	8·2	16	9·7
6	10·7	18	8·0
8	12·2	20	6·2
10	12·8		

2.6 For HCN: $I = 19·005 \times 10^{-47}$ kg m^2; $B = 1·472\,8$ cm^{-1}; for DCN: $I = 23·254 \times 10^{-47}$ kg m^2; $B = 1·203\,7$ cm^{-1}.

2.7 Vibrational frequency is about 2995 cm^{-1}.

Chapter 3

3.1 $\bar{\omega}_e = 1903·98$ cm^{-1}; $\bar{\omega}_e x_e = 13·96$ cm^{-1} (or $x_e = 7·332 \times 10^{-3}$); zero point energy, $\varepsilon_0 = 948·51$ cm^{-1}; force constant, $k = 1595·0$ kg s^{-2} (or 1595·0 N m^{-1}); dissociation energy $= 744·4$ kJ mol^{-1} (inaccurate because cubic and quartic terms in $(v + \frac{1}{2})$ are ignored and these become important at large v—see Sec. 6.1.4 for details).

3.2 Zero point energy for $(HCl + D_2) = 2937·5$ cm^{-1}, while that for $(DCl + HD)$ is 2808·5 cm^{-1}. Hence 129·0 cm$^{-1} = 1·542$ kJ mol^{-1} is evolved during the reaction.

3.3 The intensity of the hot band is about 0·36 that of the fundamental.

3.4 18·8 cm^{-1}.

3.5 (a) 1; (b) 4; (c) 3; (d) 30.

3.6 (a) about 2550 cm^{-1}; (b) about 975 cm^{-1} (taking $\nu_{CO} \approx 1100$ cm^{-1}).

Chapter 4

4.1 Raman spectroscopy; spacing $= 243·2$ cm^{-1} ($B = 60·8$ cm^{-1}).

4.2 No. The nuclear spin affects only the line intensities and in this case produces a strong, weak, strong, weak, ... alternation of intensities.

4.3 Molecule is linear (PR contour of an infra-red band) and has a centre of symmetry, hence is A—B—B—A. 3374 cm^{-1} and 3287 cm^{-1} are close to the $\equiv$CH stretching frequency (given in Table 3.4 as 3300 cm^{-1}) so molecule is acetylene, HC$\equiv$CH. Assignments:

3374 cm^{-1}:	symmetric CH stretch
3287 cm^{-1}:	asymmetric CH stretch
1973 cm^{-1};	C$\equiv$C stretch
729 and 612 cm^{-1}:	bending vibrations

4.4 Non-linear and no centre of symmetry, hence

$$\overset{\text{A}}{\diagup \diagdown}$$
$$\text{B} \qquad \text{B}.$$

3756 cm^{-1} and 3652 cm^{-1} are in the region of OH stretching frequency, so molecule is H_2O. Assignments:

3756 cm^{-1}:	asymmetric stretch
3652 cm^{-1}:	symmetric stretch
1595 cm^{-1}:	bend

4.5 Both molecules, no centre of symmetry. N_2O linear (PR bands), hence N—N—O. NO_2 non-linear, hence either

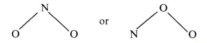

or

Chapter 5

5.1 $5331\cdot55$ cm^{-1}, $7799\cdot29$ cm^{-1} and $9139\cdot80$ cm^{-1}. Ionization energy $= 12\,186\cdot40$ cm^{-1} ($\equiv 1\cdot511$ eV).

5.2 $L = 2$; $S = \frac{3}{2}$, $J = \frac{5}{2}$. Minimum number of electrons $= 3$; possible electron configurations include: *spp, ppp, ppd*, etc. (all non-equivalent electrons).

5.3 (a) $^1S_0$, $^3S_1$; (b) $^1D_2$, $^1P_1$, $^1S_0$, $^3D_{3,2,1}$, $^3P_{2,1,0}$, $^3S_1$; (c) $^1D_2$, $^3D_{3,2,1}$; (d) $^1F_3$, $^1D_2$, $^1P_1$, $^3F_{4,3,2}$, $^3D_{3,2,1}$, $^3P_{2,1,0}$.

5.4 (a) $^1S_0$; (b) $^1D_2$, $^1S_0$, $^3P_{2,1,0}$; (c) $^1G_4$, $^1D_2$, $^1S_0$, $^3F_{4,3,2}$, $^3P_{2,1,0}$.

5.5 $^4S_1$: Total spin $= \frac{3}{2}$, hence J must be half-integral; $^2D_{7/2}$: $S = \frac{1}{2}$, $L = 2$, hence J cannot be greater than $\frac{5}{2}$; $^0P_1$: $2S + 1$ cannot be zero.

5.6 $^2P_{3/2} \rightarrow {}^2D_{3/2}$: 10 lines; $^2P_{3/2} \rightarrow {}^2D_{5/2}$: 12 lines; $^2P_{1/2} \rightarrow {}^2D_{3/2}$: 6 lines.

Chapter 6

6.1 By plotting the given data, extrapolating and taking the area under the graph, the dissociation energy is found to be approximately $12\,100$ cm^{-1} or $144\cdot7$ kJ mol^{-1}.

6.2 $490\cdot3$ kJ mol^{-1} (or $41\,000$ cm^{-1}).

6.3 Ground state: $v_{max} = 69$; $D''_e = 690\cdot0$ kJ mol^{-1}
                                 $D''_0 = 680\cdot2$ kJ mol^{-1}
    Excited state: $v_{max} = 53$; $D'_e = 581\cdot3$ kJ mol^{-1}
                                 $D'_0 = 570\cdot7$ kJ mol^{-1}

6.4 Ground state: $D''_0 = 345\cdot8$ kJ mol^{-1}
    Excited state: $D'_0 = 237\cdot4$ kJ mol^{-1}
The answers to question 6.3 are inaccurate (in this case badly so) because Eq. (6.10) relies on the use of terms up to $(v + \frac{1}{2})^2$ only; a better treatment would involve terms in $(v + \frac{1}{2})^3$ and $(v + \frac{1}{2})^4$.

6.5 Band head is in the P branch at $p = 14$ (Eq. (6.23)), and the position of the band head is at $19\,358$ cm^{-1}. The ground state has the larger internuclear distance.

Chapter 7

7.1 For H: $B_z = 0\cdot7107$ T (7107 gauss). For ^{13}C: $B_z = 2\cdot8266$ T (28 266 gauss).

7.2 Chemical shift positions are $5\cdot75\delta$ and $1\cdot05\delta$; $J_{AX} = 10$ Hz.

7.3 Eight lines of equal intensity, symmetrically disposed about the midpoint; taking the midpoint at $0\cdot0$ Hz, the lines are at: $-77\cdot5$, $-57\cdot5$, $-27\cdot5$, $-7\cdot5$, $+7\cdot5$, $+27\cdot5$, $+57\cdot5$ and $+77\cdot5$ Hz. (N.B. the J_{HH} values given are irrelevant to the ^{19}F spectrum if the chemical shifts are large.)

7.4 The BH_4 group will give rise to a $1:3:3:1$ quartet because of coupling to the CH_3; conversely the CH_3 group will show a $1:4:6:4:1$ pentet, with the same separation between the lines as exhibited in the quartet.

7.5 (a) 6; (b) 12; (c) 9; (d) 12.

7.6 Each D nucleus, having a spin of 1, splits the neighbouring H (of CHD_2) into three lines of equal intensity. Use the family tree method to show that the resultant overlapped spectrum is a $1:2:3:2:1$ pentet.

7.7 Small groups, such as $—CH_3$ or $C_6H_5—$, may often undergo internal rotation at temperatures not too far below the bulk melting point of the material. This lengthens the relaxation time of the nuclei and so sharpens the resonance frequency.

Chapter 8

8.1 Recoil velocity $= 39{\cdot}7 \text{ m s}^{-1}$; Doppler shift $= 3{\cdot}97 \times 10^{11} \text{ Hz}$.

8.2 (a) 5 lines; (b) 12 lines; (c) 12 lines.

INDEX